NONLINEAR OPTICS, QUANTUM OPTICS, AND ULTRAFAST PHENOMENA WITH X-RAYS

NONLINEAR OPTICS, QUANTUM OPTICS, AND ULTRAFAST PHENOMENA WITH X-RAYS

Physics with X-Ray Free-Electron Lasers

Edited by
BERNHARD W. ADAMS
Argonne National Laboratory

Kluwer Academic Publishers
Boston/Dordrecht/London

Distributors for North, Central and South America:
Kluwer Academic Publishers
101 Philip Drive
Assinippi Park
Norwell, Massachusetts 02061 USA
Telephone (781) 871-6600
Fax (781) 871-6528
E-Mail <kluwer@wkap.com>

Distributors for all other countries:
Kluwer Academic Publishers Group
Post Office Box 322
3300 AH Dordrecht, THE NETHERLANDS
Telephone 31 78 6576 000
Fax 31 78 6576 474
E-Mail <orderdept@wkap.nl>

Electronic Services <http://www.wkap.nl>

Library of Congress Cataloging-in-Publication

Nonlinear optics, quantum optics, and ultrafast phenomena with X-rays: physics with X-ray free-electron lasers/edited by Bernhard W. Adams.
p.cm.
Includes bibliographical references.
ISBN: 1-4020-7475-1 (acid-free paper)
1.Free electron lasers. 2.X-ray lasers. 3.Nonlinear optics. 4.Quantum optics. 5.Laser pulses, Ultrashort. I.Title: Physics with X-ray electron lasers. II.Adams, Bernhard W.

QC689.55.F75N66 2003
621.36'6—dc21

2003047593

Printed on acid-free paper.

Printed in the United States of America

Contents

List of Figures

List of Tables

Preface

It is a little more than one hundred years since the discovery of x-rays by W.C. Röntgen in 1895 and about fifty since synchrotron radiation appeared on the scene (see ch. 1, sec. 4) and set off an explosive growth of x-ray science. The laser is also about 50 years old and has probably been the cause of an even steeper progress in optical science. In both fields, science in waiting came to realization as soon as the source characteristics made it possible. Due to a number of developments, we may expect perhaps even more rapid advances and a confluence of these two fields in the decade to come: synchrotron radiation sources are beginning to provide highly coherent x-rays, and, with the advent of the x-ray free-electron laser (XFEL), x-rays with laser-like characteristics will be available.[1] On the other side, lasers are being used more and more in the production of x-rays through high harmonics (up to a few hundred eV photon energy), x-ray quantum lasers (to distinguish them from XFELs, also up to a few hundred eV) and plasma sources (up to ca. 10 keV).

Why are the three fields of nonlinear optics, quantum optics, and the physics of ultrafast phenomena with x-rays combined in this book? There is of course the simple reason that all three of them are emergent, and this is a book about future developments in x-ray physics. However, there are also other connections, with nonlinear optics being the link. Many quantum optical phenomena require nonlinear optics and the study of ultrafast phenomena usually requires high intensities, which may force nonlinear optics upon the experimentor. Furthermore, the coherent control of very short light and x-ray pulses requires nonlinear optical techniques. This book is certainly not comprehensive. The selection of material represent my view of the field and the availability of contributors.

A field that is usually associated with lasers is time-resolved studies of the dynamics of chemical reactions and of electrons in photoexcited semiconductors and metals. X-rays can provide a substantial amount of information in these studies, mainly due to their spatial resolution and element specificity.

[1] why the circuitous expression? See ch. 1, sec. 7

Efforts are currently under way to combine lasers and x-rays in time-resolved experiments from nanosecond down to picosecond timescales. With the next generation of x-rays sources, i.e., XFELs and energy-recovery linear accelerators (ERLs), the subpicosecond time domain will become available, and there are ideas for pushing the limits down to a few femtoseconds. Ultrafast science with the combination of femtosecond lasers and subpicosecond x-ray sources will probably be one of the most exciting fields in condensed matter physics of the next decade (see fig. 6.1).

The advent of XFELs will also make nonlinear optics with x-rays a topic to consider, especially when the x-rays are to be compressed in time (for ultrafast science), or in space (for small samples, down to nanocrystals, virus particles, etc.). There is, however, at least one particular case where nonlinear optical effects are observable even on an x-ray tube, and that is spontaneous parametric down conversion of x-rays (see chapter 5).

It is thus necessary for x-ray physicists to become acquainted with subjects that are already well developed for the "visible" light regime (meaning laser-generated light, i.e., far infrared to deep ultraviolet). There are also a few fields where laser physics could perhaps borrow from x-ray science. Concerning theory, the dynamical theory of diffraction and its extension to deviations from periodicity, the Takagi-Taupin theory, should be of interest to the study and design of photonic crystals. On the instrumental side, the ways of doing an experiment differ considerably between the laser and x-ray communities: A laser physicist likes to do things in a hands-on fashion, i.e., everything is in a compact setup on an optical table, and adjustments are done by hand. With x-rays, a hands-off approach is strongly recommended, and in synchrotron radiation laboratories, the art of remote controlling a hundred-meter-long beamline has been carried to perfection.

It is the intention of this book to contribute to the foreseeable confluence of the fields, both on the theoretical, and on the practical side.

About the CD that comes with this book: Most of the figures in this book were generated in color, but are printed in black-and-white for reasons of economy. You will find the color versions on the CD, many of them in several file formats, including the source for some of the graphics. There is also some additional material that is not printed in the book, such as photographs of experimental setups, a movie about dynamical diffraction, etc. Please take a look at the README file on the CD for details.

B.W.A.

Contributing Authors

Bernhard W. Adams is a visiting scientist at Argonne National Laboratory, Illinois, USA. He received his PhD from the University of Munich in 1995 and worked as a postdoctoral research fellow at DESY, Hamburg.

Philip H. Bucksbaum is Otto Laporte professor of physics at the University of Michigan and director of the Center for Frontiers in Optical Coherent and Ultrafast Science (FOCUS).

Matthew F. DeCamp is a postdoctoral fellow at the University of Michigan. He received his Ph.D. from the University of Michigan in 2002.

Eric M. Dufresne is a beamline operation scientist at the Advanced Photon Source. He received his PhD in 1995 from McGill University (Montreal) and worked as a Postdoctoral Research Fellow at the University of Michigan.

Martin E. Garcia is an assistant Professor at the Freie Universität Berlin, Germany. He received his Masters Degree in Bariloche, Argentina, his PhD at the Freie Universität Berlin, and he worked as a visiting scientist at Valladolid (Spain), Michigan (USA) and Greifswald (Germany).

Harald O. Jeschke is a postdoctoral associate at the Department of Physics & Astronomy, Rutgers University, Piscataway, NJ 08854, USA. He received his Ph.D. from the Freie Universität Berlin in 2000.

Aaron Lindenberg is at the Advanced Light Source at Lawrence Berkeley National Laboratory. He recieved his Ph.D. from the Physics Department at UC Berkeley in 2001.

David A. Reis is an assistant professor of physics and applied physics at the University of Michigan. He received his Ph.D in 1999 from the University of Rochester.

Peter Sondhauss did his Ph.D. 1999 at the Friedrich-Schiller University in Jena (Germany). At present he is a Postdoctoral Research Assistant in Justin Wark's group at the University of Oxford.

Justin S. Wark is a Professor of Physics in the Department of Physics at the University of Oxford.

Pedro Zambianchi is a postdoc at Argonne National Laboratory. He received his Ph.D. from New York State University (SUNY) at Albany in 1999.

Notation

some electromagnetic and mechanic quantities in Gaussian units					
symbol	meaning	dimensions		to SI	where
		basic	alt.		
$\mathbf{A}$	vector potential	$\frac{m^{1/2}\ell^{1/2}}{t}$	$\frac{\mathcal{E}}{q}, \frac{q}{\ell}$		20
$\mathbf{B}$	magnetic induction	$\frac{m^{1/2}}{\ell^{1/2}t}$	$\frac{q}{\ell^2}$	1 gauss = 10^{-4} T	
c	speed of light	$\frac{\ell}{t}$			4
$\mathbf{E}$	electric field	$\frac{m^{1/2}}{\ell^{1/2}t}$	$\frac{q}{\ell^2}$	$\frac{\text{1statV}}{\text{cm}} = \frac{\text{30kV}}{\text{m}}$	
$\mathcal{E}$	energy	$\frac{m\ell^2}{t}$	$\frac{q^2}{\ell}$	1 erg = 10^{-7} J	
Φ	scalar potential	$\frac{m^{1/2}\ell^{1/2}}{t}$	$\frac{\mathcal{E}}{q}, \frac{q}{\ell}$	1 statV = 300 V	20
γ	relativisticity	1			4
$\mathbf{p}$	momentum	$\frac{m\ell}{t}$			
r_e	class. el. radius	ℓ		$2.82 \cdot 10^{-13}$cm	21
ρ	charge density	$\frac{q}{\ell^3}$		$\frac{\text{statC}}{\text{cm}^3}$	
q	electric charge	$\frac{m^{1/2}\ell^{3/2}}{t}$		1 statC = $\frac{1}{3}10^{-9}$C	20

1 pF $\approx$ 1 cm, 1 statvolt/cm = 1 gauss

There are two conventions for wave numbers k, wave vectors $\mathbf{k}$ and frequencies ω. One, used for example by Jackson [1], writes wave numbers as $k = 2\pi/\lambda$ and frequencies as $\omega = 2\pi/\tau$, where λ and τ are the wavelength and the oscillation period, respectively. The phase of a wave is then $\exp(i(\mathbf{k}\cdot\mathbf{r} - \omega t))$. This is the convention used in this book. The other way of writing wave numbers and frequencies ν is $k = 1/\lambda$ and $\nu = 1/\tau$. Important works in dynamical diffraction theory use that convention, and that must be kept in mind when comparing results. There are also two possible signs of the phase of a wave. Most of this book uses $\mathbf{k}\cdot\mathbf{r} - \omega t$, except for ch. 10, where the phase is $\omega t - \mathbf{k}\cdot\mathbf{r}$.

Citations

Entire works (books, articles, etc.) are cited as [number], where the number refers to the list of references at the end of the book. A citation of the type [x].(y) means eq. (y) in work x, and [x].[y] means sec. y in work x.

Chapter 1

X-RAY SOURCES

1.802 terafurlongs per fortnight – it's not just a good idea, it's the law!

B.W. Adams
Advanced Photon Source, Argonne National Laboratory, Argonne, IL 60439, USA
adams@mail.desy.de

Abstract The development of x-ray sources and their applications is described, with special emphasis on high intensities and on short pulse duration.

1. Introduction

In a few years, the first light from an x-ray free-electron laser (XFEL) will be available for experiments. The advances in radiation characteristics, such as intensity, coherence, or reciprocal pulse duration, will probably surpass everything that x-ray science has ever experienced. In parallel, other promising concepts for the production of intense, ultrashort pulses of x-rays are under study or in development [2], such as various types of laser-driven sources and the energy-recovering linear accelerator (ERL). Before delving into the science that can be done with such a source (not to speak of the challenges it will present), let us look back at a century of x-rays and survey the current activities in the development of the next generation of x-ray sources.

2. X-Ray Tubes

The classical, and for about 50 years only, source of man-made x-rays is the x-ray tube. The radiation is emitted into a large solid angle and has both continuous and spectral line characteristics. Most x-ray tubes are operated continuously, but pulsed operation is possible. The spatial coherence properties are mainly determined by the focal spot of the electron beam on the anode and the longitudinal coherence is given by the spectral bandwidth, i.e., a few x-ray oscillations in the continuous spectrum and a few thousand in the emission lines.

3. Laser-Driven Sources

With the development of high-power short-pulse lasers, several types of laser-driven x-ray sources became possible [3]. All of them can provide x-ray pulses of very short duration, and laser pump, x-ray probe experiments enjoy the advantage of natural synchronism between the x-rays and the pump pulse from the same laser.

When an intense laser pulse is focused onto a target, x-rays can be produced by two processes, called "thermal" and "suprathermal" [4]. The former is based on heating the target material to a very high temperature, so that the thermal emission has a significant component in the x-ray region. In the latter, a laser pulse generates a plasma on a target surface and then accelerates electrons to kilo-electronvolt energies through laser-plasma interactions (see ch. 2, sec. 2.5). These electrons strike the underlying target material and generate x-rays in the same way as in an x-ray tube.

Soft x-rays, up to a few hundred eV photon energy, can be generated as high harmonics of intense laser radiation, incident on a gas. Electrons are removed from their host atoms in the gas by the strong electric field of the laser pulse and are accelerated until the electric field reverses its direction after an optical half-cycle. When such an electron returns to its host atom (ion, rather), it has again gained energy in the reversed electric field. As it traverses the Coulomb field near the nucleus, its motion becomes highly anharmonic, giving rise to the emission of short-wavelength radiation. In order to obtain a coherent effect over an extended length of the gas, the difference in the refractive indices between the fundamental and the high harmonic must be compensated. A common scheme is to use a hollow glass fiber, which is filled with a noble gas. The phase matching can be tuned through adjustment of the gas pressure [3]. Because the anharmonic part of the electron's trajectory is limited to a region close to the nucleus of the host atom, the duration of the x-rays is much less than one optical cycle of the driving laser, occurring in a single subfemtosecond burst from a single-cycle laser pulse or in a sequence of bursts if the laser pulse is longer [5].

High harmonic generation in a gas cannot (yet) produce hard x-rays. A scheme has been proposed to create very high harmonics at x-ray photon energies through the anharmonic motion of ultrarelativistic electrons in the focus of a very intense laser. Only the most powerful lasers can reach the intensity of the order of 10^{20}W/cm^2, which is necessary for that process. Again, the duration of the x-rays is limited to a brief burst or sequence of bursts, lasting much less than one optical cycle of the laser. More about relativistic electrons in electromagnetic waves can be found in ch. 2.

Finally, Thomson backscattering from relativistic electrons should be mentioned. These can be generated by laser wakefield acceleration (see ch. 2, sec. 2.5), which takes place when a laser creates a strong disturbance in a plasma, which in turn leads to very high electric fields in the longitudinal direction. The relativistic electrons can also be generated in a conventional particle accelerator, see sec. 5.

4. Synchrotrons and Storage Rings

Synchrotron radiation was discovered in 1947 [6, 7], when a bright emission was seen on the General Electric betatron. Theoretical calculations had already predicted that electrons radiate and lose energy due to the transverse acceleration required to keep them on a closed orbit, but it was only after the visual observation that the unique characteristics of synchrotron radiation were realized. Its use as a research tool really took off in the 1960s, as it was quickly found to be a very valuable contribution to a number of fields, such as crystallography, materials science, chemistry, etc. The first experiments were done at facilities intended for accelerator research and for particle physics. As the potential of synchrotron radiation became evident, more and more resources were devoted to the optimization of the radiation characteristics of the sources. Currently, the so-called third-generation sources are mature and in full "production" mode and the fourth generation is on the drawing boards, although there is still some contention as to what kind of source is to be called "fourth generation", with the x-ray free-electron laser apparently leading the field.

A synchrotron accelerates charged particles (henceforth simply "electrons") by running them on a closed orbit and sending them repeatedly through a radiofrequency resonator where they "surf" on an electromagnetic field. The resonator has two functions: it makes the electromagnetic waves that are fed into it develop the longitudinal component necessary for acceleration and it recycles the energy that is not extracted by the electrons. Dipole magnets, also called bending magnets, deflect the electrons to close the orbit, quadrupoles provide focusing and sextupoles correct for chromatic aberrations due to the electron energy spread. Because the orbit is fixed, the magnetic fields have to be ramped up as the electrons gain energy - hence the name synchrotron. The assembly of all the magnets is called the magnetic lattice and together with the resonator, it provides equilibrium positions for the electrons which run along the orbit at a velocity close to the speed of light. These positions are commonly called "buckets", and their fill of electrons is called a "bunch". The longitudinal position of an electron relative to a bucket determines the acceleration it experiences, which, in turn, counteracts any positional deviation.

A synchrotron cannot start up from zero energy. Instead, the electrons are injected at an energy E of typically many times their rest mass equivalent mc^2, whence they are brought to the final energy by ramping up the magnetic fields and thus shifting the buckets, and the electrons caught in them, to higher energies. Strictly speaking, the current third-generation synchrotron radiation sources actually are not synchrotrons at all because electrons from a booster synchrotron (see fig. 1.2) are injected at full energy and the radiofrequency resonator only compensates for the energy losses due to the emission of synchrotron radiation. Such a device is called a storage ring and is much more stable in operation than a synchrotron. Storage rings were originally built for particle physics collision experiments but the same kind of technology is used in the third-genration synchrotron radiation sources.

Acceleration of charged particles causes emission of electromagnetic waves, and thus a bending magnet is a source of synchrotron radiation. The emitted power at a given bending radius depends on the parameter $\gamma = E/(mc^2)$ to the fourth power, and therefore only electrons or positrons are used for the production of synchrotron radiation, at typically $\gamma \approx 10^4$. The emission is strongly peaked in the forward direction, i.e., tangential to the orbit (see fig. 1.1) with an opening angle of the order of γ^{-1} (see [1, 8] for details). One should keep in mind, though, that this single-electron emission angle must be convoluted with the angular distribution of electrons within one bunch. The first generation of synchrotron radiation sources consisted of machines built for particle physics experiments and the emission from bending magnets was used only in the so-called "parasitic mode". In the second generation, special magnetic

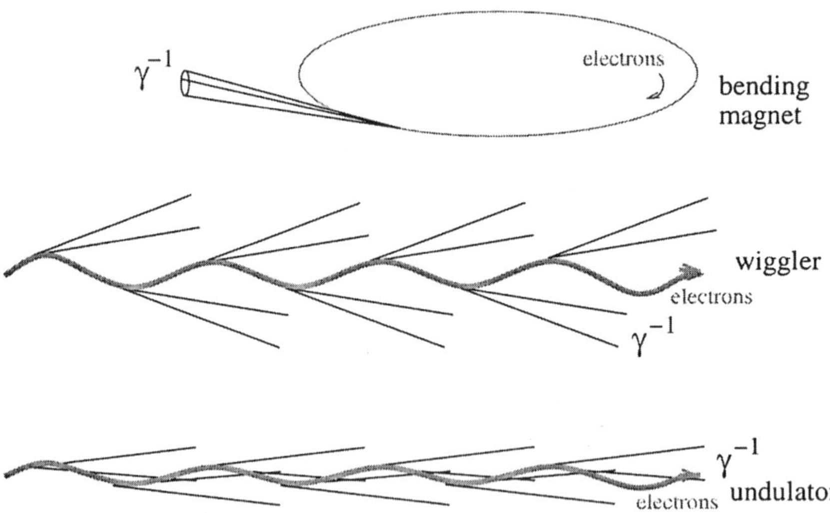

Figure 1.1: Bending magnet, wiggler, and undulator emission.

structures, called wigglers and undulators, were inserted into storage ring particle colliders. The generic name for both types is "insertion device" and both consist of alternatingly poled magnets in a periodic array with a period λ_u (the undulator wavelength). They force the electrons to take an oscillatory detour, thus effectively reducing their longitudinal velocity v_u to a little less than the vacuum speed c of light. In an undulator, the angles of the trajectory undulations are smaller than the $1/\gamma$ emission cones of the electrons and interference occurs, for each electron separately, between the waves emitted in all the bends. This gives rise to distinct peaks in the emission spectrum. Along the undulator axis, the wavelengths of these peaks are given by odd integer fractions of the slippage over one undulator period, between the light, going at velocity c, and the electrons, going at velocity v_u. The width of the interference line is given by the number of undulator poles and the energy spread of the electron beam. A wiggler induces the electrons to much larger angular deviations from the straight path and the intensity from the bends adds up incoherently.

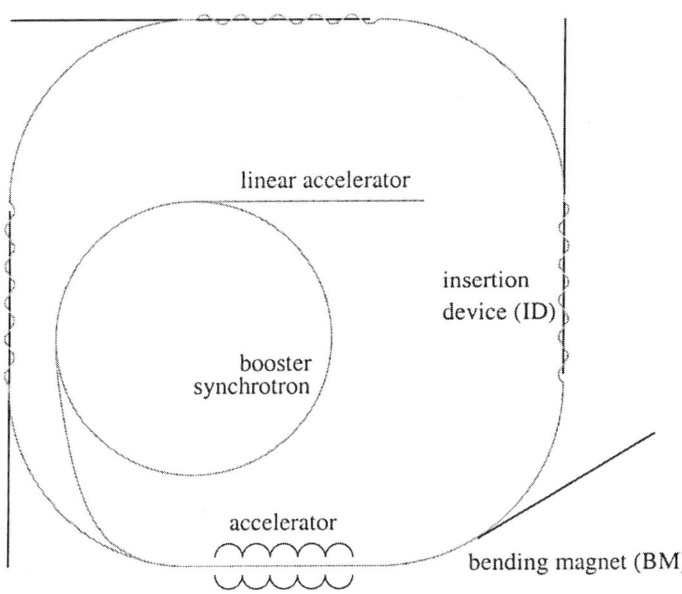

Figure 1.2: Schematic of a 3rd-generation storage ring facility, showing a linear accelerator, which injects at a few 100 MeV into the booster synchrotron which brings the electrons to several GeV (having to ramp the magnetic field up in the process). The electrons are transferred to the storage ring at full final energy, which is maintained against losses through synchrotron radiation by an accelerator structure. Radiation from bending magnets and from insertion devices is used. For the latter, the ring contains dedicated straight sections. In reality, there are many more straights than just the four shown here, and the machine looks much more like a ring.

The third generation consists of storage rings that were built specifically with the purpose of providing synchtrotron radiation and could be optimized accordingly. This applies mostly to the electron beam optics. The third generation is currently represented by three storage rings for hard x-rays (the European Synchrotron Radiation Facility (ESRF), the Advanced Photon Source (APS) in the U.S. and the Super Photon ring-8GeV (SPring8) in Japan) and a few smaller ones for soft x-rays (the Advanced Light Source (ALS) at Berkeley, BESSY II in Berlin, etc.). Most beam lines at 3rd-generation sources are equipped with undulators because these provide a more collimated beam than wigglers and because the need for monochromatic radiation, which is required for most experiments, are best met by the line spectrum of an undulator.

The ESRF has 992 buckets, spaced 2.82 ns from each other, giving a circulation period of 2.8 μs, and the APS has 1296 buckets, spaced 2.84 ns, giving a circulation period of 3.68 μs. Typical fill patterns [9, 10] are contiguous 2/3 of the buckets full, 1/3 empty at the ESRF, or the 23+1 pattern at the APS with 24 buckets selected at 153 ns intervals, 23 of which are filled and one left empty (see ch. 6, figs. 6.4 and 6.5). Most facilities also have a single-bunch mode, which gives long dark times between flashes of x-rays, corresponding to the ring's circumference. This mode is needed mostly in time-of-flight spectroscopy and in nuclear resonant time domain techniques.[1]

The electrons in a bunch can be thought of as a hot gas that is confined by the magnetic structure of the ring and ponderomotive potentials in the microwave resonators that replenish the particle energy. After a few thousand round trips, any phase space distribution within that gas cloud will relax to an equilibrium, which is given solely by the nominal particle energy and the magnetic electron optics. It is that equilibrium phase space distribution that determines the emission characteristics of the storage ring, such as pulse duration and transverse emittance. The spectral width of the undulator lines also has a contribution from the energy spread which becomes dominant in undulators with more than about 100 poles.

5. Pulse Slicing and Ultrafast Thomson Scattering

The equilibrium length of electron bunches in a typical 3rd-generation storage ring is several millimeters to centimeters, corresponding to a duration of the emitted x-ray pulses of tens to hundreds of picoseconds. At the APS, running in a standard operating mode, the value is about 70 ps. Two techniques for obtaining short x-ray pulses from a storage ring are the pulse-slicing technique and the Thomson scattering technique. Both were developed at Berkeley National Laboratory [11, 12, 13].

In the pulse-slicing technique, an ultrashort laser pulse interacts with the electrons as they traverse an undulator and transfers energy to them, in an inverse free-electron laser process [14]. Corresponding to the short laser pulse, only the electrons in a short slice in the bunch are affected in this way. Having a slightly different energy than the other electrons, they take a different trajectory

[1] With ^{57}Fe, the most commonly used Mössbauer isotope, a few 100 ns are usually sufficient, but there are other, more long-lived Mössbauer nuclear species.

in the next energy-dispersive section (i.e., bending magnet) of the storage ring. Their synchrotron emission in following bending magnets or undulators goes therefore in a different direction and can be separated from the main beam. The duration of these x-ray pulses is given by the laser pulse and dispersive effects in the storage ring. About 300 fs have been demonstrated [12]. Because the laser has a much lower repetition rate than the storage ring and also because only a small fraction of the electrons in one bunch contribute to the ultrashort emission, the intensity is much less than that of the main beam.

The Thomson scattering technique also relies on a short-pulse laser to produce short x-ray pulses. A pulse of laser light is directed at roughly a right angle towards a bunch of relativistic electrons. Thomson (or, Compton) scattered photons are concentrated in the forward direction of the electrons due to relativistic effects. The resulting x-ray beam has a duration given by the intersection of the laser pulse with the electron beam.

6. Energy-Recovering Linacs

With recent advances in the technology of superconducting microwave resonators and with laser-driven photocathode electron guns, a type of machine becomes feasible that can provide a much more brilliant beam than a storage ring while being rather similar to one in all operational aspects - the energy recovering linear accelerator (ERL) [15]. The idea is to generate a bunch of electrons with a much tighter phase space distribution than the equilibrium distribution of the ring, accelerate it to several GeV in a superconducting high-field-gradient linear accelerator and send it on just one or very few round trips in a machine that has the same kind of magnetic lattice and insertion devices as a conventional storage ring. There is only very little growth in the electron phase space volume during these few round trips, and therefore the synchrotron radiation has an exceedingly high brilliance and can have a very short pulse duration (100 fs versus 100 ps in a storage ring).

Instead of dumping the electrons and incurring a large energy bill,[2] the beam is brought back to the linear accelerator 180 degree off-phase to the field present there. It is thus decelerated instead of accelerated and feeds its energy back into the superconducting resonator. The resonant cavities in the linear accelerator have thus the additional function of storing the energy that is extracted from one bunch and reusing it.

In passing, we should take note of a particular source, which is close to coming on line (hopefully in the spring of 2003) and will have a couple of year's worth of operating time until the Linear Coherent Light Source (LCLS) [16]

---

[2]Take a typical 3rd-generation source like the APS, 7 GeV and 100 mA and dump that - the power is 700 MW!

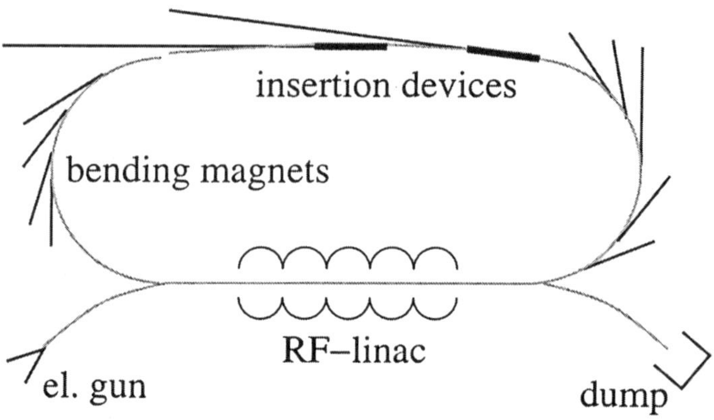

Figure 1.3: Schematic of an ERL. Electron bunches from a low-emittance electron gun are pre-accelerated to a few hundred MeV and injected into the main linear accelerator, where they are brought to the full energy of several GeV. After one pass through the recirculator where synchrotron radiation is produced, the electrons enter the linear accelerator off-phase by π. They leave it with a few hundred MeV and are dumped.

XFEL will be built: The Sub-Picosecond Photon Source (SPPS) at the Stanford Linear Accelerator is a non-recirculating linear-accelerator-driven undulator which will produce x-ray pulses of about 80 fs duration. Because it is not meant to be an FEL, the energy spread of the electrons is almost not an issue (see below) and therefore the electron bunches can be compressed to even less length than they will have in the LCLS. After passage through the undulator, the electrons are dumped without any energy recuperation. This is feasible only because of the low repetition rate of SPPS.

7. X-Ray Free-Electron Lasers

Currently, two projects for free-electron lasers (FELs) that should be capable of producing hard x-rays are in advanced stages of planning, one (TESLA, [17]) near Hamburg, and the other (LCLS [16]) at Stanford. Several laboratories are also pursuing soft x-ray FEL projects (with no claim at comprehensiveness: DESY [18], BESSY [19], MIT-Bates [20, 21], SPring8 [22], but these are outside the scope of this book. The synchrotron radiation community has grown used to a tremendous pace of progress in relevant radiation parameters,[3] but the XFELs will bring an unprecedented increase in brilliance and other relevant parameters. In particular, the x-rays will be generated in pulses of less than 200 fs duration (see end of sec. 7.1 of this chapter), making them very attractive in the

[3]Advocates of synchrotron radiation research like to point out that the growth in brilliance over the past thirty years eclipses Moore's law of the semiconductor industry [23], which states that the number of transistors on the current leading generation of computer chip doubles every 18 months.

study of ultrafast phenomena (see ch. 6, fig. 6.1). For this reason, and also due to the physics of FEL emission, which is based on collective motion and radiation of large numbers of electrons, one might claim that the XFEL is the ultimate accelerator-based x-ray source. That collective radiation, shown schematically

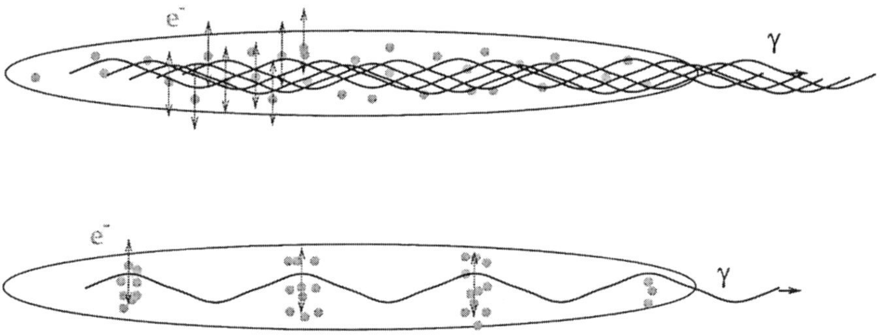

Figure 1.4: Emission from randomly distributed (top) and microbunched (bottom) charges. The intensity radiated by the collective motion of charges scales as the square of the charge in one microbunch.

in fig. 1.4, as opposed to the randomly phased emission from single electrons in an undulator at a synchrotron, is due to a periodic density modulation of the electron bunch at the radiated wavelength. It is called microbunching and is the central term of FEL physics.

The physics of free-electron lasers differs from that of other lasers in a number of ways. First and foremost, the amplifying medium is no ordinary matter, but rather a beam of (usually, and certainly in the case of an XFEL) ultrarelativistic electrons. Also, an FEL is not inherently a quantum device. Only in the special case that the energy spread of the electrons is less than the photon recoil energy does it make sense to use a quantum description of the gain process. Finally, other than most visible-light lasers, the planned XFELs have no resonator.

7.1 The Physics of the FEL Process

The two defining features of any laser are stimulation and amplified emission. Other than in a quantum laser, these two processes are separated from each other in a classical FEL (but see discussion of the quantum FEL below, sec. 7.3). The basic ingredients of any FEL are a relativistic electron beam of sufficiently high quality (i.e., a high and sharply concentrated density in transverse and longitudinal phase space) and an undulator. When viewed in the laboratory frame, FEL radiation appears at the resonance wavelength of the undulator. In the rest frame of the electron bunch [24], the periodic magnetic field of the undulator is relativistically compressed and appears as an electromagnetic wave moving towards the electrons. Conversely, an electromagnetic wave coming from behind suffers a relativistic red shift, and, if its frequency is the undulator fundamental in the lab frame, then it equals that of the undulator "wave" in the electron frame. The resulting standing wave (see fig. 1.5) creates a ponderomotive potential for the electrons, which induces them to a damped oscillatory motion towards the minima, shown schematically in longitudinal phase space in fig. 1.6.

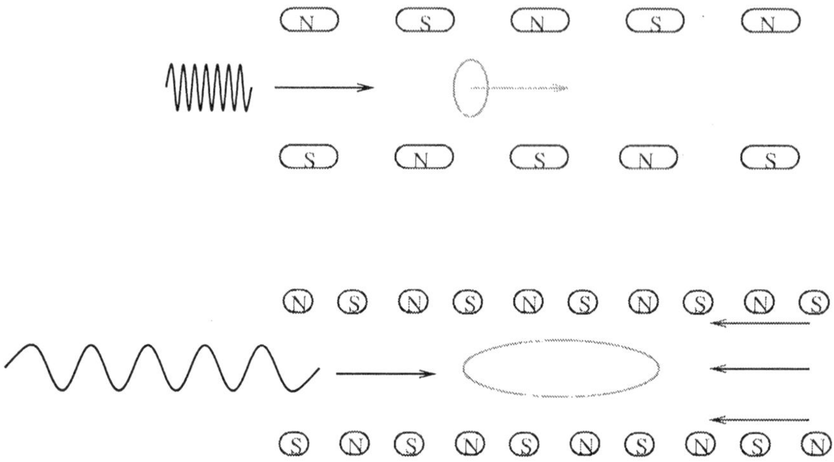

Figure 1.5: Ponderomotive potential in an FEL, lab frame vs. particle frame. Rest frame (top): electron bunch in an undulator and radiation catching up with it from behind. Particle frame (bottom): the same situation in the electron rest frame. The relativistically compressed undulator structure appears as a wave from the right, which interferes with the red-shifted radiation from the left, forming a standing wave pattern, which leads to a ponderomotive potential for the electrons.

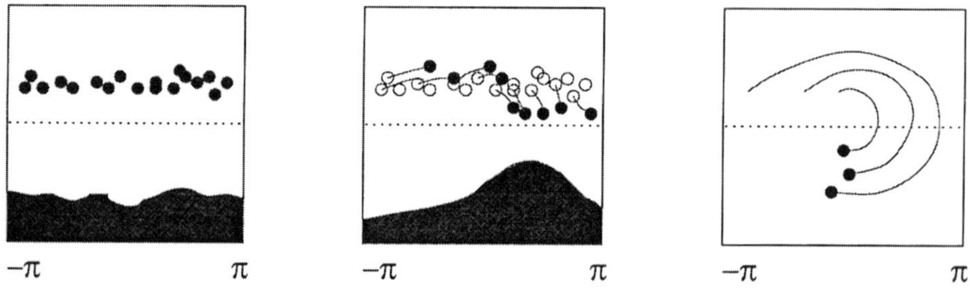

Figure 1.6: Longitudinal phase space evolution in an FEL due to the ponderomotive potential. Dots denote electrons occupying positions in phase space, lines are phase space trajectories. Left: Initial statistical distribution in longitudinal phase, slightly above the energy of undulator resonance. Middle: depending on the phase, different trajectories are taken, resulting in a bunching in the projection upon the phase. Right: Maximum bunching in phase: from here on debunching sets in. This set of graphs is inspired by figs. 1-9 to 1-11 in [25]. In the lower halves of the left and middle graphs, a projection of the electron density onto the spatial coordinate is indicated. All three graphs are just cartoons of the physics, done without any calculation.

Due to the concentration of electrons in the potential minima, the charge density within the bunch becomes modulated at the undulator emission wavelength (as seen from the lab frame) and radiates collectively as shown in fig. 1.4. This leads in turn to a stronger ponderomotive potential, and so forth. As the electrons get concentrated in space (or equivalently the phase along the longitudinal direction), they spread out in longitudinal momentum. At some point along the process, the microbunching is maximal and further evolution of the system leads to overshooting within the ponderomotive potential and debunching in space. By Madey's theorem [26], gain in an FEL is proportional to the derivative of the spontaneous undulator emission over the photon energy (see, for example, [27] for a demonstrative calculation). Maximum gain does therefore occur not right at the center of the undulator lines but at slightly longer wavelengths. The energy spread of the electrons traversing the undulator must be considerably smaller than the width of the gain curve, or the FEL will not work.

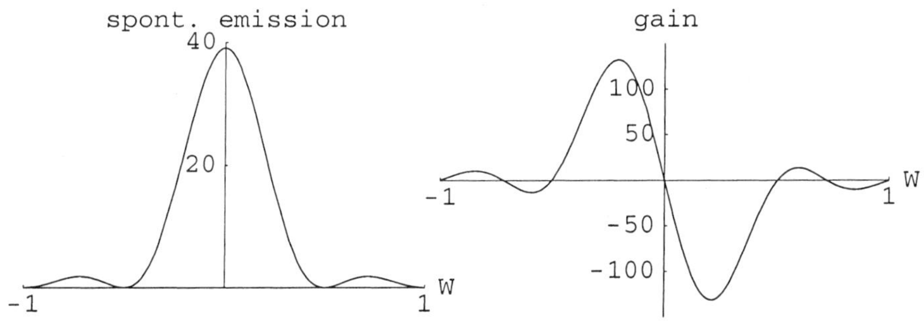

Figure 1.7: Left: Spectrum of spontaneous undulator emission over the detuning Ω from the photon energy of the nominal undulator emission. Right: FEL gain curve is the derivative of the spontaneous emission curve.

For lack of high-reflectance, normal-incidence x-ray mirrors, it is all but impossible to construct a resonator to enclose the gain medium. The lasing action must therefore occur at extremely high gain in a single pass of the electrons through the undulator. This process is commonly called self-amplified spontaneous emission (SASE) and can be described within classical electromagnetism. As a bunch of electrons of several ten GeV particle energy enters an undulator, the statistical distribution of charges within it leads to spontaneous emission of x-rays at the undulator lines. This spontaneous emission occurs preferentially into modes of the electromagnetic field which are well matched to the particular statistical charge distribution within the bunch because of a collective radiation phenomenon, called superradiance: the electric field that the electrons exert work on in the emission process [8] is generated not only by themselves (as with single emitting electrons), but also by all the other ones. Depending on the phase of an electron relative to the others, it will radiate more or less than a single electron would. Superradiance is usually discussed in terms of multiplet states of many-atom systems and their parities [28], but the concept is applicable here, too. The spontaneously generated radiation creates a ponderomotive potential, which leads to microbunching and gain, as described above. Once the amplified radiated field is stronger than that due to the initial statistical charge distribution, the FEL operates as a simple linear amplifier. The crossover from spontaneous superradiant emission to amplified emission separates the so-called lethargy and gain regions (see fig. 1.8). The surviving

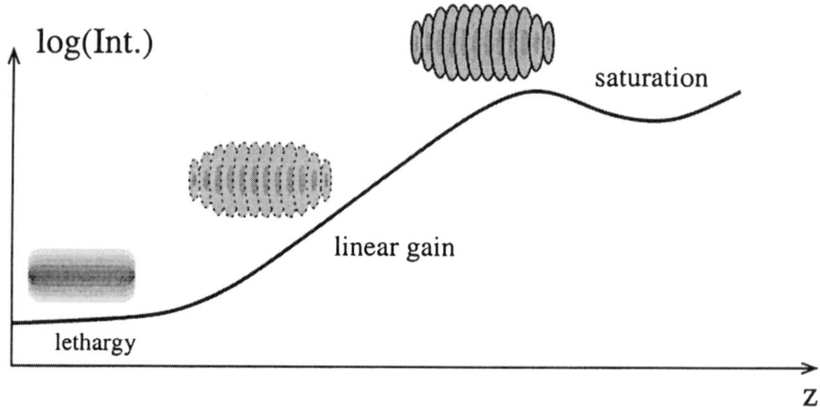

Figure 1.8: Intensity growth in a SASE FEL along the length z of the undulator, showing lethargy, exponential growth in the linear amplification region and saturation. The increasing microbunching in the electron bunch is indicated. Figure created after the model of fig. 2.2.1 in [29].

modes experience initially a linear gain, i.e., after passage of a certain length of undulator, the amplitude of a given electromagnetic wave is multiplied by a factor that depends on frequency and phase of the wave but not on the position along the undulator. After the microbunching has become as complete as possible with a given initial energy distribution in the bunch, the electrons begin to overshoot in the ponderomotive potential, causing them to extract energy out of the electromagnetic field. This extraction is not complete but rather leads to an oscillation of the intensity around a saturation value, as the electrons continue downward in the undulator.

Model calculations show that the production of hard x-rays of around 10 keV photon energy in a SASE FEL process requires an extremely high current of electrons (ca. 3000 A) in an electron bunch of 10 to 50 GeV particle energy and with emittance characteristics that are at the limit of present day technology. It seems that these reqirements can be met only with a laser-driven photocatode electron gun feeding a linear accelerator. However, even a photocathode electron gun cannot directly provide the high current necessary for x-ray free-electron lasing. The electron bunches therefore need to be compressed in length, so that the instantaneous current at a given bunch charge (of the order of 1 nC) is increased to the required level. Bunch compression works by inducing an energy "chirp" in the bunch through proper phasing of the accelerator radiofrequency, and then sending the bunches through a dispersive section where more energetic particles from the bunch tail go a shorter path and thus catch up with those towards the bunch head. This corresponds to a rotation in phase space.

7.2 Hard X-Ray FEL Facilities in Planning

Currently (as of 2002), there are two projects for hard x-ray FELs in advanced stages of development: the LCLS [16] in Stanford, CA, USA and TESLA [17] in Hamburg, Germany. The linear accelerator of the LCLS is the Stanford Linear Accelerator (SLAC) was built in 1962 for the purposes of particle physics. About one-third of its length will be used to bring electrons from a laser-driven photocathode electron gun to an energy of up to 15 GeV, whereupon they enter a 121-m-long undulator for the SASE process. Five experimental stations are planned to be arrayed along the x-ray beam. The SLAC linear accelerator consists of microwave resonators made of copper, operating at room temperature with a repetition rate of 120/s.

TESLA is a completely new project. The linear accelerator will be made of superconducting niobium resonators. Because absorption of the microwave power in the resonator walls is very small (nominally zero), it is possible to keep the full field strength in them for an extended time. Rather than working with single electron bunches, it will accelerate electron bunch trains of 1 ms duration and 200 ns bunch-to-bunch spacing at a repetition rate of 10/s. The bunches will be distributed among a number (five, at the present stage of planning) of beam lines, each with one or two undulators, and serving tens of experimental stations in a dedicated user facility.

Both facilities will be using several stages of bunch compression to achieve the high beam current in bunches of less than 100 μm in length, corresponding to durations of 100 fs to 200 fs. This makes the SASE XFELs very attractive for the study of ultrafast phenomena (see ch. 6, fig. 6.1). It is expected that the x-rays will be fully transversely coherent and have a longitudinal coherence of the order of 0.1 fs, corresponding to the gain bandwidth of about 10^{-3}, relative. Because the coherence time will be very much shorter than the bunch length, each x-ray pulse will be composed of a unique pattern of many short intensity spikes.

7.3 The Quantum FEL

When the principle of the free-electron laser was first worked out, the theory was developed within the framework of quantum physics (see also [30] for a discussion of quantum effects in "ordinary" synchrotron radiation). A "usual" laser with an atomic or molecular gain medium has two energy levels associated with the actual lasing transition. At least one more level is required to separate the pumping and lasing processes. A generic example is the three-level laser shown in fig. 1.9. The pump process takes atoms in the gain medium from the

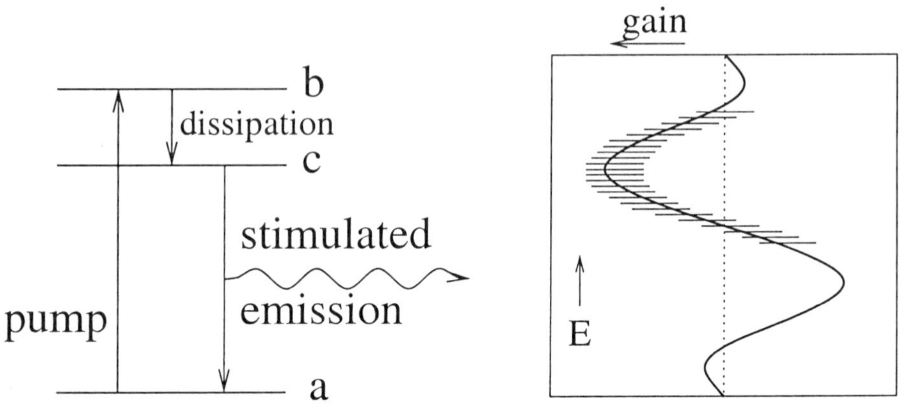

Figure 1.9: Levels in an atomic/molecular laser and an FEL. Left: A generic atomic/molecular three-level laser (the ruby laser works like this). Right: Many recoil energy levels within the gain curve in a free-electron laser (see text).

lower state (a) to the upper state (b). Spontaneous emission and/or nonradiative transfer causes relaxation from (b) to the upper laser level (c), whence stimulated emission takes the atoms back to (a), and so forth. In thermodynamic language, a laser is a machine that transforms a flow of pump energy and entropy (transition (a) to (b) in the example) into a low-entropy flow of energy in the laser beam ((c) to (a)) and a high-entropy flow of energy out of a dissipative process ((b) to (c)).

In a free-electron laser, the role of atomic or molecular energy levels is taken by states of the electron bunch with particular kinetic energies,[4] separated by the photon recoil energy. Under real-world conditions, there is a multitude of such levels and they overlap strongly because of the energy spread of the electrons within the bunch and the undulator linewidth (given by the number of undulator poles within one gain length). A quantum free-electron laser must have a natural emission linewidth that is less than the recoil energy [31, 32], requiring (1) a low electron energy spread, (2) a large number of undulator poles, probably achievable only with an optical undulator, i.e., a high-power visible-light laser beam, and (3) a low gain to keep the gain length long. None of these conditions are fulfilled in a linear-accelerator-driven SASE XFEL and classical theory (as outlined above) is fully adequate.

[4]More precisely: states of subvolumes of the bunch which are coherently microbunched over the slippage length between electrons and electromagnetic waves within a gain length.

7.4 Lasing without Inversion

There have been proposals [27, 33, 34, 35, 36, 37] to apply the idea of lasing without inversion (LWI) to free-electron lasers. These ideas apply to low-gain FELs, as will become evident below, and are therfore not directly transferable to single-pass x-ray FELs, which rely on high-gain operation. It is nevertheless possible that they could be relevant to the lethargy region of a SASE FEL, and for that reason they shall be mentioned here.

Whether lasing necessarily requires inversion or not depends on the definition of the term, that is, which accounting rules are to be used for determining level occupations. In LWI, the inversion does not simply appear on a projection of state occupation onto the energy scale. It is, however, present in a more subtle way with respect to the occupation of states that can participate in the lasing transition versus absorption processes. There are several variants of LWI, but the most common ones involve three energy levels,[5] two of which (a) and (b) are much closer to each other than they are with respect to the third one (c). A popular configuration is the so-called lambda with (a) and (b) at the feet of the letter Λ and (c) at the tip. A small number of atoms in an ensemble is excited to the state (c) and the majority is in the lower states (a) and (b). An auxiliary laser (or longer wavelength) field drives states (a) and (b) into a coherent superposition, which gives rise to a periodically (at a frequency corresponding to the spacing between (a) and (b)) appearing destructive interference of transitions from (a/b) to (c). Another superposition of (a) and (b), which is in quadrature to the one forced by the auxiliary field, is wide open to be occupied by transitions from (c) and permits thus stimulated emission. The system is therefore not inverted if one simply counts the number of atoms in states (a) and (b) versus the ones in (c). It *is*, however, inverted with respect to atoms in state (c) versus atoms in the superposition of (a) and (b) that is in quadrature to the auxiliary field.

The proposals for LWI in a free-electron laser work on an analogy of that scheme by inducing an energy-dependent phase shift on electrons in a bunch as they traverse a magnetic chicane between two undulator sections. The phase shift is designed to lead to destructive interference for absorption by electrons of lower energy (i.e., being on the absorptive side of the FEL gain curve for a particular photon energy) in the two undulator sections and to lead to constructive interference for electrons of higher energy. This requires of course roughly equal degrees of microbunching in the two undulator sections and is therefore applicable only to low-gain free-electron lasers, which the SASE FEL is not. However, LWI might be useful to reduce the lethargy length in SASE FELs, i.e.,for the part of the undulator where strong gain has not set in.

---

[5] not to be confused with the three levels shown in fig. 1.9

8. Comparison of Sources

Nonlinear optics and the science of ultrafast phenomena require high intensities and in most cases also a high degree of coherence. When comparing x-ray sources for their usefulness in these fields, the figure of merit is therefore the peak brilliance, defined as the number of photons per second, per milliradian-square, per milimeter-square and per 0.1% energy bandwidth *within the duration of an x-ray flash from one bunch.* For many other applications, whose efficiency scales linearly with the number of photons or number of coherent photons, the figure of merit would be the average brilliance, i.e., the peak brilliance multiplied with the duty cycle (bunch length divided by bunch-to-bunch spacing). Figure 1.10 shows a comparison of the peak brilliance for existing

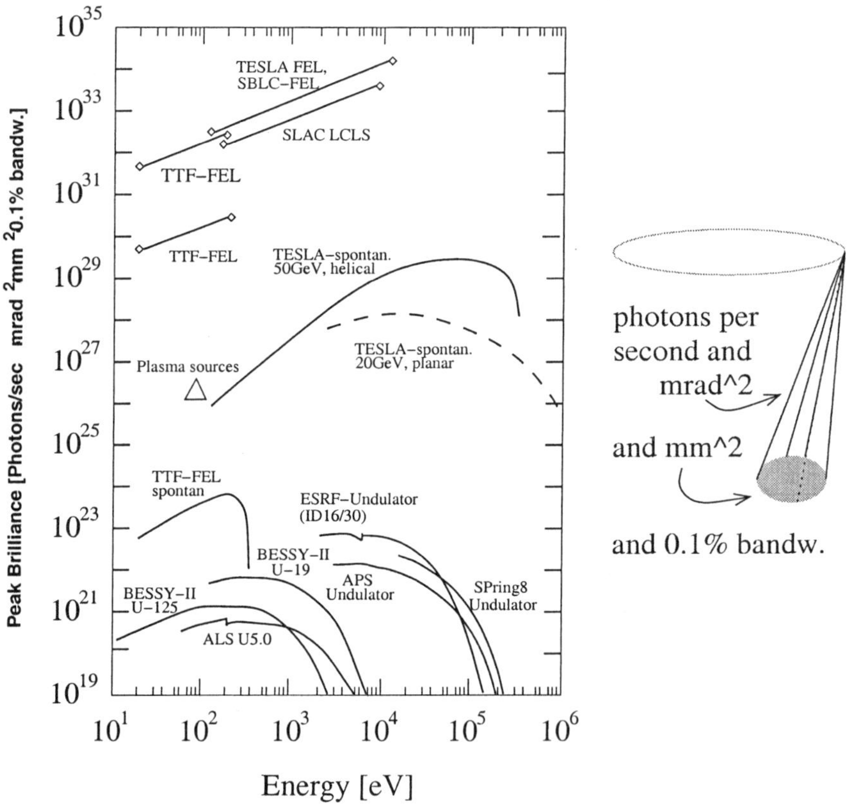

Figure 1.10: Left: brilliance comparison chart, taken from [29] and redrawn for graphic quality. Right: explanation of the term "brilliance" (the area indicated by the ellipse is to be taken at the undulator exit).

x-ray sources (all in the lower part of the graph) versus the planned linear-accelerator-driven sources. A distinction is drawn between the TESLA FEL and TESLA spontaneous radiation, because the latter, due to statistical emission of radiation (fig. 1.4, upper part), is orders of magnitude below the brilliance of collective FEL emission (fig. 1.4, lower part). However, at the current level of technology, the SASE effect is expected to work only with photon energies up to 10 to 15 keV. At higher photon energies, only spontaneous undulator radiation is available, but, for several reasons, that still has a much higher brilliance than the output of the current 3rd-generation undulators. Linear accelerators can provide electron bunches with much less emittance than those in a storage ring (see this chapter, sec. 4), and the FEL driver linacs will make full use of that advantage. Linear accelerators can also output much shorter electron bunches than those that circulate in a storage ring (again, sec. 4). This feature favors the peak, but not the time-averaged brilliance. Finally, both, TESLA and SLAC (which will drive the LCLS) have rather generous particle energy margins. If the TESLA FEL will be joined with the TESLA collider (which will bring electrons and positrons up to 250 GeV, each), then it is no big problem to send up to 50 GeV electrons into an undulator (for comparison: ESRF operates at 6 GeV, the APS at 7 GeV, SPring8 at 8 GeV, and Petra at DESY at 12 GeV). Missing in the chart in fig. 1.10 are the new kids on the block, the ERLs (see sec. 6). They will have a much lower transverse and pulse duration than the current storage rings, and their peak brilliance will be a few orders of magnitude below the spontaneous emission from TESLA, because the individual bunch charge is less. Also, there is no plan to operate one at 20 to 50 GeV, and therefore they will provide very few high-energy photons. The time-average brilliance will be higher than the spontaneous emission from TESLA because they will run at very high repetition rates.

Chapter 2

NONLINEAR OPTICS OF FREE ELECTRONS

Tell me, O Octopus, I begs,
Is those things arms, or is they legs?
I marvel at thee, Octopus;
If I were thou, I'd call me us.
– Ogden Nash

B.W. Adams
Advanced Photon Source, Argonne National Laboratory, Argonne, IL 60439, USA
adams@mail.desy.de

Abstract The free-electron model is well suited to the study of linear and nonlinear x-ray scattering. Here, the relativistic motion of free electrons in electromagnetic fields is described within the Hamilton-Jacobi formalism and nonlinear scattering of electromagnetic waves is interpreted as the re-radiation from these charges in anharmonic motion.

Preface to Chapters 2 to 5

The following chapters 2, 3, 4 and 5 are meant as a unit, with later chapters drawing upon results of earlier ones. The first two deal with theory, with this one preparing the ground for the dynamical diffraction theory in ch. 3. Chapter 4 is a collection of proposals for novel x-ray optical elements, based on the some of the theory developed in ch. 3, and ch. 5 shows experimental results whose interpretation requires results from this chapter and ch. 3.

The dynamical theory of x-ray diffraction is usually developed on the basis of macroscopic electromagnetism, using the electric displacement **D** and the linear electric susceptibility. Both extensions of the theory in ch. 3 would encounter problems with that macroscopic approach (explained in detail in ch. 3), which is the reason for the microscopic formulation, and thus this chapter 2.

Although spontanous parametric down conversion is a quantum optical effect, it is described here in a semiclassical manner, with a calculation of the optical nonlinearity based upon the Hamilton-Jacobi formalism and introduction of vacuum fluctuations as a driver of the effect. This approach is sufficient at the present level of experimental evidence (see ch. 5).

1. Introduction

X-rays, like all electromagnetic waves, are scattered as they induce charges to an oscillatory motion which leads to re-emission (but see remarks on p. 24). Generally, this motion and the radiated field are related to the incident fields in some nonlinear way, which may be due to anharmonic potentials or the intrinsically nonlinear coupling of charged particles to the electromagnetic field. A relatively simple model case is that of free electrons(bound electrons and resonances are treated in ch. 11). It is particularly applicable in the context of this book—matter in strong laser fields and x-rays diffracting in a sample of light elements: A strong laser field can strip the outer electrons off an atom, and, once they are freed, they contribute strongly to the scattering of light. X-rays, on the other hand, come at photon energies far above the binding energies of all electrons in light elements and the majority of electrons in heavier elements. Except for a photon energy right at an x-ray absorption edge, the free-electron description of the interaction of x-rays with matter is therefore well justified. Whereas a consideration of the depth of the binding potential versus the photon energy provides a basis for the free electron picture, it is the anharmonicity of the potential that needs to be considered for nonlinear optical effects. This is discussed below.

What is the origin of the nonlinear optical response of electrons? For a qualitative understanding, we might start with the Lorentz equation for the acceleration of a particle of charge q and mass m in an electromagnetic field. Expressed in terms of the scalar and vector potentials Φ and $\mathbf{A}$, it reads [1].(6.29), [1].(6.31), [1].(6.113):

$$\dot{\mathbf{v}} = \frac{q}{mc}\left(-c\nabla\Phi - \dot{\mathbf{A}} + \mathbf{v}\times(\nabla\times\mathbf{A})\right), \tag{2.1}$$

In the transverse gauge, Φ is due to the static charge distribution and $\mathbf{A}$ represents an incident electromagnetic wave, in the absence of which the particle rests at a position $\mathbf{r}$ in a potential minimum, i.e., $\mathbf{v} = \mathbf{0}$ and $(\nabla\Phi)(\mathbf{r}) = \mathbf{0}$. For small amplitudes $\mathbf{A}$, the particle performs a harmonic oscillation $\mathbf{x}(\mathbf{r}, t)$ about $\mathbf{r}$. As the wave and the resulting motion become stronger, the particle's motion becomes anharmonic due to both, anharmonicities in Φ and the Lorentz force term $(q/mc)\mathbf{v}\times(\nabla\times\mathbf{A})$. The latter induces a longitudinal component in the motion, and, in the case of a single plane wave, a figure-eight type trajectory results, as shown in fig. 2.1. In a weak electromagnetic wave (say, in the output of a laser pointer), the term $\mathbf{v}/c$ is very small and the figure-eight looks more like a straight line. The opposite is true for an intense few-cycle laser pulse. The particle is then accelerated first in the transverse direction along $\mathbf{E}$, but as it acquires a relativistic energy within a fraction of a single optical cycle, it gets deflected in the direction along the wave vector by the magnetic field component in the wave. This longitudinal motion in the magnetic field

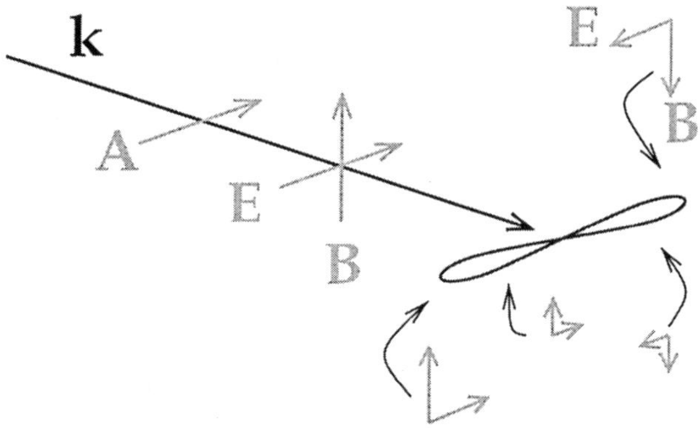

Figure 2.1: The figure-eight motion of a free charge in a plane electromagnetic wave. In the figure to the left is shown the vector potential $\mathbf{A}$ and at the right the corresponding fields $\mathbf{E}$ and $\mathbf{B}$. The fields $\mathbf{B}$ and $\mathbf{E}$ are strongest at the turning points of the motion, and $\mathbf{A}$ is strongest in the center. The longitudinal motion, due to the $\mathbf{A}^2$ term in eq. (2.11) (or the $\mathbf{v} \times \mathbf{B}$ term in the Lorentz equation), has twice the frequency of the transverse motion: the charge moves forward on both turning points and backward on the long stretches.

gives rise to another deflection into the transverse direction and so on. Actually, the distinction between electric and magnetic field components should only be taken as an aid to intuition, because it breaks down in the context of rapidly changing relativistic particle velocities. Consequently, the sections to follow will be based upon the relativistic Hamilton-Jacobi formalism.

An estimate of the nonlinearity and the relative importance of an anharmonic potential of outer-shell electrons versus the Lorentz force can be found as follows: For a small amplitude $\mathbf{A}$, oscillating at a frequency ω, and neglecting for now the scalar potential, eq. (2.1) can be integrated to give $\mathbf{v} = -(q/mc)\mathbf{A}$ and $\mathbf{x} = -i(q/mc\omega)\mathbf{A}$. The expansion $\nabla\Phi(\mathbf{x}) \approx \mathbf{x}\cdot\nabla\nabla\Phi + (\mathbf{x}, \mathbf{x}) : \nabla\nabla\nabla\Phi + \ldots$, together with the Poisson equation $\nabla^2\Phi = 4\pi\rho$ for the charge density ρ, gives then for an electron of charge $q = -e$:

$$|\nabla\Phi| \approx \underbrace{\frac{4\pi\rho r_e}{ek}|\mathbf{A}|}_{|\nabla\Phi|_{harm.}} + \underbrace{\frac{4\pi r_e^2|\nabla\rho|}{e^2k^2}|\mathbf{A}|^2}_{|\nabla\Phi|_{anh.}} + \ldots, \tag{2.2}$$

where $r_e = e^2/mc^2 = 2.82 \cdot 10^{-13}$ cm is the classical electron radius and $k = \omega/c$. Factors of i and -1 are disregarded in this estimate. The use of the Poisson equation for this order-of-magnitude estimate implies the assumption that all tensor components of $\nabla\nabla\Phi$ are of about the same magnitude.

Using $|\nabla \times \mathbf{A}| = k|\mathbf{A}|$, the Lorentz force term becomes:

$$|\mathbf{v} \times (\nabla \times \mathbf{A})| \approx \frac{r_e kc}{e}|\mathbf{A}|^2 \tag{2.3}$$

and the ratio of the Lorentz force term to the anharmonic potential is

$$\frac{|\mathbf{v} \times (\nabla \times \mathbf{A})|}{c|\nabla\Phi|_{anh.}} = \frac{ek^3}{4\pi r_e|\nabla\rho|} \tag{2.4}$$

With a typical valence electron density ρ/e of 1 per Å^{-3} (it is 1.07 for diamond, 0.7 for Si, 1.4 for Ge, ...), which, for outer-shell electrons, varies substantially within roughly 1Å (the Bohr radius is 0.5 Å), the denominator becomes $e * 3.5 \cdot 10^{26}$ m^{-3}. For a visible-light wavelength of, say, 633 nm, the numerator takes on the value of $e * k^3 = e * 10^{21}$ m^{-3}, which is less than the denominator, but the opposite is true for an x-ray wavelength of 1Å, which makes the numerator $e*k^3 = e*2.5 \cdot 10^{32}$ m^{-3} be much larger than the denominator. In conclusion, nonlinear optics of visible light in condensed matter is dominated by anharmonic potentials in the chemical bonds, but nonlinear optics of x-rays in matter made of light elements is dominated by the Lorentz force. This justifies the use of the free-electron model for the study of nonlinear optics of x-rays, at least for light elements where there are no tightly bound electrons. Nonlinear x-ray optics with tightly bound electrons, including x-ray resonances, is treated in ch. 11. At very high laser intensities, anharmonicities within chemical bonds become irrelevant, too, and optical nonlinearities are due to collective plasma effects and the Lorentz-force-type nonlinear response of free electrons.

Two more nonlinear effects are not explicitly contained in eq. (2.1): as the electron oscillates, it samples the fields at $\mathbf{r}+\mathbf{x}$ and the current it generates must also be attributed to that point. Within the second order, these two effects appear only in the contribution of $(q/mc)\dot{\mathbf{A}}$ to $\dot{\mathbf{v}}$ and cancel each other out (see also ch. 3, sec. 4.2): One contribution, $\mathbf{J}^{(2a)}$, to the second-order current density at point $\mathbf{r}$ comes from the first-order velocity at $\mathbf{r}+\mathbf{x}$ (where the electron actually is), obtained by integration of the first-order acceleration $\dot{\mathbf{v}}^{(1)} = (q/mc)\dot{\mathbf{A}}$ to $\mathbf{v}^{(1)} = (q/mc)\mathbf{A}$ with $\mathbf{A}(\mathbf{r}+\mathbf{x})$ inserted. Another contribution, $\mathbf{J}^{(2b)}$, is obtained by back-tracking the first order current density $-e\eta(\mathbf{r})\mathbf{v}^{(1)}$ from $\mathbf{r}+\mathbf{x}$ (where the current is generated) to $\mathbf{r}$. Taylor expansions of $\mathbf{J}^{(2a)}$ and $\mathbf{J}^{(2b)}$ yield opposite terms of equal magnitude. Beyond the second order, there is a contribution from electron sampling the magnetic field at $\mathbf{r}+\mathbf{x}$, which remains without opposing terms. Finally, there is a nonlinearity due to the spatial inhomogeneity of the charge density (see p. 78)

Nonlinear optical effects with x-rays are much weaker than with visible light, because, for a given electric field strength, the more rapid reversal of the electron's motion within an optical half-cycle suppresses both, the oscillation amplitude (which enters the anharmonic potential) and the oscillation velocity (which enters the Lorentz force term). However, whereas the terms based upon Φ go like ω^{-4} (in eq. (2.2), $|\nabla\Phi|_{anh.} \sim |\mathbf{A}|^2/k^2 \sim |\mathbf{E}|^2\omega^{-4}$), the Lorentz force goes only like ω^{-1} (in eq. (2.3), $|\mathbf{v} \times (\nabla \times \mathbf{A})| \sim k|\mathbf{A}|^2 \sim |\mathbf{E}|^2\omega^{-1}$) and therefore dominates at x-ray frequencies.

An electron in an electromagnetic wave reaches relativistic velocities [38] if the dimensionless vector potential $\mathcal{A} = q\mathbf{A}/mc^2 > 1$. At a photon energy of 1 eV, $|\mathcal{A}| = 1$ corresponds to a power level of $9 \cdot 10^{17}$ W cm^{-2}, which is well within reach of current high-power lasers [39, 40, 41]. For x-ray photons of 10^4 eV, the power density must be 10^8 times as high, and that is beyond any third-generation x-ray source and also beyond x-ray FELs with current focusing capabilities (see ch. 1, sec. 8). There is one exception, though, namely spontaneous parametric down conversion. It is (barely) measurable because it is driven by vacuum fluctuations of the electromagnetic field whose virtual power density is exceedingly high at x-ray frequencies (see ch. 5). Relativistic free-electron effects can also be expected in the combination of a strong laser beam and an x-ray.

Due to its different character (see above), nonlinear optics of x-rays cannot be described in the same way as that of visible light, which is commonly based upon a phenomenological description with a series expansion of the electrical polarization [42].(1.2), [43].(1-2):

$$\mathbf{P} = \chi^{(1)} \cdot \mathbf{E} + \chi^{(2)} : \mathbf{EE} + \chi^{(3)} \vdots\, \mathbf{EEE} + \ldots, \tag{2.5}$$

where the n-th order susceptibility, $\chi^{(n)}$ is a tensor of rank $n+1$ and generally contains any anisotropies of the material. All even-order susceptibilities $(\chi^{(2)}, \chi^{(4)}, \ldots)$ vanish if the medium has an inversion symmetry, which is the case for inversion symmetric crystals and for amorphous media. The common nonlinear optical crystals (LBO, KTP, etc.) have noncentrosymmetric unit cells.

In cases where free electrons do play a role, one considers usually only a single plane wave (packet), and the longitudinal motion of the electrons has no direct effect on the radiated fields [43].[1-2]. It does, however, appear indirectly because a longitudinal velocity of an electron gives rise to a Lorentz force in the transverse direction. The motion resulting from that effect is a third-order effect and is polarized like the incident electric field. It is therefore possible to expand the nonlinear optical response of free electrons as in eq. (2.5), with only the odd-order susceptibilities $\chi^{(1)}, \chi^{(3)}, \ldots$ present.

With x-rays propagating in a crystal, diffraction plays an important role, and often more than a single plane wave is present. A longitudinal motion of an electron relative to one of them generally has a transverse component relative to at least some of the others (see fig. 2.2). Therefore, a $\chi^{(2)}$-like contribution appears, which cannot be expressed in terms of only the electrical field (because the direction of $\mathbf{B}$ is not determined by that of $\mathbf{E}$ alone). One might think of augmenting eq. (2.5) with magnetic field terms, but a better way is to base the formalism on the vector potential and induced currents, instead of the electrical field and induced polarization. This is pursued in the following sections and in ch. 3, sec. 4.

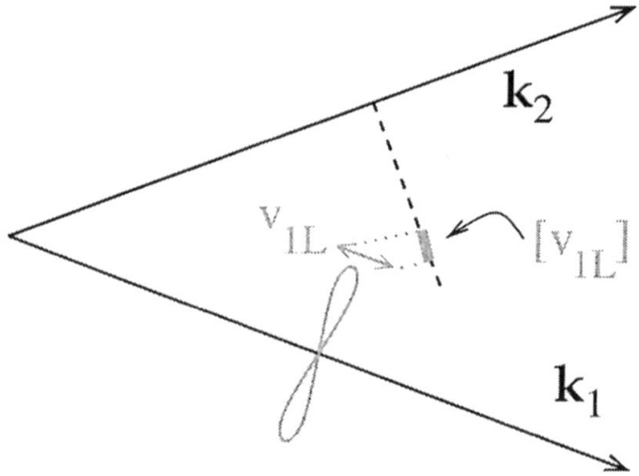

Figure 2.2: Longitudinal and transverse current: The longitudinal part $\mathbf{v}_{1L}$ of the figure-eight motion in fig. 2.1 does not contribute to the transverse current if only a single wave (plane or curved), here with wave vector $\mathbf{k}_1$ is present. There is however a nonvanishing projection $[\mathbf{v}_{1L}]_2$ of $\mathbf{v}_{1L}$ onto a direction transverse to another wave $\mathbf{k}_2$.

2. Relativistic Electrons in Electromagnetic Waves

The use of classical electrodynamics and the model of free electrons are quite appropriate for the description of coherent x-ray scattering from a large number of electrons. Although the energy of a typical x-ray photon is a sizable part of the electron rest mass, it is still very much less than the combined rest mass of all electrons within a typical coherence volume of x-rays from a synchrotron radiation source (of the order of 10^{14} electrons in a typical x-ray coherence volume of 1 μm (longitudinal), times $(10\ \mu\text{m})^2$ (transverse)). For small recoil energies of 10^{-3} eV and below, the relevant reaction mass is even larger (by a factor of 4000 to 5000) because the nuclei are coupled to the electrons on time scales reciprocal to the phonon energies.

Solutions for a single plane wave or wave packet have been known for a long time [44, 45, 46]. The quantum version of the plane-wave solution is the Gordon-Volkov solution [47] of the Schrödinger/Dirac equation. Analytic solutions are also possible for multiple plane waves with parallel propagation directions, for two crossing plane waves in a particular configuration [48, 49, 50], or for one running wave and one static field. An approximate solution for two crossing plane waves can be found by use of a gauge transformation, which decouples the effects of the waves in first order (sec. 2.4). In all cases, only a single electron is considered and interactions between electrons are neglected. Also not included in the considerations are self-interactions, i.e., the modification of the electromagnetic field by the electron itself.

Although the emphasis of this book is on nonlinear optics, the topic of electron acceleration shall also be touched upon briefly. The Lawson-Woodward theorem [51, 52] states that there can be no net acceleration of an electron in a plane wave packet, that is, after the passage of the wave packet, the electron is displaced in the longitudinal direction, but all acceleration has cancelled out, and the momentum is the same as before. A great deal of work has been devoted to bypassing this theorem, and one must therefore keep its limitations in mind. It applies to (1) classical physics of (2) a free charge in (3) an electromagnetic plane wave (4) packet in (5) an unbounded vacuum. If any of these conditions are not met, it is possible to accelerate electrons in electromagnetic waves. This is discussed in more detail in sec. 2.5, see also [30] for a discussion of quantum effects in synchrotron radiation. Here, only brief mention shall be made of counterpoints: (1) Compton scattering (not at all negligible with x-rays) and the quantum statistical nature of emission from an accelerated charge, (2) electrons bound to atoms, (3) the electric field in a focus does have a longitudinal component, (4) one can (at least theoretically) devise a rectified electromagnetic pulse (see sec. 2.5), and (5) dielectrics, RF resonators, etc., modify the electromagntic field, giving it a non-plane-wave character.

For the purposes of laser acceleration of electrons, a detailed knowledge of the electron's motion in the fields is not really necessary, and the lack of solutions for the three-dimensional fields required can be ameliorated by use of an approximation, using the ponderomotive potential in terms of the pulse envelopes (see sec. 2.5).

2.1 Single Plane Wave Packet

In order to introduce the concepts and the notation, we will first take a look at the well-known case of a relativistic spinless particle in an electromagnetic plane wave (packet). The vector potential of the wave packet is

$$A_\mu(\eta) = (A^0)_\mu P(\eta) e^{i\eta} + cc. \qquad \eta = k_\nu x^\nu, \tag{2.6}$$

where $(A^0)_\nu$ is a constant four-vector, which can be written as a constant real-valued four-vector times a phase factor $\exp(i\phi)$ which is independent of x_ν. $P(\eta)$ is the pulse envelope, the wave vector k_ν is a null-vector $k_\nu k^\nu = 0$ and the Lorentz condition $k_\nu A^\nu = 0$ is fulfilled. The metric $(1, -1, -1, -1)$, is used here. With this, the Hamilton-Jacobi equation for the particle is [38, 46, 53]:

$$(S_{,\nu} - qA_\nu)(S^{,\nu} - qA^\nu) = m^2. \tag{2.7}$$

The comma means a partial derivative, i.e., $S_{,\nu} = \partial S / \partial x^\nu$. A solution of eq. (2.7) can be written as

$$S(x_\nu) = p^\nu x_\nu + F(\eta), \tag{2.8}$$

where p^ν is the particle's 4-momentum *before* interaction with the field,[1] Upon insertion into (2.7), we get

$$(p_\mu - qA_\mu)(p^\mu - qA^\mu) - m^2 + 2F'k_\mu(p^\mu - qA^\mu) + (F')^2 k_\mu k^\mu = 0. \quad (2.9)$$

With $k_\mu A^\mu = 0$ and $k_\mu k^\mu = 0$, a first-order linear ordinary differential equation for $F(\eta)$ results. Using also $p_\nu p^\nu = m^2$, the solution is

$$F(\eta) = q\int^\eta \frac{p_\mu A^\mu}{k_\alpha p^\alpha} d\eta' - q^2 \int^\eta \frac{A_\mu A^\mu}{2k_\alpha p^\alpha} d\eta'. \quad (2.10)$$

The particle's position X_μ can be determined by equating the position $(X_0)_\nu$ without fields to $(\partial S/\partial p^\nu)$

$$X_\nu = (X_0)_\nu - \frac{q}{k_\alpha p^\alpha}\int^\eta A_\nu d\eta' - \frac{q^2 k_\nu}{(k_\alpha p^\alpha)^2}\int^\eta \frac{p^\beta A_\beta}{q} - \frac{A^\beta A_\beta}{2} d\eta'. \quad (2.11)$$

This is an implicit expression because X_ν is contained in η in the integral bounds. Therefore, the particle's motion is anharmonic beyond the 2nd order.

The canonical momentum is found by $(P_c)_\nu = S_{,\nu}$. In the Lorentz gauge, the kinematical momentum [1].(12.11) is $(P_k)_\nu = (P_c)_\nu - qA_\nu$, and thus

$$(P_k)_\nu = -qA_\nu + p_\nu + qk_\nu \frac{p^\beta A_\beta}{k^\alpha p_\alpha} - q^2 k_\nu \frac{A^\beta A_\beta}{2k^\alpha p_\alpha}. \quad (2.12)$$

Equation (2.12) is not implicit in the way of eq. (2.11), but in another way it is: when evaluated at some point r_μ in space-time (which enters through η as argument to A_ν), it gives the momentum of the particle in r_μ if it happens to be there. That, however, is determined by the initial particle distribution and the trajectory, as given in eq. (2.11).

In (1+3)-notation, $A_\mu = (\Phi, \mathbf{A})$, $k_\mu = (\omega/c, \mathbf{k})$ and $p_\mu = (m^{(0)}c, \mathbf{p}_0)$, where $m^{(0)}$ is the relativistic mass $\gamma^{(0)}m$ at the velocity that the particle had before interaction with the field. With these definitions, eq. (2.12) becomes:

$$\begin{aligned} mc &= -\frac{q}{c}\Phi + m^{(0)}c + \frac{q\omega}{c^2}\frac{\Phi m^{(0)}c - \mathbf{p}_0\cdot\mathbf{A}}{\omega m^{(0)} - \mathbf{k}\cdot\mathbf{p}} - \frac{q^2\omega}{2c^3}\frac{\Phi^2 - \mathbf{A}^2}{\omega m^{(0)} - \mathbf{k}\cdot\mathbf{p}} \\ \mathbf{p} &= -\frac{q}{c}\mathbf{A} + \mathbf{p}_0 + \frac{q}{c}\mathbf{k}\frac{\Phi m^{(0)}c - \mathbf{p}_0\cdot\mathbf{A}}{\omega m^{(0)} - \mathbf{k}\cdot\mathbf{p}} - \frac{q^2}{c^2}\mathbf{k}\frac{\Phi^2 - \mathbf{A}^2}{\omega m^{(0)} - \mathbf{k}\cdot\mathbf{p}}. \end{aligned} \quad (2.13)$$

[1] More generally, instead of p^ν, a 4-momentum of the form $p^\nu + \sigma k^\nu$ with any scalar σ could be inserted, and $F(\eta)$ would then be made to compensate for the term σk_μ, see [38].

The charged particle describes a figure-eight motion, centered on a point that moves with $\mathbf{v}_0$. This is shown in fig. 2.1 for $\mathbf{v}_0 = \mathbf{0}$ (i.e., $p_\mu = (mc, 0, 0, 0)$):

With the aid of eqs. (2.11) and (2.12), it is easy to see that the effect of a plane wave packet on a free electron is a net longitudinal displacement due to the A-square term in eq. (2.11). There is however no net acceleration because eq. (2.12) reduces to $(P_k)_\nu = p_\nu$ for times after the passage of the wave packet.

2.2 Multiple Parallel Plane Waves

The vector potential is now a sum of plane waves

$$A_\mu = \sum_j (A_j)_\mu = \sum_j (A_j^0)_\mu e^{i\eta_j}, \qquad \eta_j = (k_j)_\nu x^\nu, \tag{2.14}$$

all of which have parallel wave vectors, i.e., $(k_j)_\nu = \sigma_{j\ell}(k_\ell)_\nu$ for some scalar $\sigma_{j\ell}$ (no summation over j or ℓ). The sum in eq. (2.14) contains terms in complex conjugate pairs because the vector potential is real-valued: for each summation index i, there is exactly one index ℓ with $(A_\ell)_\mu = (A_j^*)_\mu$. The number of plane waves in the sum is unlimited, and an integral could be used instead of the sum to synthesize wave packets—as long as all wave vectors are parallel. Because of the scalar relation between the wave vectors, it is possible to express a solution of eq. (2.7) with a single parameter η in the form of eq. (2.8) and thus to obtain a linear ordinary differential equation for $F(\eta)$. The only difference to eq. (2.10) is that $p_\nu A^\nu$ and $A_\nu A^\nu$ now contain many interfering terms.

In the particular case of two parallel waves with crossed polarizations of the electric fields in a charge-free environment (apart from the particular particle whose motion is being considered), one may alternatively keep both parameters η_1 and η_2 and write a solution of eq. (2.7) as $S(x_\nu) = p^\nu x_\nu + F_1(\eta_1) + F_2(\eta_2)$. Because of the crossed polarizations, there are no terms in $A_\nu A^\nu$, containing $\exp(i(\eta_1 \pm \eta_2))$. One may therefore find separate solutions for F_1 and F_2, both of the form of eq. (2.10). In other words, for two parallel waves in crossed polarizations, the motion of the charged particle is simply a superposition of the independent motions in single plane waves.

2.3 Multiple Plane Waves, Nonrelativistic Approximation

With multiple non-parallel waves having phases η_j, nonlinear terms arise from products of partial derivatives of the type $(\partial F/\partial\eta_i)(\partial F/\eta_j)(k_i)_\nu(k_j)^\nu$, where $(k_i)_\nu(k_j)^\nu$ do not vanish due to the non-parallelity. However, no conceivable laboratory-based x-ray source can bring electrons to relativistic oscillatory velocites, as it is possible in the focus of a high-power laser. When dealing with x-rays only, the function F in eqs. (2.8) and (2.21) is therefore much smaller than $p_\nu x^\nu$ (which is dominated by $p_0 x^0$), and a nonrelativistic approximation can be employed by neglecting the nonlinear terms.

The vector potential is expressed in n pairs of complex conjugate plane waves, as in eq. (2.14) but now with the waves not necessarily being parallel. In keeping with the nonrelativistic approximation, the transverse gauge will be used here and in ch. 3, sec. 4, which builds upon the results derived here.

We can anticipate that $(p_\nu - qA_\nu)(p^\nu - qA^\nu)$ in eq. (2.7) will lead to terms which depend on η_j and on $\eta_j + \eta_\ell$. With the nonlinear terms neglected, a suitable approximation of S is

$$S = p_\nu x^\nu + \sum_j F_j(\eta_j) + \sum_{j\ell}{}' F_{j\ell}(\eta_j + \eta_\ell), \tag{2.15}$$

where the prime on the sum over j and ℓ means that terms with $\eta_j + \eta_\ell = 0$ are to be omitted from the summation and $F_{j\ell}$ is taken as a function of a single independent variable $\eta_j + \eta_\ell$. Upon insertion into eq. (2.7), using $p_\nu p^\nu = m^2$ and omission of the nonlinear terms, one gets

$$\begin{aligned} 0 = & -2q\sum_j p_\nu (A_j)^\nu + q^2 \sum_j (A_j^0)_\nu (A_j^{0*})^\nu + 2\sum_j F_j'(k_j)_\nu \,(p^\nu - qA^\nu) + \\ & q^2 \sum_{j\ell}{}'(A_j)_\nu (A_\ell)^\nu + 2\sum_{j\ell}{}'(F_{j\ell})'[(k_j)_\nu + (k_j)_\nu]\,(p^\nu - qA^\nu)\,, \end{aligned} \tag{2.16}$$

where $(A_j)_\nu = (A_j^0)_\nu \exp(i\eta_j)$ by definition in eq. (2.14). As above, the prime on $\sum_{j\ell}{}'(A_j)_\nu (A_\ell)^\nu$ signifies omission of the terms with $\eta_j + \eta_\ell = 0$ because these are contained within the sum $\sum_j (A_j^0)_\nu (A_j^{0*})^\nu$ in the first line of eq. (2.16).

The terms $(k_j)_\nu A^\nu$ can be made to vanish for all j only in special cases (see secs. 2.2 and 2.4). However, these terms may be neglected in a nonrelativistic approximation because then the fields are so weak that $q(k_j)_\nu A^\nu$ is much smaller than $(k_j)_\nu p^\nu$. The two would be of the same magnitude if the energy of the charged particle in the field would approach the rest mass. Doing so, one may split eq. (2.16) into separate ordinary differential equations for each F_j and $F_{j\ell}$ with solutions

$$\begin{aligned} F_j(\eta_j) &= \frac{q\, p_\nu (A_j^0)^\nu}{(k_j)_\nu p^\nu} \int^{\eta_j} e^{i\eta_j'}\, d\eta_j' - \frac{q^2}{2}\frac{(A_j^0)_\nu (A_j^{0*})^\nu}{(k_j)_\nu p^\nu} \int^{\eta_j} d\eta_j' \\ F_{j\ell}(\eta_j + \eta_\ell) &= -\frac{q^2}{2}\frac{(A_j^0)_\nu (A_\ell^0)^\nu}{((k_j)_\nu + (k_\ell)_\nu)p^\nu} \int^{\eta_j+\eta_\ell} e^{i(\eta_j'+\eta_\ell')}\, d(\eta_j'+\eta_\ell'). \end{aligned} \tag{2.17}$$

Position and momentum of the particle can be found as in eqs. (2.11) and (2.12). The latter will be needed later, in ch. 3, sec. 4, and is therefore spelled out here:

$$(P_k)_\nu = -qA_\nu + p_\nu + q\sum_j (k_j)_\nu \frac{p_\beta (A_j^0)^\beta e^{i\eta_j}}{(k_j)_\alpha p^\alpha} - \frac{q^2}{2}\sum_j (k_j)_\nu \frac{(A_j^0)_\beta (A_j^{0*})^\beta}{(k_j)_\alpha p^\alpha}$$
$$-\frac{q^2}{2}{\sum_{j\ell}}' \left[(k_j)_\nu + (k_\ell)_\nu\right] \frac{(A_j^0)_\beta (A_\ell^0)^\beta}{[(k_j)_\alpha + (k_\ell)_\alpha] p^\alpha} e^{i(\eta_j+\eta_\ell)}, \qquad (2.18)$$

where the first two terms represent a superposition of an oscillatory motion due to the electric field and the initial momentum, the third one tells the longitudinal motion due to the Lorentz force, the fourth term is a momentum in the fields that is coupled to the electron, and the fifth term gives the nonlinear coupling of the fields to the electron. In nonrelativistic notation, eq. (2.18) is:

$$\mathbf{p}_k = -\frac{q}{c}\mathbf{A} + \mathbf{p}_0 + \frac{q}{c}\sum_j \mathbf{k}_j \frac{\left[\mathbf{p}_0 \cdot \mathbf{A}_j^0 - mc\Phi_j^0\right] e^{i\eta_j}}{m\omega_j - \mathbf{p}_0 \cdot \mathbf{k}_j} \qquad (2.19)$$
$$-\frac{q^2}{2c^2}\sum_j \mathbf{k}_j \frac{|\mathbf{A}_j^0|^2}{m\omega_j - \mathbf{p}_0 \cdot \mathbf{k}_j} - \frac{q^2}{2c^2}{\sum_{j\ell}}' (\mathbf{k}_j + \mathbf{k}_\ell) \frac{(\mathbf{A}_j^0 \cdot \mathbf{A}_\ell^0 - \Phi_j^0 \Phi_\ell^0) e^{i(\eta_j+\eta_\ell)}}{m(\omega_j+\omega_\ell) - \mathbf{p}_0 \cdot (\mathbf{k}_j + \mathbf{k}_\ell)},$$

with $p_\nu = (mc, \mathbf{p}_0)$, $(k_j)_\nu = (\omega_j, \mathbf{k}_j)$ and $(A_j^0)_\nu = (\Phi_j^0, \mathbf{A}_j^0)$ and the superscript 0 on Φ_j^0 and $\mathbf{A}_j^0$ should not be taken as a summation index.

2.4 Relativistic Electrons in Two Plane Wave Packets

The vector potential is now the sum of the two wave packets

$$(A_j)_\mu(\eta_j) = (A_j^0)_\mu P_j(\eta_j) e^{i\eta_j} + cc., \quad \eta_j = (k_j)_\nu x^\nu, \quad (A_j)_\nu (k_j)^\nu = 0, \qquad (2.20)$$

with generally non-parallel wave vectors $(k_1)_\nu$ and $(k_2)_\nu$ ("parallel" meaning that a scalar σ exists, so that $(k_2)_\nu = \sigma(k_1)_\nu$). The special case of parallel wave vectors is discussed below.

A solution of eq. (2.7) is now sought with

$$S(\eta_1, \eta_2) = p_\nu x^\nu + F(\eta_1, \eta_2). \qquad (2.21)$$

Insertion of eq. (2.21) into (2.7) yields an equation similar to (2.9), except for the presence of partial derivatives of F, a nonlinearity due to $(k_1)_\nu (k_2)^\nu \neq 0$, and the fact that the coefficients of $\partial F/\partial\eta_j$ need not necessarily be constant, as in eq. (2.9) by virtue of the Lorentz condition $k_\mu A^\mu = 0$. However, the freedom within the Lorentz class of gauges can be exploited to make these coefficients constant by finding scalar functions $G_j, j = 1, 2$ that preserve the Lorentz condition $(k_j)_\nu (A_j)^\nu$, i.e., fulfill $(G_j)_{;\nu}^{;\nu}$ (see [1].(6.42)), and transform the fields

$(A_j)_\nu \rightarrow (A_j^{\supset})_\nu = (A_j)_\nu + (G_j)_{,\nu}$, so that $(k_1)_\nu (A_2^{\supset})^\nu = (k_2)_\nu (A_1^{\supset})^\nu = 0$. Although still nonlinear, the differential equation is now simplified to some degree which permits an iterative solution with a spectral representation of $F(\eta_1, \eta_2)$. In a geometrical interpretation, the use of the gauge freedom is analogous to finding the part $\mathbf{a}_\perp = \mathbf{a} - \mathbf{b}(\mathbf{a} \cdot \mathbf{b}/|\mathbf{b}|^2)$ of a vector $\mathbf{a}$ that is orthogonal to another vector $\mathbf{b}$.

The functions $G_j, j = 1, 2$ are:

$$G_j = -\frac{(k_{o(j)})_\alpha (A_j^0)^\alpha}{(k_1)_\beta (k_2)^\beta} \int_{-\infty}^{\eta_j} P_j(\eta_j') e^{i\eta_j'} + cc., \qquad \begin{array}{l} o(1) = 2 \\ o(2) = 1. \end{array} \tag{2.22}$$

With the new potential explicitly written down,

$$(A_j^{\supset})_\nu = \left((A_j^0)_\nu - (k_j)_\nu \frac{(k_{o(j)})_\alpha (A_j^0)^\alpha}{(k_1)_\beta (k_2)^\beta} \right) P_j(\eta_j) e^{i\eta_j} + cc., \tag{2.23}$$

it is evident that, in addition to preserving the Lorentz condition, the gauge also makes $(k_1)_\nu (A_2^{\supset})^\nu$ and $(k_2)_\nu (A_1^{\supset})^\nu$ vanish. Although $(k_1)_\nu (k_2)^\nu$ appears in the denominator, the case of parallel wave vectors need not be excluded because the Lorentz condition makes the numerator $(k_{o(j)})_\alpha (A_j^0)^\alpha = 0$, as well. Numerator and denominator tend to zero to the same order, as the two waves are made more and more parallel to each other (meaning $(k_2)^\beta = \sigma (k_1)^\beta \rightarrow 0$ without $(k_2)^0 \rightarrow 0$ or $(k_2)^0 \rightarrow 0$) and the gauge term (i.e., the fraction within the parentheses) in eq. (2.23) stays finite. Although unnecessary for parallel waves, application of the gauge transformation is possible and does introduce a change to the vector potentials within the gauge freedom. The above procedure cannot be applied to more than two non-parallel wave packets because that would exceed the available gauge freedom.

The contents of the parentheses in eq. (2.23) are now abbreviated $(A_j^{0\supset})_\mu$, so that the gauge-transformed potentials can be written as

$$(A_j^{\supset})_\mu (\eta_j) = (A_j^{0\supset})_\mu P_j(\eta_j) e^{i\eta_j} + cc., \qquad \begin{array}{r} (A_j^{\supset})_\nu (k_j)^\nu = 0 \\ (A_j^{\supset})_\nu (k_{o(j)})^\nu = 0, \end{array} \tag{2.24}$$

in the same way as eq. (2.20), but now with an additional transversality property.

Insertion of eq. (2.21) into eq. (2.7), and making use of $(k_j)_\nu (A_\ell^{\supset})^\nu$, $j, \ell = 1, 2$, and of $p_\nu p^\nu = m^2$ gives then

$$-2qp^\mu (A^{\supset})_\mu + q^2 (A^{\supset})^\mu (A^{\supset})_\mu + 2p^\mu \left(\frac{\partial F}{\partial \eta_1} (k_1)_\mu + \frac{\partial F}{\partial \eta_2} (k_2)_\mu \right) + \frac{\partial F}{\partial \eta_1} \frac{\partial F}{\partial \eta_1} (k_1)_\mu (k_2)^\mu = 0, \tag{2.25}$$

where $(A^{\supset})_\nu = (A_1^{\supset})_\nu + (A_2^{\supset})_\nu$.

The wave packets are now written in terms of a spectral representation

$$P_j(\eta_j)e^{i\eta_j} + cc. = \int_{-\infty}^{\infty} \mathcal{P}_j(u)e^{iu\eta_j}du. \tag{2.26}$$

For computational purposes, a discretization

$$P_1(\eta_j)e^{i\eta_j} + cc. \approx \sum_{u=-M}^{M-1} (\mathcal{P}_j)_u e^{i(u/U)\eta_j} \tag{2.27}$$

is adopted, and likewise for $j = 2$, with v, V and N, instead of u, U and M. The sums must of course consist of finite numbers $2M$ and $2N$ of terms, M/U and N/V being the cutoff frequencies.

Correspondingly, F is written as

$$F(\eta_1, \eta_2) = \sum_{u,v=0}^{\pm\infty} c_{uv}e^{i((u/U)\eta_1+(v/V)\eta_2)}. \tag{2.28}$$

Insertion of this series into eq. (2.25) gives a system of equations, which can be sorted into lines, according to the exponents $i((u/U)\eta_1 + (v/V)\eta_2)$.

The following abbreviations for $j = 1, 2$ simplify the notation:

$$\begin{aligned}
(\mathfrak{A}_j)_u &= q^2(A_j^0)^\nu(A_j^0)_\nu\sum_s(\mathcal{P}_j)_s(\mathcal{P}_j)_{u-s} - 2qp^\nu(A_j^{0\supset})_\nu(\mathcal{P}_j)_u \\
\mathfrak{C}_{u,v} &= 2q^2(A_1^{0\supset})^\nu(A_2^{0\supset})_\nu(\mathcal{P}_1)_u(\mathcal{P}_2)_v \\
\mathfrak{K}_j &= 2p^\nu(k_j)_\nu \qquad\qquad \mathfrak{I} = (k_1)^\nu(k_2)_\nu
\end{aligned} \tag{2.29}$$

With the exponential factored out, the line in the system of equations, belonging to $\exp(i((u/U)\eta_1 + (v/V)\eta_2)$ is:

$$\begin{aligned}
&\delta_{v,0}(\mathfrak{A}_1)_u + \delta_{u,0}(\mathfrak{A}_2)_v + \mathfrak{C}_{u,v} + \\
&i(\frac{u}{U}\mathfrak{K}_1 + \frac{v}{V}\mathfrak{K}_2)c_{u,v} - \mathfrak{I}\sum_{m,n}\frac{m}{U}\frac{n}{V}c_{(m,v-n)}c_{(u-m,n)} = 0.
\end{aligned} \tag{2.30}$$

The set of equations (2.30) can now be solved iteratively. To obtain the coefficients $c_{u,v}^{(\ell)}$ in generation ℓ, the $c_{u,v}^{(\ell-1)}$ of the preceding generation $\ell - 1$ are inserted into the double sum, whereupon each line (for indices u, v) in eq. (2.30) can be solved for $c^{(\ell)}_{(u,v)}$. The iteration is started by setting $c^{(0)}_{u,v}$ all equal to zero. Direct calculation of the double sum over m, n for each pair u, v is a formidable task. However, using the fact that the double sum is a convolution, a much more efficient computation is possible in the Fourier domain by use of the Fast Fourier Transform (FFT). Having performed the iteration, one may calculate position and canonical momentum by taking the appropriate gradients of S (see [38]).

2.4.1 Discussion. Because the electromagnetic gauge has one degree of freedom, the above procedure is applicable only to the case of two plane wave packets. There are, however, two potentially very interesting applications to that model: laser acceleration of electrons by two intersecting laser beams and laser diagnostics of a free-electron laser. The latter case is well matched to the model, with relativistic electrons subjected to an intense electromagnetic plane wave from behind (the FEL radiation) and another intense electromagnetic wave from the side (the diagnostic laser).

When comparing the iterative solution in terms of a spectral expansion to an explicit calculation of the electron's trajectory with the Lorentz equation, the two methods can be seen as complementary. The latter is useful if the initial position and momentum are precisely known, and one wants to calculate the trajectory. The spectral expansion gives a more overall picture for undetermined initial positions and immediately yields information on the re-radiated electromagnetic waves, particularly for the harmonics. It should however be noted that the back action of the radiated fields on the electron is not taken into account, and that can be rather significant at relativistic electron velocities.

2.5 Laser Acceleration of Electrons

A particle accelerator has the purpose of increasing the kinetic energy of electrons (or other charged particles) and providing them to an experiment. In order to obtain a net increase in the kinetic energy between the entrance to and the exit from the device, the integral of the electric field over the particle trajectory must be nonzero. A plane electromagnetic wave packet (eq. (2.12)) in vacuum generally does not fulfill that condition. Its longitudinal component is zero, and the transverse component cancels out. After the passage of such a wave packet, the electron is displaced and has the same momentum it had before. An idea to "rectify" the electric field in a plane wave packet and thus obtain a net acceleration from a plane wave is being discussed in the literature [54]. The vector potential of such a rectified pulse would not return to the value it had before the arrival of the pulse but would rather have a rounded step-like shape. A practical realization would be challenging.

Many novel concepts for the acceleration of electrons and other charged particles have been developed that work by either making an electromagnetic wave develop a longitudinal component or by combining several waves or a wave and a static magnetic field [55, 56, 57, 58], so that one provides a longitudinal component relative to the other. The traditional way of making an electromagnetic wave develop a longitudinal component for the purposes of particle acceleration is to slow down its phase velocity by injecting it into an RF resonator (see ch. 1, sec. 4). The achievable accelerating field gradient is limited by the properties of the materials in the resonator walls. This limit can be overcome by the use of high-power lasers in concepts, such as the plasma wave accelerators, the laser focus accelerator, the inverse free-electron laser, etc.

One type of plasma wave accelerator [59, 60, 61, 62, 63] is the plasma wakefield accelerator, which works by having an electromagnetic wave packet excite a density wave in a plasma. Electrons can "surf" on that wave and be accelerated by the longitudinal electric field (which exists because the density wave propagates at less than the speed of light). In the plasma beat wave accelerator, the density wave is created by two copropagating laser beams whose frequencies differ by the plasma frequency.

The electromagnetic field in a focus has a longitudinal component [64], which can be used for the purposes of acceleration [65]. It can be described as a superposition of an infinite number of plane waves, but it does not take that many. There are several proposals for combining two electromagnetic waves for the purposes of electron acceleration, and, in most of them, such as the inverse free-electron laser [14, 66, 67], one wave is actually a static (in the lab frame) magnetic field. More complicated concepts, such as a sequence of crossed foci [68], have also been proposed. In order to overcome the limitation of a short interaction length, the use of Bessel beams has been proposed [69].

The relativistic motion of an electron in a superposition of plane waves of arbitrary directions is very complicated, and no analytic solution exists. A series expansion for the two-wave case is shown in sec. 2.4, and a nonrelativistic approximation for multiple plane waves is given in sec. 2.3.

However, for the purposes of electron acceleration, a detailed knowledge of the motion in the fields is not necessary—only the momentum after passage of the waves is of interest. An approximate solution can then be found [70, 71, 72] for the case of an arbitrary long-envelope electromagnetic field by introduction of the ponderomotive potential $U = (q^2/mc^2)A_\nu A^\nu$. When averaged over many cycles of the electromagnetic fields, the electron's motion $\overline{x}^\nu, \overline{p}^\nu$ is described by

$$\frac{d\overline{x}^\nu}{d\tau} = \frac{\overline{p}^n u}{m}, \qquad \frac{dp^\nu}{d\tau} = \frac{1}{2}\frac{\partial U}{\partial x_\nu}, \tag{2.31}$$

where τ is the relativistic proper time of the electron. When comparing to, for example, eq. (4) in [72], please note that U is defined as a dimensionless quantity there and has the dimensions of an energy here.

Chapter 3

DYNAMICAL DIFFRACTION

A dozen, a gross, and a score,
Plus three times the square root of four,
Divided by seven,
Plus five times eleven,
Equals nine squared plus zero, no more.
– Jon Saxton

B.W. Adams
Advanced Photon Source, Argonne National Laboratory, Argonne, IL 60439, USA
adams@mail.desy.de

Abstract The theoretical basis for coherent x-ray diffraction is the dynamical diffraction theory. It is introduced here on the basis of microscopic electromagnetism, using the vector potential, charge densities and currents, instead of the electric field and susceptibilities, as is usually done. Based upon this approach, the theory is extended in two ways. The first is an explicit inclusion of time dependence by a combination of Kato's eikonal theory with the Takagi theory. Order-of-magnitude estimates can be obtained directly from that theory without running a computer simulation. The results will be needed in ch. 4. The other extension of the dynamical theory is the inclusion of nonlinear optical effects. These results will be needed in ch. 5.

1. Introduction

The dynamical theory of diffraction deals with coherent multiple scattering phenomena. Its name, and that of the kinematical theory that deals with single scattering cases, stems from an analogy with classical mechanics: Whereas kinematics simply describes the motion of masses in space and time, dynamics incorporates interactions between them and with external potentials. Likewise, the kinematical theory gives only the directions of diffracting waves in crystals and the dynamical theory describes the evolution of coupled wave amplitudes in a scattering potential. The perfect crystal theory originated in the works of Ewald and von Laue and is almost a century old. It was first formulated for x-rays but was later adapted to particles, such as neutrons and electrons. With the advent of optical communication technology, a similar theory was developed independently for visible-light photons. There is however one important difference between the x-ray and visible light cases: A so-called photonic crystal can exhibit a photonic band gap, i.e., a range of photon energies within which no propagating mode exists *for any direction.* Because of the relatively weak interaction of x-rays with matter, a similar effect (i.e., Bragg reflection) occurs only within a narrow angular range. Electrons with their strong interaction with matter do of course experience band gaps in crystals. Section 2 gives a brief introduction to the dynamical theory of x-ray diffraction. It is not meant to be comprehensive and covers only those subjects that are necessary groundwork for the two extensions covered later in the chapter, namely the time-dependent eikonal Takagi-Taupin theory and the nonlinear dynamical diffraction theory. For the same reason, the introduction to the dynamical theory given here differs from the standard approach. As argued below and in secs. 3 and 4, both extensions require the use of microscopic electromagnetism in terms of the electron density and the vector potential, instead of the electric susceptibility and the electric field. A brief outline of the macroscopic formulation is given in app. 5.1 and more details can be found in the literature [73, 74]. There is a downside to the microscopic approach, namely the lack of a natural way to include photoelectric absorption within the context of classical electrodynamics. In order to satisfy the requirements of both, nonlinear optics and photoelectric absorption, the matter would have to be quantized [75]. In keeping with the main thrust of this book and to avoid lengthy derivations, we go here with the microscopic approach and model photoelectric absorption phenomenologically.

In the nineteen-sixtees, gemetric-optical approaches of the dynamical diffraction theory for disturbed crystals were formulated by P. Pennig and D. Polder [76] and by N. Kato and U. Bonse through the introduction of an eikonal. Kato's theory [74, 77, 78, 79] refers to the so-called Laue case (see p. 47) and Bonse [80, 81] treated the so-called Bragg case (see p. 47). In parallel, S. Takagi [82, 83] and D. Taupin [84] developed a perturbational wave-optical approach that leads to a system of partial first order differential equations. The latter is commonly used for the interpretation of x-ray diffractive analysis of crystal defects and semiconductor device structures, and for the design of precise x-ray optical elements in synchrotron radiation facilities. That theory is modified in sec. 3 in two ways: It is synthesized with the eikonal theory [85, 86] and time dependence is included explicitly (see also ch. 10), permitting the treatment of x-ray propagation under the conditions of a rapidly changing structure factor, such as in the presence of optical phonons induced by a short-pulse laser. This becomes particularly important if the time scale of these changes approaches the reciprocal energy bandwidth of x-ray reflections (typically sub-picoseconds) upon which novel x-ray optical elements might be based (see ch. 4 and [86]). This is also the time scale at which the macroscopic approach to dynamical diffraction may fail because an expression of the electrical polarization in terms of the susceptibility times the electric field strength does not account for the motion of charges due to the other influences that cause the structure factor to change (such as optical phonons). With a simple approximation used here, the microscopic and macroscopic approaches are equivalent in this respect. However, if done more carefully, the microscopic formulation predicts frequency shifts not seen with macroscopic electromagnetism (see app. 5.6.2).

The way the eikonal function is introduced here differs from Kato's. It is used here as a means to minimize the derivatives in the differential equations and can be determined in a numerical variational procedure. On a qualitative level, it makes the concept of the dispersion surface available within the Takagi theory and permits simple qualitative and order of magnitude estimates of the influence of a spatial or temporal disturbance on the diffracting wave fields. Computer simulations on the basis of the minimized derivatives in the differential equations can be done more precisely. Other than in Kato's theory, it is not necessary for the disturbance to be small: Deviations from the local dispersion surface, as approximated through the eikonal function, are picked up by the Takagi-type differential equations. This remaining deviation can be interpreted as the driving force for a change of the diffracting wave fields. Therefore, the theory introduced here is capable of qualitative predictions due to its geomtric-optical character, and can be used for quantitative calculations due to its wave-optical character. It can give more precise results (see p. 56) than the standard Takagi theory because, instead of the full influence of the crystal's scattering power on the x-rays, only their deviation from geometric-optical behavior need to be treated in a wave-optical calculation.

The last section deals with an extension of dynamical x-ray diffraction to nonlinear optics. With x-rays far off any resonances, the magnetic components of the electromagnetic waves match the electric components in importance for nonlinear effects (see discussion on p. 21). Other than with visible light [43, 87], a formulation in terms of nonlinear electric susceptibilities would then be insufficient. The theory is therefore based on microscopic electromagnetism, using the scalar and vector potentials, charge densities and currents.

2. Linear Perfect Crystal Theory

By use of the transverse gauge $\nabla \cdot \mathbf{A} = 0$, Maxwell's equations are equivalently expressed by the following two equations, one each for the scalar potential Φ ([1].(6.44)) and the vector potential $\mathbf{A}$ ([1].(6.52)).

$$\nabla^2\Phi = -4\pi\rho \tag{3.1}$$

$$\nabla^2\mathbf{A} - \frac{1}{c^2}\frac{\partial^2\mathbf{A}}{\partial t^2} = -\frac{4\pi}{c}\mathbf{J} + \frac{1}{c}\nabla\frac{\partial\Phi}{\partial t} = -\frac{4\pi}{c}\mathbf{J}_t, \tag{3.2}$$

where $\mathbf{J}$ and $\mathbf{J}_t$ are the current density and the transverse current density (see [1].[6.5] for details):

$$\mathbf{J}_t(\mathbf{r}) = \frac{1}{4\pi}\nabla\times\nabla\times\int\frac{\mathbf{J}(\mathbf{r}')}{|\mathbf{r}'-\mathbf{r}|}d\mathbf{r}'. \tag{3.3}$$

In order to find the current density $\mathbf{J}$, and thus $\mathbf{J}_t$, a model must be developed for the action of the electromagnetic fields upon the charges. In a quantum physical picture, the equilibrium charge density is established by balancing the electrostatic potential of the nuclei against a kinetic energy due to the uncertainty relation and the multiparticle effects of screening and exchange interaction. In a purely classical picture, it is not possible to determine both, the electron density $\eta(\mathbf{r})$ and the density $\nu(\mathbf{r})$ of nuclear charge carriers (protons), from first principles. Instead, one must prescribe a density of positive background charges and build a model that maintains an equilibrium electron density in a consistent manner. The common model of placing the electrons in harmonic potentials is good for resonant cases but not if multiple Fourier components of the periodic electron density come into play. Here, the emphasis is on the diffraction of x-rays in crystals of light elements whose electrons have binding energies well below the x-ray photon energy. This case is served better by a "jellium"-type model (see fig. 3.1), in which the total charge density $\rho(\mathbf{r})$ in the absence of electromagnetic waves is split into a highly mobile electronic charge density $-e\eta(\mathbf{r})$ and a sluggish positive background charge, which does not reflect the actual distribution of nuclei but rather mirrors the electronic charge density as $e\eta(\mathbf{r})$. By this cancellation of charges, the charge density $\rho(\mathbf{r})$ vanishes everywhere in the absence of electromagnetic waves, and the electrons are in equilibrium. Electromagnetic waves induce a differential motion of the

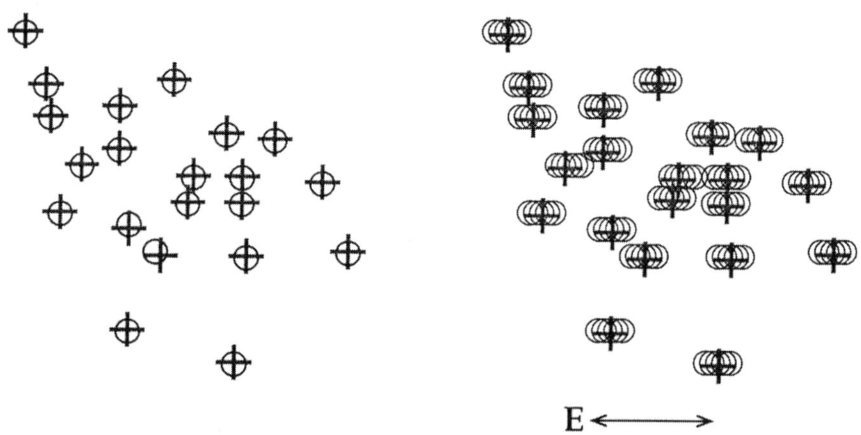

Figure 3.1: The jellium model of charges. Left, equilibrium case, electronic charge (circles) is compensated exactly by a positive background charge. Right, the presence of an electromagnetic wave ($\mathbf{E}$ field shown) makes the electrons oscillate while the positive charges stay put.

electronic charge density relative to the background charge, giving rise to a current and a weak (see app. 5.2) electrostatic potential.

The background charge density $e\eta$ contributes much less to scattering than $-e\eta$ of the electrons because it is tied to the heavy nuclei. It will simply be assumed static. The linear nuclear scattering contribution[1] is however stronger than the nonlinear scattering from electrons and will therefore come up again (and be dismissed) when nonlinear effects are discussed in sec. 4. Similarly, the electron's spin contributes far less than its charge to each of the (linear or nonlinear) orders of interaction with the electromagnetic field.

2.1 Perfect Lattice, Fourier and Bloch Sums

The spatial periodicity of a crystal suggests solving the wave equation (3.2) in terms of Fourier series and Bloch waves. In crystallographic terminology, a linear operator $\mathbf{h}$ that maps a real-space vector $\mathbf{r}$ to the scalar phase $\mathbf{h} \cdot \mathbf{r}$ is a reciprocal lattice vector and a member of reciprocal space. The electron density of the crystal in the absence of electromagnetic waves can now be expressed as a Fourier series[2]

$$\eta(\mathbf{r}) = \sum_{\mathbf{h}} \tilde{\eta}_{\mathbf{h}} e^{i\mathbf{h}\cdot\mathbf{r}}. \tag{3.4}$$

It is real-valued, and therefore (i) $\tilde{\eta}^*_{\mathbf{h}} = \tilde{\eta}_{-\mathbf{h}}$. If the electron density is inversion symmetric relative to the origin, then $\eta(\mathbf{r}) = \eta(-\mathbf{r})$, and thus also (ii) $\tilde{\eta}_{\mathbf{h}} = \tilde{\eta}_{-\mathbf{h}}$ for all $\mathbf{h}$.[3] The quantities $F_H = V\tilde{\eta}_H$ are called "structure factors" in crystallographic terminology, where V is the unit cell volume.

The vector potential is real-valued and can be written as a sum of two complex conjugate Bloch waves $\mathbf{A}(\mathbf{r},t) = \mathbf{A}^{(+)}(\mathbf{r},t) + \mathbf{A}^{(-)}(\mathbf{r},t)$, with the waves in $\mathbf{A}^{(+)}(\mathbf{r},t)$ and $\mathbf{A}^{(-)}(\mathbf{r},t)$ running forward and backward in space and time, respectively.

$$\mathbf{A}^{(+)}(\mathbf{r},t) = \sum_{\mathbf{h}} \tilde{\mathbf{A}}^{(+)}_{\mathbf{h}} e^{i(\mathbf{k}_{\mathbf{h}}\cdot\mathbf{r}-\omega t)} \qquad \mathbf{k}_{\mathbf{h}} = \mathbf{k}_{\mathbf{0}} + \mathbf{h}, \tag{3.5}$$

where $\tilde{\mathbf{A}}^{(+)}_{\mathbf{h}}$ and $\tilde{\mathbf{A}}^{(-)}_{-\mathbf{h}}$ are complex conjugates of each other. The wave vector $\mathbf{k}_{\mathbf{0}}$ will henceforth be called the "Bloch base vector". Similarly to eq. (3.5), the scalar potential Φ is written as $\Phi^{(+)}(\mathbf{r},t) + \Phi^{(-)}(\mathbf{r},t)$, with

$$\Phi^{(+)}(\mathbf{r},t) = \sum_{\mathbf{h}} \tilde{\Phi}^{(+)}_{\mathbf{h}} e^{i(\mathbf{k}_{\mathbf{h}}\cdot\mathbf{r}-\omega t+\varphi_{\mathbf{h}})}, \tag{3.6}$$

where $\varphi_{\mathbf{h}}$ allows for a phase shift between the vector and scalar potentials. $\tilde{\Phi}^{(+)}_{\mathbf{h}}$ and $\tilde{\Phi}^{(-)}_{-\mathbf{h}}$ are complex conjugates of each other and $\varphi_{-\mathbf{h}} = -\varphi_{\mathbf{h}}$.

The current density $\mathbf{J} = -e\eta\mathbf{v}$ is given by the electronic charge density $-e\eta$ and velocity $\mathbf{v}$. In the absence of absorption, $\mathbf{v}$ can be determined by use of

[1] Meaning scattering of electromagnetic waves from the nuclei moving as units, not Mössbauser scattering.

[2] Please see the remarks on the notation of wave vectors, etc. on p. xxiii.

[3] For a given $\mathbf{h}$, it is always possible to make $\tilde{\eta}_{\mathbf{h}} = \tilde{\eta}_{-\mathbf{h}}$ by a judicious choice of the origin. The structure is inversion-symmetric if this occurs at the same origin for all $\mathbf{h}$.

the Lorentz equation, or directly by eq. (2.19). Absorption can be modeled phenomenologically by the introduction of a phase μ and multiplying the velocity, as determined by eq. 2.19, with a factor $\exp(i\mu)$. Another possibility for the phenomenological modeling of absorption would be to make η complex-valued, with the possibility of differentiating between absorption due to the Fourier components of $\tilde{\eta}_{\mathbf{h}}$, but that would stretch the jellium model too far.

The current density $\mathbf{J}(\mathbf{r},t)$ may be split as $\mathbf{J} = \mathbf{J}^{(+)} + \mathbf{J}^{(-)}$ into two parts, $\mathbf{J}^{(+)}$ due to $\mathbf{A}^{(+)}$ and $\Phi^{(+)}$, and $\mathbf{J}^{(-)}$ due to $\mathbf{A}^{(-)}$ and $\Phi^{(-)}$. The electrons are assumed to be initially at rest, i.e., $\mathbf{p}_0 = \mathbf{0}$, and thus the linear part of eq. (2.19) gives:

$$\begin{aligned} \mathbf{J}^{(+)} &= -r_e\, c \sum_{\mathbf{h}} \tilde{\eta}_{\mathbf{h}} e^{i\mathbf{h}\cdot\mathbf{r}} \sum_{\mathbf{h}'} \left(\tilde{\mathbf{A}}^{(+)}_{\mathbf{h}'} + \frac{\mathbf{k}_{\mathbf{h}'}}{k} \tilde{\Phi}^{(+)}_{\mathbf{h}'} e^{i\varphi_{\mathbf{h}'}} \right) e^{i(\mathbf{k}_{\mathbf{h}'}\cdot\mathbf{r}-\omega t)} e^{i\mu} \\ \mathbf{J}^{(-)} &= -r_e\, c \sum_{\mathbf{h}} \tilde{\eta}_{\mathbf{h}} e^{i\mathbf{h}\cdot\mathbf{r}} \sum_{\mathbf{h}'} \left(\tilde{\mathbf{A}}^{(-)}_{-\mathbf{h}'} - \frac{\mathbf{k}_{\mathbf{h}'}}{k} \tilde{\Phi}^{(-)}_{-\mathbf{h}'} e^{-i\varphi_{\mathbf{h}'}} \right) e^{-i(\mathbf{k}_{\mathbf{h}'}\cdot\mathbf{r}-\omega t)} e^{i\mu}, \end{aligned} \tag{3.7}$$

where $r_e = e^2/mc^2 = 2.82 \cdot 10^{-15}$ m is the classical electron radius and $k = \omega/c$ is the wave number. If $\mu = 0$, then $\mathbf{J}^{(+)}$ and $\mathbf{J}^{(-)}$ are complex conjugates of each other. Otherwise, $\mathbf{J}$ is complex-valued, and therefore phase-shifted relative to $\mathbf{A}$. The scalar potential $\Phi_{\mathbf{h}}$ in eq. (3.7) gives rise to a longitudinal current, which can be expressed in terms of the vector potential $\mathbf{A}_{\mathbf{h}}$ and is smaller than the transverse current, directly due to $\mathbf{A}_{\mathbf{h}}$, by a factor of about 10^{-6}. It shall be ignored here but is considered in app. 5.2, and also in the context of nonlinear scattering in sec. 4.

2.2 The System of Linear Equations

To make the exponents of the current and the fields themselves comparable, the index transformations $\mathbf{h} + \mathbf{h}' \to \mathbf{h}$, $\mathbf{h}' \to \mathbf{h}'$ (and thus $\mathbf{h} \to \mathbf{h} - \mathbf{h}'$) for $\mathbf{J}^{(+)}$ and $\mathbf{h}' - \mathbf{h} \to \mathbf{h}$, $\mathbf{h}' \to \mathbf{h}'$ (and thus $\mathbf{h} \to \mathbf{h}' - \mathbf{h}$) for $\mathbf{J}^{(-)}$ are applied. This gives

$$\mathbf{J}^{(+)} = -r_e c \sum_{\mathbf{h}\mathbf{h}'} \tilde{\eta}_{\mathbf{h}-\mathbf{h}'} \tilde{\mathbf{A}}^{(+)}_{\mathbf{h}'} e^{i(\mathbf{k}_{\mathbf{h}}\cdot\mathbf{r}-\omega t)} e^{i\mu}. \tag{3.8}$$

Upon insertion of eq. (3.8) into (3.3) and use of (see appendix 5.6.3) $\int \exp(i\mathbf{k}\cdot\mathbf{r}')/|\mathbf{r}'-\mathbf{r}|d\mathbf{r}' = (4\pi/|\mathbf{k}|^2)\exp(i\mathbf{k}\cdot\mathbf{r})$, the transverse current density becomes

$$\mathbf{J}_t^{(+)} = -r_e c \sum_{\mathbf{h},\mathbf{h}'} \tilde{\eta}_{\mathbf{h}-\mathbf{h}'} \left[\tilde{\mathbf{A}}^{(+)}_{\mathbf{h}'} \right]_{\mathbf{h}} e^{i(\mathbf{k}_{\mathbf{h}}\cdot\mathbf{r}-\omega t)} e^{i\mu} \tag{3.9}$$

with

$$\left[\tilde{\mathbf{A}}^{(+)}_{\mathbf{h}'} \right]_{\mathbf{h}} = \frac{\mathbf{k}_{\mathbf{h}} \times \tilde{\mathbf{A}}^{(+)}_{\mathbf{h}'} \times \mathbf{k}_{\mathbf{h}}}{|\mathbf{k}_{\mathbf{h}}|^2}. \tag{3.10}$$

At this point, some clarification may be required about the meaning of eq. (3.9). Application of the transverse current formula (3.3) does, of course, remove the longitudinal contributions due to the induced scalar potential Φ (which is ignored here, anyway). However, there is more to it than that. The individual contributions to the sum in eq. (3.8) are not transverse because they have $\tilde{\mathbf{A}}^{(+)}_{\mathbf{h}'}$, i.e., the amplitude for $\mathbf{k}_{\mathbf{h}'}$, associated with $\exp(i\mathbf{k}_{\mathbf{h}} \cdot \mathbf{r} - \omega t)$. Application of eq. (3.3) redistributes the Bloch wave components of the current and makes them individually transverse.

We can now insert eqs. (3.5) and (3.9) into (3.2) and consider each of the Fourier-Bloch components with an exponent $\exp(\pm i(\mathbf{k}_{\mathbf{h}} \cdot \mathbf{r} - \omega t))$ separately. This gives for each $\mathbf{h}$:

$$\begin{aligned} (k^2 - \mathbf{k_h}^2)\tilde{\mathbf{A}}^{(+)}_{\mathbf{h}} &- 4\pi r_e e^{i\mu} \sum_{\mathbf{h}'} \tilde{\eta}_{\mathbf{h}-\mathbf{h}'} \left[\tilde{\mathbf{A}}^{(+)}_{\mathbf{h}'}\right]_{\mathbf{h}} = 0 \\ (k^2 - \mathbf{k_h}^2)\tilde{\mathbf{A}}^{(-)}_{-\mathbf{h}} &- 4\pi r_e e^{i\mu} \sum_{\mathbf{h}'} \tilde{\eta}_{\mathbf{h}'-\mathbf{h}} \left[\tilde{\mathbf{A}}^{(-)}_{-\mathbf{h}'}\right]_{\mathbf{h}} = 0. \end{aligned} \tag{3.11}$$

The deviation of the wave vector $\mathbf{k_h}$ from k in eq. (3.11) is due to the scattering contributions of all Bloch component amplitudes $\left[\tilde{\mathbf{A}}^{(\pm)}_{\pm\mathbf{h}'}\right]_{\mathbf{h}}$ contained in the sum over $\mathbf{h}'$, into the one with amplitude $\tilde{\mathbf{A}}^{(\pm)}_{\pm\mathbf{h}}$. The term in the sum with $\mathbf{h}' = \mathbf{h}$ describes refraction due to the average electron density given by $\tilde{\eta}_0$. To get an idea of the deviation $k - |\mathbf{k_h}|$, we can approximate $k^2 - \mathbf{k_h}^2 \approx 2k(k - |\mathbf{k}_0|)$. With k of the order of $10^{10}\mathrm{m}^{-1}$ to $10^{11}\mathrm{m}^{-1}$ for typical x-rays, $r_e = 2.82 \cdot 10^{-15}\mathrm{m}$ and η of the order of $10^{29}m^{-3}$ to $10^{31}m^{-3}$ (it is $7 \cdot 10^{29}m^{-3}$ for Si), the deviation is of the order of $10^5 m^{-1}$, or ca. 10^{-5} of k itself. Likewise, the average refractive index (i.e., without diffraction effects) differs by that much from unity, see eq. (3.12) in sec 2.3.

2.3 The Dispersion Surface

Equation (3.11) defines a system of linear equations to be solved for the wave fields inside the crystal. Far off the Bragg condition $|\mathbf{k}_0| \approx |\mathbf{k}_0 + \mathbf{h}|$ for all $\mathbf{h}$, the diffracted waves have negligibly small amplitudes and all that is left is $k^2 - 4\pi r_e \tilde{\eta}_0 = \mathbf{k}_0^2$, giving the refractive index (see also app. 5.1):

$$n = \sqrt{1 - \frac{4\pi r_e e^{i\mu}}{k^2}\tilde{\eta}_0} \approx 1 - \frac{2\pi r_e e^{i\mu}}{k^2}\tilde{\eta}_0 =: 1 - \delta(k) - i\beta(k), \tag{3.12}$$

where δ is the refractive index decrement and β is the absorption index.

If the Bragg condition $|\mathbf{k_0}| \approx |\mathbf{k_0} + \mathbf{H}_n|$ is fulfilled for a number N of reciprocal lattice vectors[4] $\mathbf{H}_n, n = 1 \ldots N$, then eq. (3.11) is a system of linear equations with $N + 1$ lines for $N + 1$ field amplitudes $\tilde{\mathbf{A}}_\mathbf{0}^{(+)}, \tilde{\mathbf{A}}_{\mathbf{H}_1}^{(+)} \ldots \tilde{\mathbf{A}}_{\mathbf{H}_N}^{(+)}$. Nontrivial solutions (i.e., not all $\tilde{\mathbf{A}}_\mathbf{h} = \mathbf{0}$) are found if the determinant of coefficients vanishes, leading to an algebraic equation for $(k, \mathbf{k_0})$.

The following discussion refers to the two-wave case, i.e., $N = 1$ and only two amplitudes $\tilde{\mathbf{A}}_\mathbf{0}^{(+)}$ and $\tilde{\mathbf{A}}_\mathbf{H}^{(+)}$ are of appreciable magnitude. The case of more than two waves is addressed again at the end of this section. With two waves, eq. (3.11) is reduced to two lines, and the determinant is:

$$\det(k, \mathbf{k_0}) = ({k_r}^2 - \mathbf{k}_0^2)({k_r}^2 - (\mathbf{k_0} + \mathbf{H})^2) - (4\pi r_e)^2 e^{2i\mu} C^2 \tilde{\eta}_\mathbf{H} \tilde{\eta}_{-\mathbf{H}}, \quad (3.13)$$

where $k_r = k\sqrt{1 - 4\pi r_e \exp(i\mu)\tilde{\eta}_0 / k^2}$ is the refraction-corrected wave number and $C = (\tilde{\mathbf{A}}_\mathbf{0} \cdot \tilde{\mathbf{A}}_\mathbf{H})/(|\tilde{\mathbf{A}}_\mathbf{0}||\tilde{\mathbf{A}}_\mathbf{H}|)$ is a polarization factor, which equals 1 or $\cos 2\Theta$ for σ and π polarization, respectively. The symbols σ and π stand for 'senkrecht' and 'parallel', German for perpendicular and parallel, and denote the electrical polarization relative to the scattering plane defined by $\mathbf{k_0}$ and $\mathbf{H}$.

Figure 3.2 shows a surface in (k_x, k_y, k), with $k_z = 0$, that represents solutions of $\det(k, \mathbf{k_0}) = 0$ with *real-valued* $\mathbf{k_0}$. Complex solutions are discussed below and in sec. 2.5, see also [88]. Using the approximation $|\mathbf{k_0}| \approx k \approx |\mathbf{k_0} + \mathbf{H}|$, and abbreviating $|\mathbf{k_0}| - k_r = \xi_0$ and $|\mathbf{k_0} + \mathbf{H}| - k_r = \xi_h$, eq. (3.13) can be simplified to define the so-called dispersion hyperbola (more generally, dispersion surface)

$$\xi_0 \xi_h = \frac{4\pi^2 r_e^2 C^2 \tilde{\eta}_\mathbf{H} \tilde{\eta}_{-\mathbf{H}}}{k^2}. \quad (3.14)$$

Appendix 5.1 gives a comparison with conventional notation [73]. The origin of the coordinates ξ_0 and ξ_h is at the intersection Q (terminology of [73]) of the two straight lines in fig. 3.2, upper right part, and ξ_0 and ξ_h are measured along the directions of $\mathbf{k_0}$ and $\mathbf{k_h}$. A point on a dispersion hyperbola for which amplitudes $\tilde{\mathbf{A}}_\mathbf{0}^{(+)}$ and $\tilde{\mathbf{A}}_\mathbf{H}^{(+)}$ are solutions of the corresponding system of linear equations is commonly called the tie point for these amplitudes. Its coordinates ξ_0 and ξ_h (see app. 5.6.1) are related to the amplitudes by:

$$\frac{\Re \xi_0}{\Re \xi_h} \approx \frac{\left|\tilde{\mathbf{A}}_\mathbf{H}^{(+)}\right|^2}{\left|\tilde{\mathbf{A}}_\mathbf{0}^{(+)}\right|^2}, \quad (3.15)$$

where $\Re$ means the real part and "$\approx$" becomes "$=$" in the absence of absorption.

[4]Please note the terminology: Lower case $\mathbf{h}$ is used in the function of a summation index and uppercase $\mathbf{H}$ is used to denote particular reciprocal lattice vectors.

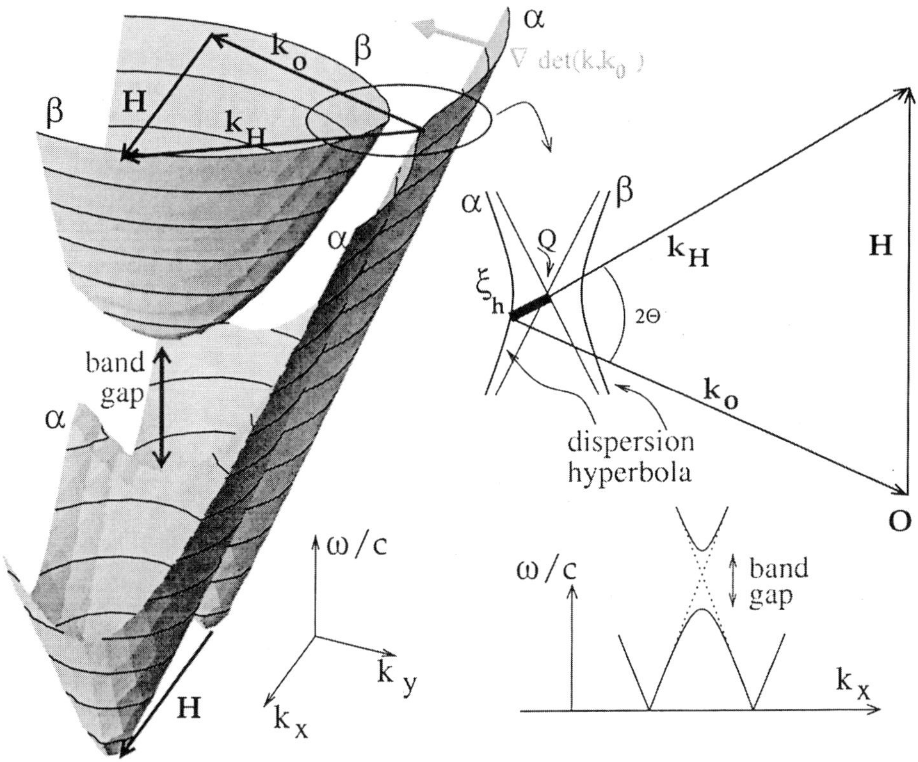

Figure 3.2: The dispersion surface $\det(k, \mathbf{k}_O) = 0$ of the two-wave case, shown in momentum-energy-space. The cones represent the free-space photon dispersion $|\mathbf{k}_O|c = \omega$ and $|\mathbf{k}_O + \mathbf{H}|c = \omega$. Close to their intersection, diffraction couples the waves and a splitting into the α and β branches occurs. A closeup of the intersection region in a cut at constant ω is shown in the upper right part of the figure. The dispersion surface in this cut can be approximated by a hyperbola, its asymptotes being circles about the points **0** and **H**. In the lower right corner, a cut containing the energy and the k_x axes is shown. This is the familiar representation of a band structure with a direct band gap. The figure should actually be continued periodically along k_x. For the meaning of $\nabla \det(k, \mathbf{k}_O)$, see text. The gap is heavily exaggerated.

The two branches (see fig. 3.2) of the hyperbola with real-valued (ξ_0, ξ_h) are commonly called α, or "1" (for $\xi_0, \xi_h > 0$) and β, or "2" (for $\xi_0, \xi_h < 0$).

In the presence of absorption, i.e., for $\mu \neq 0$, eq. (3.13) has complex-valued solutions $\mathbf{k}_O$, but even if $\mu = 0$, complex-valued $\mathbf{k}_O$ are admissible (see sec. 2.5). Because the reciprocal lattice vectors are real-valued, all wave vectors (i.e., $\mathbf{k}_O$ and $\mathbf{k}_H$ in the two-wave case) must have the same imaginary part, henceforth called $\mathcal{K}$. A complex-valued wave vector requires the presence of a crystal surface, to be discussed further in sec. 2.5.

Because of the relation $\mathbf{k}_H = \mathbf{k}_O + \mathbf{H}$, the two waves form a standing wave pattern whose periodicity is that of the lattice planes belonging to $\mathbf{H}$. Its phase, given by the relative phases of $\tilde{\mathbf{A}}_0^{(+)}$ vs. $\tilde{\mathbf{A}}_H^{(+)}$, depends on the choice of the coordinate origin but in any case, the standing waves belonging to the α and β branches differ by a phase of close to π (exactly so in the limit of a vanishing imaginary wave vector $\mathcal{K}$). This can be seen as follows: Suppose first that $\mathcal{K} = \mathbf{0}$, implying that the absorption phase μ is zero. Equation (3.11) for $\mathbf{h} = \mathbf{H}$ then requires that the relative phase between $\tilde{\mathbf{A}}_0^{(+)}$ and $\tilde{\mathbf{A}}_\mathbf{H}^{(+)}$ be the opposite of that of $\tilde{\eta}_\mathbf{H}$ (which depends on the coordinate origin), or that much plus a phase of π. The former case belongs to the β, and the latter, with a larger value of ${\mathbf{k_H}}^2$, to the α branch. It is also obvious from eq. (3.11) that, as one changes the ratio $|\tilde{\mathbf{A}}_\mathbf{H}^{(+)}|/|\tilde{\mathbf{A}}_0^{(+)}|$ by moving along either one of the two branches, the relative phase between $\tilde{\mathbf{A}}_\mathbf{H}^{(+)}$ and $\tilde{\mathbf{A}}_0^{(+)}$ remains unchanged. Neither this statement, nor the exact value of π between the α and β branches, remain strictly valid if $\mathcal{K} \neq \mathbf{0}$ is admitted because then, the phase of the factor $(k^2 - {\mathbf{k_h}}^2)$ in eq. (3.11) is affected as the length of the real part of $\mathbf{k_0}$ is changed. Going back to $\mathcal{K} = \mathbf{0}$, one may see from eq. (3.11) that $|\mathbf{k_0}|$ and $|\mathbf{k_H}|$ take extremal values on the α and β branches: a change δ in the relative phase between $\tilde{\mathbf{A}}_\mathbf{H}^{(+)}$ and $\tilde{\mathbf{A}}_0^{(+)}$ from the value taken on either branch induces a first-order growth (as $\sin\delta$) in the imaginary part $\mathcal{K}$, and a second-order change (as $\cos\delta$) in the real parts of the wave vectors. This is going to be important in sec. 3. Corresponding to the maximal length of the wave vectors on the α branch is a maximal effective refractive index due to a position of the standing waves that avoids the electron density on the lattice planes. Conversely, the antinodes of the standing waves for the β branch are at the maxima of the electron density, making the effective refractive index, and thus the lengths of $\mathbf{k_0}$ and $\mathbf{k_H}$, minimal.

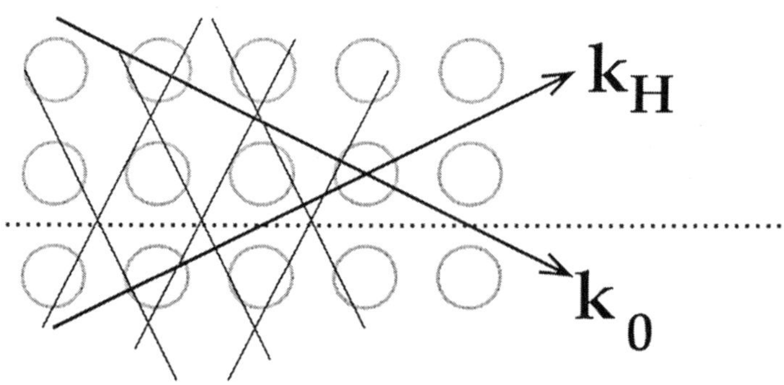

Figure 3.3: Standing waves in dynamical diffraction. Shown are atoms (circles) in a lattice, two waves with wave vectors $\mathbf{k_0}$ and $\mathbf{k_h}$ and the crests of the electric field (solid lines perpendicular to $\mathbf{k_0}$ and $\mathbf{k_h}$). Interference makes the maxima (along the dotted horizontal line) of the standing waves lie between the lattice planes on which the atoms sit. This is the situation for the α branch.

Instead of obtaining the extremal lengths of the wave vectors as a result of dynamical diffraction theory, one may also use it as a starting point [89], motivated by Fermat's principle of extremal optical path length, and derive the dispersion surface from there. Interestingly, Fermat's principle appears here truly as a principle of minimum (the usual case) and maximum (a very rare case) optical path length *simultaneously*.

The positions of the intensity maxima of the standing waves have an effect not only on the effective refractive index but also on the absorption. A mode on the α branch suffers anomalously little of it because it avoids the electrons, particularly the highly concentrated core electrons whose binding energies are of the order of magnitude of x-ray photons. This is the so-called Borrmann effect [90]. It is most pronounced for σ-polarized x-rays near the symmetry center of the α branch because there, the interference cancellation at the lattice planes is perfect.[5] The part of the α branch that can sustain x-ray transmission without appreciable absorption losses shrinks around the symmetry point, as the propagation length is increased. For a mode on the β branch, interference tends to increase absorption over the value for nondiffracting conditions.

Similarly to the two-wave case, a dispersion surface with $N + 1$ branches can be found if $N + 1 > 2$ waves are involved. Because electromagnetic waves permit two orthogonal polarization directions for a given wave vector, $N + 1$ is generally twice as large as the number of wave vectors. With the exception of the two-wave case, a separation into two decoupled sets of polarization states (i.e., σ and π in the two-wave case) is generally impossible. Important are the six-wave case with three wave vectors, and multiple wave cases for back-reflection in cubic crystal structures [91]. The polynomial in $(k, \mathbf{k_0})$ obtained from the determinant condition will then be of a higher order, resulting in a correspondingly higher number of branches of the dispersion surface.

For each particular point $(k, \mathbf{k_0})$ on the dispersion surface, a nontrivial set of solution amplitudes $\{\tilde{\mathbf{A}}_\mathbf{h}\}$ for eq. (3.11) exists. Conversely, a different set of amplitudes is a solution at another (or no other) point on the dispersion surface (see sec. 3.6). However, any set $\{\tilde{\mathbf{A}}_{\mathbf{h}_j}, j = 1\ldots N\}$ of complex-valued field amplitudes of an N-wave case can be written as a sum $\sum_n \{\tilde{\mathbf{A}}_{\mathbf{h}_j}, j = 1\ldots N\}_n$ of N sets of field amplitudes, where each set belongs to one particular tie point.

2.4 Phase and Group Velocity, Beam Direction

In a dispersive medium, there is generally a difference between the phase velocity $\mathbf{k}(\omega/k^2)$ and the group velocity $\nabla_\mathbf{k}\omega$ of electromagnetic waves. In some applications, it becomes necessary to also consider higher derivatives, such as the dispersion $\nabla_\mathbf{k}\nabla_\mathbf{k}\omega$ of the group velocity. Furthermore, the phase gradient of a wave and the direction of energy flow do generally go in different directions, if the optical medium is not isotropic. The dispersion surface is a

[5] But even so, an electron density that is not highly concentrated on the lattice planes (such as in the odd-odd-odd reflections in the diamond structure, for example Si (111)), leaves a significant amount of absorption

very powerful tool to visualize and determine all these properties, especially so for diffractive optics (i.e., here x-ray crystal optics). Having multiple branches and being nonconvex, it is a bit more complicated than for purely refractive optics, but it works in basically the same way (see for example [92]). The dispersion surface shown in fig. 3.2 corresponds to Fresnel's ellipsoid in the refractive optics of anisotropic media.

A particular point P on the dispersion surface in fig. 3.2 gives the phase velocities $\omega/|\mathbf{k_0}|$ and $\omega/|\mathbf{k_h}|$ of the two waves with wave vectors $\mathbf{k_0}$ and $\mathbf{k_H}$. Other properties, such as the group velocity and the propagation direction in space and time of a wave packet synthesized from Fourier components taken from the vicinity of P, can be determined from the gradient $\nabla \det(k, \mathbf{k_0})$, taken on the domain of real-valued $(k, \mathbf{k_0})$ (shown in fig. 3.2). The projection of $\nabla \det(k, \mathbf{k_0})$ onto the momentum coordinates gives the direction of the Poynting vector (see [93, 94, 95], app. 5.5 and fig. 3.4, left-hand side) and a projection onto a plane containing the photon energy and momentum (fig. 3.4, right-hand side) gives the propagation in space and time. Taking the slope of the direction perpendicular to that projection gives the group velocity.

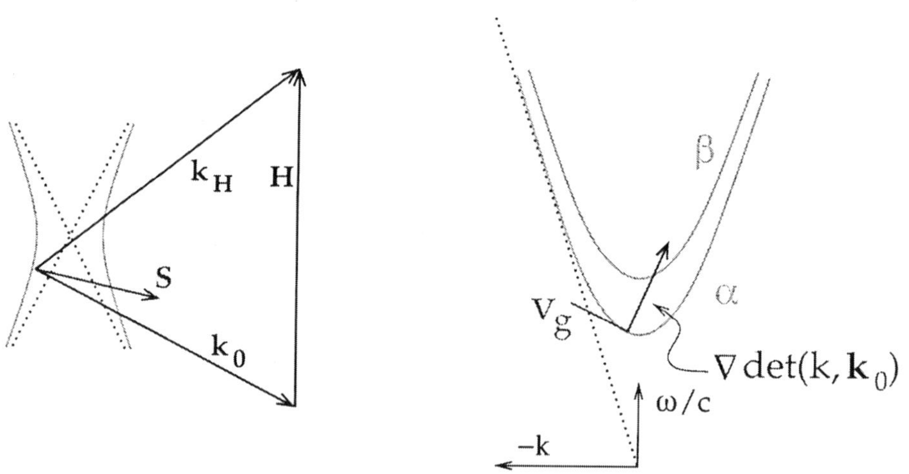

Figure 3.4: 2D-projections of the dispersion surface. Left: A projection of fig. 3.2 onto the (k_x, k_y) plane. The projection $\mathbf{S}$ of $\nabla \det(k, \mathbf{k_0})$ onto the (k_x, k_y) plane inherits the property of $\nabla \det(k, \mathbf{k_0})$ being perpendicular to the dispersion surface and is perpendicular to the dispersion hyperbola in the above graph. Right: The group velocity $\mathbf{v}_g$ is given by the slope of the dispersion surface.

A striking demonstration of the dependence of the beam direction of a diffracting beam on the detuning from the Bragg angle can seen in fig. 3.5. More material can be found on the CD in the directory dxd/Poynting. There are still images and also a video clip of the luminescent trace of a diffracting x-ray beam inside a diamond crystal.

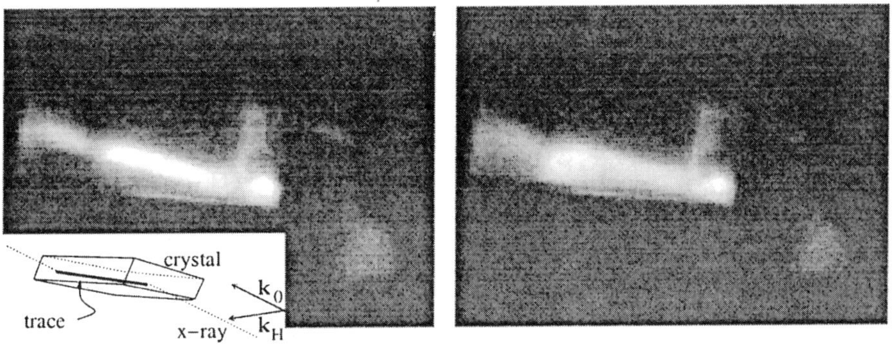

Figure 3.5: Photos of the Poynting vector in dynamical diffraction. An x-ray beam diffracting inside a diamond crystal leaves a luminescent trace which is seen by a TV camera in the beam time described in ch. 5, sec. 2.7.3. Left: With a large detuning, the beam direction is along $\mathbf{k_O}$. Right: With a range of angles, close to the Bragg angle, beams in a fan between $\mathbf{k_O}$ and $\mathbf{k_H}$ are excited. The insert shows a sketch for orientation. Compare also with fig. 5.23. The photos, and particularly the video clip, on the CD show the effect in more detail.

2.5 Extinction and Boundary Conditions

Our considerations so far have dealt only with the waves inside a perfect crystal of infinite extent. In this idealized case, solutions $\mathbf{k_O}$ of eq. (3.13) must be real-valued because, otherwise, the amplitudes would grow to infinity in some direction. An interruption of the crystal can however put a break to such a growth, and therefore complex-valued $\mathbf{k_O}$ are admissible in the presence of a surface. Absorption can be the cause of imaginary parts of the wave vectors, but, even in the absence of absorption, complex solutions of eq. (3.13) are possible. These can be found between[6] the α and β branches of the dispersion surface in fig. 3.2 (but not outside of the double-cone structure, or inside of the β-"bowl").

Because the waves usually originate outside of the crystal and must leave the crystal to be detected,[7] the Bloch waves inside the crystal must be matched continuously across the crystal boundaries to the simple plane waves outside. It follows that any difference of wave vectors across a boundary must be parallel to the surface normal vector. A generalization to this rule will be discussed in sec. 3.9. The wave vector matching across the boundary is demonstrated in fig. 3.6 for the so-called Bragg and Laue cases. The graph on the left-hand side of fig. 3.6 shows the asymmetric Bragg case. If no part of the real-valued dispersion surface can be reached by adding a vector parallel to the surface normal to the incident wave vector, then the Bloch wave inside the crystal has a base wave vector $\mathbf{k_O}$ with a large imaginary part. Continuity across the

---

[6]Figure 3.2 shows only the real parts of $(k, \mathbf{k_O})$. The determinant $\det(k, \mathbf{k_O})$ is nonzero off the dispersion surface in the figure if $(k, \mathbf{k_O})$ is restricted to the real domain. It can however vanish if complex-valued $(k, \mathbf{k_O})$ are admitted.

[7]There are exceptions, such as Kossel lines [96, 97], which are a dynamical diffraction phenomenon with an internal source, or the standing wave technique [98], which has a "detector" internal to the crystal.

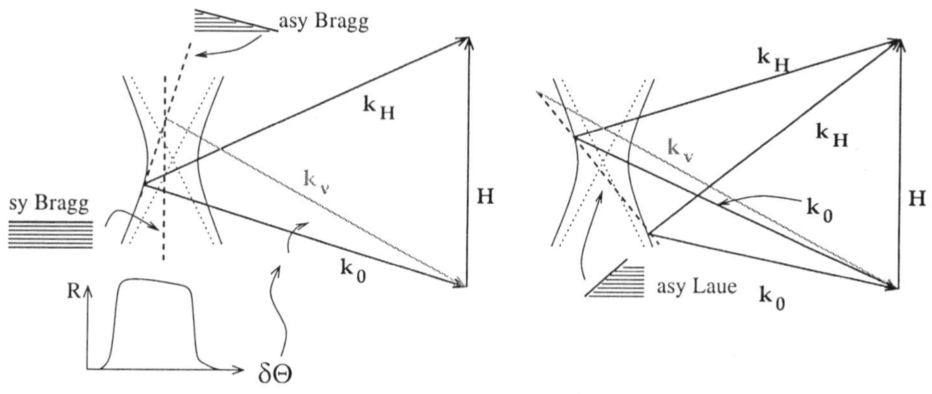

Figure 3.6: Bragg (left) and Laue (right) types of boundary conditions with vacuum wave vector $\mathbf{k}_v$ and Bloch component wave vectors $\mathbf{k}_0$ and $\mathbf{k}_H$. Shown is the dispersion surface of the two-wave case and the surface normals and real-space surfaces and lattice planes for an asymmetric (asy Bragg) and a symmetric Bragg (sy Bragg) case and for an asymmetric Laue (asy Laue) case. Also shown on the left-hand side is the so-called rocking curve of the Bragg reflectivity over the incidence angle. See also fig. 6.11 for the reflectivity curve of a real reflection (Si (333) at 10.367 keV). The slight slope of the flat top is an absorption effect.

boundary demands that the imaginary part be parallel to the surface normal, and therefore there is extinction of the Bloch waves along that direction, just as with total reflection. For a slightly different direction of the incident wave vector, the surface normal can hit either the α or the β branch, and a propagating mode is excited. The angular width of total reflection is called the "rocking width".

For the symmetrical Bragg case, the rocking width can be determined in a geometric argument: Using eq. (3.11) and the approximation $(k^2-\mathbf{k}_h^2) \approx 2k\xi_h$, the difference of the wave numbers on the apices of the dispersion surface is $\Delta k = 4\pi r_e|\tilde{\eta}_\mathrm{H}|C/k$, and a geometrical projection gives for the difference wave vector $\Delta\mathbf{k}$ between these points

$$|\Delta\mathbf{k}| = \frac{4\pi r_e|\tilde{\eta}_\mathrm{H}|C}{k\cos\Theta}, \tag{3.16}$$

which, by simple geometry, is related to the angular width $\Delta\Theta$ of the total reflection range by $k\Delta\Theta\sin\Theta = |\Delta\mathbf{k}|$. Therefore, the rocking width is:

$$\Delta\Theta = \frac{8\pi r_e C|\tilde{\eta}_\mathrm{H}|}{k^2\sin 2\Theta} \tag{3.17}$$

On the right-hand side of fig. 3.6, the surface normal intersects both branches of the real-valued dispersion surface for any incidence angle. Therefore two modes are excited, and an interesting interference phenomenon occurs. Whenever the waves with wave vectors $\mathbf{k}_0$ on the α and β branches are in phase, then the ones with wave vectors $\mathbf{k}_\mathrm{H}$ are out of phase by π, and vice versa.

Because there is a wave vector difference $\Delta\mathbf{k}$ between the α and the β mode, this phase difference changes upon propagation into the crystal, and the period of this so-called pendellösung is reciprocal to $\Delta\mathbf{k}$. Pendellösung is German for "pendulum solution" and is meant as an analogy to the problem of two coupled pendula of slightly different lengths.

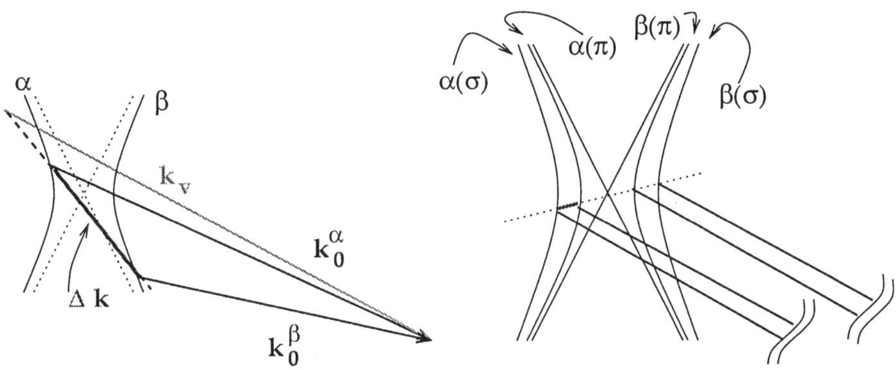

Figure 3.7: Pendellösung: Left: A difference $\Delta\mathbf{k}$ between tie points on the α and β branches is the cause of a beat-type interference upon propagation. Right: With both, σ and π polarized radiation present, the α and β branches both split into pairs and additional pendellösung periods appear. The σ-π beat results in a periodic change of the helicity.

The maximum pendellösung length $L_\mathbf{H}$ corresponds to the minimum $\Delta|\mathbf{k}|$ in eq. (3.16) between the apices of the two branches of the dispersion surface and therefore gives a measure of the scattering power of the crystal for a given reciprocal lattice vector.

$$L_\mathbf{H} = \frac{2\pi}{|\Delta\mathbf{k}|} = \frac{k\cos\Theta}{2r_e\tilde{\eta}_\mathbf{H}C} = \frac{\lambda}{\Delta\Theta\sin\Theta} \tag{3.18}$$

The expression for $L_\mathbf{H}$ in the commonly used notation of[73] is given in app. 5.1, eq. (3.78), compare also [74], p. 97 (where 2δ has the meaning of $\Delta\Theta$ here). A characteristic pendellösung time $\tau_\mathbf{H}$ can be defined as the time it takes the x-rays to traverse $L_\mathbf{H}$ at their group velocity, which is $c\cos\Theta$ at the symmetry points of the dispersion surface (i.e., for equally large $|\tilde{\mathbf{A}}_0|$ and $|\tilde{\mathbf{A}}_\mathbf{H}|$).

$$\tau_\mathbf{H} = \frac{k}{2cr_e\tilde{\eta}_\mathbf{H}C} \tag{3.19}$$

Equation (3.13) shows that the dispersion surface depends, among other things, on the polarization of the x-rays. Figure 3.7, on the right-hand side, shows the α and β branches for both polarization states and the difference wave vector between $\alpha(\sigma)$ and $\alpha(\pi)$. The wave vector difference between the polarization states results in a helicity pendellösung, which is routinely used in synchrotron radiation laboratories to generate circularly polarized x-rays [99].

3. Extended Takagi-Taupin Theory

The dynamical theory shall now be extended to include disturbances of a perfect crystal lattice by static strain or acoustic phonons, deformations of the unit cells by optical phonons, and direct electronic changes of the structure factor. The latter two will be important in ch. 4. Time dependence will be included explicitly (see also ch. 10 for a somewhat different approach). The theory is a combination of geometrical optics and wave optics because an eikonal function will be introduced and a wave-optical calculation picks up what is left over from the geometric-optical approximation. With these dual properties, the range from adiabatic change to abrupt transitions can be covered.

For static or quasistatic disturbances (criteria (i) to (iii) below), the theory might be developed on the basis of macroscopic electromagnetism. If, however, the disturbance of the diffracting crystal occurs on a time scale that is comparable to that of the x-ray scattering itself (see below), then the motion of charges due to the disturbance becomes relevant. This motion is not accounted for in a formulation of the theory in which the electric polarization is proportional to the fields by way of the susceptibility χ (see also discussion in app. 5.3). The same may also be seen in a formal argument: Spatial derivatives of χ are contained explicitly in eq. (3.67), but time derivatives are not. Spatial and temporal disturbances of χ are therefore treated in fundamentally different ways, and the theory might be inappropriate for problems in which a disturbance occurs within the x-ray scattering time which is equivalently given by any one of (i) the reciprocal energy bandwidth of an x-ray Bragg reflection, (ii) the time, given by eq. (3.19), for the x-rays to traverse one pendellösung period, or (iii) the x-ray oscillation frequency times the electric susceptibility. Depending on the scattering strength in the respective reflection, this x-ray time scale is between a few femtoseconds and picoseconds.

It is then better to employ a microscopic description. The one developed in this chapter has of course its own shortcomings—it is in particular not adequate for x-ray resonant cases. But with the emphasis on coherent manipulation of diffracting x-rays (ch. 4), crystals of light elements with absorption edges well below the x-ray energy are more important, anyway. A full treatment of resonant x-ray diffraction, with the resonant atoms oscillating at frequencies of the order of the reciprocal bandwidth of the x-ray reflection can be treated neither on the basis of macroscopic electromagnetism nor on the basis of the jellium model used here and is outside the scope of this book.

The problems of static strain and lattice defects are usually treated by the Takagi-Taupin theory [82, 83, 84] (henceforth the TTT). In order to accommodate the variable diffractive properties across the crystal, a modulated Bloch wave is defined, i.e., instead of having constant Bloch component wave amplitudes as in eq. (3.5), a slight spatial variability $\tilde{\mathbf{E}}_{\mathbf{h}}(\mathbf{r})$ is admitted. The wave vector $\mathbf{k_0}$ is chosen as deemed most suitable for a particular problem and then kept constant throughout. Usually [83].(9), the refraction-corrected wave vector $\mathbf{k}_r$ of a plane wave incident on the crystal is used. Even if it is chosen to match a tie point in a locally approximated perfect crystal at any point within the disturbed crystal, it will not fulfill eq. (3.11) at other points due to the change of the Bloch component amplitudes during propagation (see fig. 3.8). The TTT

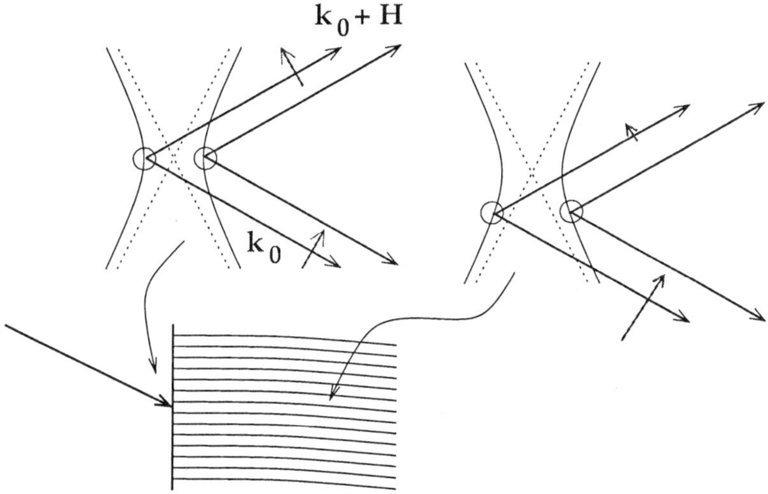

Figure 3.8: Motion of the tie points in response to changing field amplitudes in a distorted crystal.

does not contain the concept of a dispersion surface and thus makes it difficult to visualize the x-ray diffraction in the same way as for a perfect crystal. There is however a possibility to define a locally approximated dispersion surface by the introduction of an eikonal function [85, 86]. This facilitates an intuitive understanding, permits qualitative statements without having to run a computer simulation, and can make computer simulations more accurate than those based on the TTT. The same technique can also be applied to time-dependent phenomena. The purpose of the eikonal function is to take up as much as possible of the spatial and temporal dependence of the field amplitudes into a factor common to all of them that comprise one mode of dynamical diffraction. Only the remaining variability of the field amplitudes needs to be determined by solving a differential equation. How large that remainder is depends of course on the details of the problem, and in particular on the strength of a disturbance, the propagation length to be covered, and the stability of a solution of the differential equations derived below with respect to a disturbance. The latter point is illustrated in an example in sec. 3.8.2 and all of them are discussed in sec. 3.10. Interestingly, a Takagi theory of magnetic x-ray diffraction in disturbed crystals is also based upon microscopic electromagnetism [75].

3.1 Disturbed Lattice, Fourier and Bloch Sums

Acoustic phonons and static strain in a crystal lattice can be expressed by a vector field $\mathbf{u}(\mathbf{r}, t)$, which represents the displacement of the lattice from $\mathbf{r}$ to $\mathbf{r} + \mathbf{u}$. Technically, the disturbed lattice is now a manifold over the space that can be parametrized by the coordinates of the positions $\mathbf{r}$ of the same lattice in the undisturbed space. One can now define a local reciprocal lattice [83] as the tangent space in point $\mathbf{r}$ in which the local reciprocal lattice vectors $\mathbf{h}^\circ$ (see fig. 3.14) are associated with the global reciprocal lattice vectors $\mathbf{h}$ of the perfect lattice by:

$$\mathbf{h}^\circ = \nabla(\mathbf{h} \cdot (\mathbf{r} - \mathbf{u}(\mathbf{r}))) = \mathbf{h} - \nabla \mathbf{u} \cdot \mathbf{h}. \tag{3.20}$$

If told (at variance with the facts) that the lattice is perfectly periodic and asked to determine the reciprocal lattice from the electron density in the immediate vicinity of $\mathbf{r}$, the reciprocal lattice vectors in eq. (3.20) are what one would come up with (with an accuracy given by the available size of the vicinity of $\mathbf{r}$). Take, for example, the curved lattice planes shown in fig. 3.14, with $\mathbf{u}$ going against the direction of $\mathbf{H}$ (shown in the figure). As one goes to the right, $\mathbf{u} \cdot \mathbf{H}$ becomes more negative. Therefore, $\nabla \mathbf{u} \cdot \mathbf{H}$ is directed to the left, and $\mathbf{H} - \nabla \mathbf{u} \cdot \mathbf{H}$ is rotated clockwise with respect to $\mathbf{H}$ (shown in the figure).

It is also possible that the electron density is modulated on the background of an undisturbed lattice. i.e., there is no displacement $\mathbf{u}$ and the reciprocal lattice vectors remain unchanged. In contrast to the above displacive disturbance of the crystal, this one might be called a structure factor disturbance. In the most general case, when both types of disturbance are present, the Fourier sum eq. (3.4) for the electron density must be modified in both, the Fourier amplitudes, and the exponents: The latter must read $i\mathbf{h} \cdot (\mathbf{r} - \mathbf{u})$ instead of $i\mathbf{h} \cdot \mathbf{r}$ to express a static or acoustic phonon-induced lattice deformation. Optical phonons, excitons, electronic excitation, etc., modify the electron density on the background of an existing lattice, and their effect can be taken into account by making the Fourier amplitudes depend on space and time as $\tilde{\eta}(\mathbf{r}, t)$. The expression (3.4) is therefore replaced with:

$$\eta(\mathbf{r}, t) = \sum_{\mathbf{h}} \tilde{\eta}_{\mathbf{h}}(\mathbf{r}, t) e^{i\mathbf{h} \cdot (\mathbf{r} - \mathbf{u})}. \tag{3.21}$$

It is convenient for the further development of the theory to model $\eta(\mathbf{r}, t)$ as a sum of distorted Bloch waves, representing optical phonons, excitons, etc.,

$$\eta(\mathbf{r}, t) = \sum_m \sum_{\mathbf{h}} \tilde{\eta}_{m,\mathbf{h}} e^{i\mathfrak{s}_m} e^{i\mathbf{h} \cdot (\mathbf{r} - \mathbf{u})}, \qquad \mathfrak{s}_m = (\mathfrak{k}_m \cdot (\mathbf{r} - \mathbf{u}) - \varpi_m t), \tag{3.22}$$

where $\mathfrak{k}_m$ and ϖ_m are the base vector and the frequency of the m-th Bloch wave with $\mathfrak{k}_{-m} = -\mathfrak{k}_m$ and $\varpi_{-m} = -\varpi_m$. The terms with index $m = 0$ represent the electron density of a perfect, undisturbed crystal, i.e., $\mathfrak{k}_0 = \mathbf{0}$ and $\varpi_0 = 0$. Please note that the phase $\mathfrak{s}_m$ is defined relative to the deformed lattice, i.e., contains $\mathfrak{k}_m \cdot (\mathbf{r} - \mathbf{u})$, instead of $\mathfrak{k}_m \cdot \mathbf{r}$. The reason is that a modification of the electron density within the unit cells (i.e., an optical phonon, or a modification in the band/bond-structure, see ch. 4, fig. 4.1) is 'carried along' with an overall lattice distortion $\mathbf{u}$. The Fourier components $\tilde{\eta}_{m,\mathbf{h}}$ can be naturally grouped into sets of four, each containing indices $\pm m$ and $\pm\mathbf{h}$ (except of course for $m = 0$ and $\mathbf{h}$). Several statements can be made about the members of each set (see

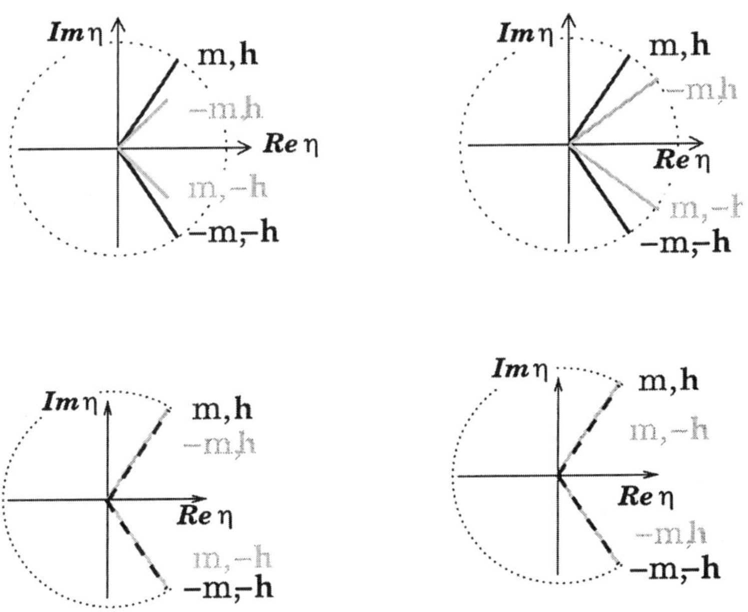

Figure 3.9: Possible and impossible combinations of $\tilde{\eta}_{m,\mathbf{h}}$. The one in the upper left-hand graph is excluded by definition (see (2) in the text). The upper right-hand graph shows the most general possible case with $\tilde{\eta}_{\pm m,\pm\mathbf{h}}$ all having the same modulus and coming in complex conjugate pairs and the two lower ones show special cases (3) amnd (4) in the text.

fig. 3.9): (1), any disturbance to the lattice is a real-valued change of the real-valued electron density. Therefore, $\tilde{\eta}_{m,\mathbf{h}}$ and $\tilde{\eta}_{-m,-\mathbf{h}}$ are complex conjugates of each other. (2), $\tilde{\eta}_{m,\mathbf{h}}$ and $\tilde{\eta}_{-m,\mathbf{h}}$ must have the same modulus, excluding the configuration in the upper left-hand graph of fig. 3.9, because otherwise, the reciprocal lattice would be changed: Take, for example, the extreme case of $\tilde{\eta}_{m,\mathbf{h}} \neq 0$ and $\tilde{\eta}_{-m,\mathbf{h}} = 0$, i.e., only two Fourier components, $\tilde{\eta}_{m,\mathbf{h}}$ and $\tilde{\eta}_{-m,-\mathbf{h}}$, are present with exponents $\pm i((\mathbf{h} + \mathfrak{k}_m) \cdot (\mathbf{r} - \mathbf{u}) - \varpi_m t)$, and the lattice periodicity is actually given by the reciprocal vector $\mathbf{h} + \mathfrak{k}_m$. This is however excluded by definition because lattice parameter changes are to be described by $\mathbf{u}$. The most general possible case is therefore the one shown in the upper right-hand graph in fig. 3.9. If (3) $\tilde{\eta}_{m,\mathbf{h}} = \tilde{\eta}_{-m,\mathbf{h}}$ for a given $\mathbf{h}$ and all

m (lower left-hand graph in fig. 3.9), then the disturbance does not introduce any inversion-asymmetry relative to $\mathbf{h}$ beyond what may already be present in the crystal in the sense that the coordinate origin can then be chosen to make $\tilde{\eta}_{m,\mathbf{h}}$ real-valued for all m. If this coordinate origin is independent of $\mathbf{h}$, then the crystal is inversion-symmetric relative to it and the disturbance does not change that property. (4) If $\tilde{\eta}_{0,\mathbf{h}} = \tilde{\eta}_{0,-\mathbf{h}}$ for all $\mathbf{h}$ (lower right-hand graph in fig. 3.9), then the structure is inversion-symmetric in the absence of the disturbance and that property is preserved under the disturbance only if $\tilde{\eta}_{m,\mathbf{h}} = \tilde{\eta}_{-m,\mathbf{h}}$ and $\tilde{\eta}_{m,\mathbf{h}} = \tilde{\eta}_{m,-\mathbf{h}}$ for all $m, \mathbf{h}$.

Equation (3.22) is a very general expression, permitting a multitude of wave vectors and frequencies of a disturbance, for each reciproal latice vector separately. In many practical cases, such as scattering in the presence of laser-generated optical phonons, it will be sufficient to keep only terms with $m = 0, \pm 1$ for each $\mathbf{h}$ in eq. (3.22), with $\tilde{\eta}_{0,\mathbf{h}}$ representing the respective Fourier component of the electron density in the undisturbed crystal, and $\tilde{\eta}_{\pm 1,\mathbf{h}}$ and being the amplitudes of a single plane wave with wave vectors $\mathfrak{k}_1$ and $\mathfrak{k}_{-1} = -\mathfrak{k}_1$.

$$\eta(\mathbf{r}, t) = \sum_{\mathbf{h}} \left(\tilde{\eta}_{0,\mathbf{h}} + \left(\tilde{\eta}_{1,\mathbf{h}} e^{i\mathfrak{s}_1} + \tilde{\eta}_{-1,\mathbf{h}} e^{-i\mathfrak{s}_1}\right)\right) e^{i\mathbf{h}\cdot(\mathbf{r}-\mathbf{u})}. \tag{3.23}$$

Generally, a spatial and—depending on the type of the problem—also a temporal dependence of the Bloch wave amplitudes will have to be admitted as $\tilde{\mathbf{A}}(\mathbf{r}, t)$. The above-mentioned eikonal factor $\exp(i\varphi(\mathbf{r}, t))$ applies to all of the Bloch sum, in the same way as $\exp(i(\mathbf{k}_0 \cdot \mathbf{r} - \omega t))$ in eq. (3.5). The optical path length along the wave vector $\mathbf{k}_0 + \nabla\varphi$ is given by $\psi = (\mathbf{k}_0 \cdot \mathbf{r} + \varphi)/k$, *and it is that quantity that is usually called the eikonal.* To distinguish φ from it, φ will be called "eikonal function" here. A more detailed discussion of the eikonal is given in sec. 3.4. In the usual definition, the eikonal ψ has the dimension of a length [100]. This is not very appropriate here because time dependence of the crystal diffractive properties will be treated on a same footing with spatial dependence.

In the theory of dynamical diffraction for a perfect crystal, several Bloch base vectors $\mathbf{k}_0$ are required to describe the full diffraction problem on several branches of the dispersion surface. Here, $\mathbf{k}_0 + \nabla\varphi$ takes the role of $\mathbf{k}_0$ in sec. 2 and $\omega + \dot{\varphi}$ corresponds to ω there. The Bloch base vector $\mathbf{k}_0$ and frequency ω will now be taken to be unique (as in the original TTT), but a number of eikonal functions φ_n will be admitted, corresponding to the number N of branches of the dispersion surface in the related perfect-crystal diffraction problem. The positive-directed part of the modulated multiple Bloch wave is then (compare with eq. (3.5)):

$$\mathbf{A}^{(+)}(\mathbf{r}, t) = \sum_{n,\mathbf{h}} \left(\tilde{\mathbf{A}}_n\right)_{\mathbf{h}} (\mathbf{r}, t) e^{iS_{n,\mathbf{h}}(\mathbf{r},t)} \tag{3.24}$$

with

$$S_{n,\mathbf{h}}(\mathbf{r},t) = (\mathbf{k_0}\cdot\mathbf{r} + \varphi_n(\mathbf{r},t) - \omega t) + \mathbf{h}\cdot(\mathbf{r}-\mathbf{u}). \tag{3.25}$$

Other than in sec. 4, linear superposition of the x-rays is valid within the scope of this section, and the index n will be suppressed in most of the following text.

Please note that only $\mathbf{h}$, but not $\mathbf{k_0}$, is multiplied with $(\mathbf{r}-\mathbf{u})$. This is different from eq. (3.22) and is a *choice*, reflecting a particular point of view, namely that the Bloch wave is "labeled" with $\mathbf{k_0}$, which is determined by the simple plane wave incident on the crystal. The derivations might just as well be carried out with $\mathbf{k_0}\cdot(\mathbf{r}-\mathbf{u})$, and $\nabla\varphi$ would make up for the difference (see below).

3.2 The System of Differential Equations

Just as in the derivation of eq. (3.8), we can now combine eqs. (3.22) and (3.24) and use the index transformation $\mathbf{h}+\mathbf{h}'\to\mathbf{h}$ and $\mathbf{h}'\to\mathbf{h}'$ to obtain the contribution of the fields to the current. In order to reduce the formula clutter, only the positive-directed waves and currents $\tilde{\mathbf{A}}_{\mathbf{h}}^{(+)}$ and $\mathbf{J}^{(+)}$ will be considered, and the $(+)$ will be suppressed. Also, $\mathbf{r}$ and t will henceforth be omitted as arguments to $\tilde{\eta}$, $\tilde{\mathbf{A}}_{\mathbf{h}'}^{(+)}$, $S_{\mathbf{h}}(\mathbf{r},t)$ and $\mathfrak{s}_m(\mathbf{r},t)$.

$$\mathbf{J}(\mathbf{r},t) = -r_e c \sum_{\mathbf{h},\mathbf{h}',m} \tilde{\eta}_{m,\mathbf{h}-\mathbf{h}'}\tilde{\mathbf{A}}_{\mathbf{h}'} e^{i(S_{\mathbf{h}}+\mathfrak{s}_m)}. \tag{3.26}$$

In order to express the integral in eq. (3.3) for the transverse current, we use Taylor expansions up to first order in $\mathbf{r}'$, centered at $\mathbf{r}$ for $\mathbf{h}\cdot\mathbf{u}(\mathbf{r}',t)$, $\varphi(\mathbf{r}',t)$, and $\tilde{\mathbf{A}}_{\mathbf{h}'}$. In this simple approximation, the results are actually equivalent to those obtained from macroscopic electromagnetism (see app. 5.6.2).

$$\mathbf{J}_t = -r_e c \sum_{\mathbf{h},\mathbf{h}',m} \tilde{\eta}_{m,\mathbf{h}-\mathbf{h}'}\left[\tilde{\mathbf{A}}_{\mathbf{h}'}\right]_{\mathbf{h}} e^{i(S_{\mathbf{h}}+\mathfrak{s}_m)}, \tag{3.27}$$

where, using the abbreviation $\mathbf{k}_{\mathbf{h}^\circ} = \mathbf{k_0}+\mathbf{h}-\nabla\mathbf{u}\cdot\mathbf{h}$,

$$\left[\tilde{\mathbf{A}}_{\mathbf{h}'}\right]_{\mathbf{h}}(\mathbf{r},t) = \frac{(\mathbf{k}_{\mathbf{h}^\circ}+\mathfrak{k}_m+\nabla\varphi)\times\tilde{\mathbf{A}}_{\mathbf{h}'}\times(\mathbf{k}_{\mathbf{h}^\circ}+\mathfrak{k}_m+\nabla\varphi)}{|\mathbf{k}_{\mathbf{h}^\circ}+\mathfrak{k}_m+\nabla\varphi|^2}. \tag{3.28}$$

Upon insertion of eqs. (3.24) and (3.27) into (3.2) and neglect of second-order derivatives of the field amplitudes $\tilde{\mathbf{A}}$ and products of first-order derivatives of φ and $\mathbf{u}$ with first-order derivatives of $\tilde{\mathbf{A}}$:

$$2i\left(\frac{k}{c}-\frac{\dot\varphi}{c^2}+\frac{\dot{\mathbf{u}}\cdot\mathbf{h}}{c^2}\right)\frac{\partial\tilde{\mathbf{A}}_{\mathbf{h}}}{\partial t} - 2i\left(\mathbf{k}_{\mathbf{h}^\circ}+\nabla\varphi\right)\cdot\nabla\tilde{\mathbf{A}}_{\mathbf{h}} =$$
$$\left[\left(k-\frac{\dot\varphi}{c}+\frac{\dot{\mathbf{u}}\cdot\mathbf{h}}{c}\right)^2-(\mathbf{k}_{\mathbf{h}^\circ}+\nabla\varphi)^2\right]\tilde{\mathbf{A}}_{\mathbf{h}} - 4\pi r_e\sum_{m,\mathbf{h}'}\tilde{\eta}_{m,\mathbf{h}-\mathbf{h}'}e^{i\mathfrak{s}_m}\left[\tilde{\mathbf{A}}_{\mathbf{h}'}\right]_{\mathbf{h}} \tag{3.29}$$

For any given wave vector $\mathbf{k}_0$ and any given function $\varphi(\mathbf{r}, t)$, this is a system of first-order partial differential equations. *To preclude any misunderstanding: The function $\varphi(\mathbf{r}, t)$ is not sought as a solution of eq. (3.29). It is an arbitrary function, available (details below) to tune some of the properties of eq. (3.29), which is then solved for the field amplitudes.*

The magnitude of the right-hand side—and thus the magnitude of the derivatives of $\tilde{\mathbf{A}}_{\mathrm{h}}$ on the left-hand side—depends on the choice of $\mathbf{k}_0$ and $\varphi(\mathbf{r}, t)$. In a time-independent problem, $\dot{\mathbf{u}} \cdot \mathbf{h} = 0$ and φ can be made independent of t. The right-hand side of eq. (3.29) is then essentially the same as in eq. (3.11), and the solutions $(\mathbf{k}_{\mathrm{h}^\circ} + \nabla\varphi)$ of the determinant condition provide the local dispersion surfaces (as defined at the beginning of sec. 3.1). Tie points are now located on the local dispersion surfaces. If, for a given set $\{\tilde{\mathbf{A}}_{\mathrm{h}}\}$ of amplitudes, the right-hand side of eq. (3.29) does not vanish, then the wave vector $(\mathbf{k}_{\mathrm{h}^\circ} + \nabla\varphi)$ represents no tie point (one need not exist for the given set $\{\tilde{\mathbf{A}}_{\mathrm{h}}\}$, but the field amplitudes can always be split into modes, for which tie points do exist *locally* (see p. 45)). In such a case, eq. (3.29) makes $i\nabla\tilde{\mathbf{A}}_{\mathrm{h}}$ pick up the balance, as shown graphically in in fig. 3.10 (see also the discussion in sec.3.9 for the equivalence of $\nabla\varphi\tilde{\mathbf{A}}_{\mathrm{h}}$ and $i\nabla\tilde{\mathbf{A}}_{\mathrm{h}}$). A similar argument holds for the

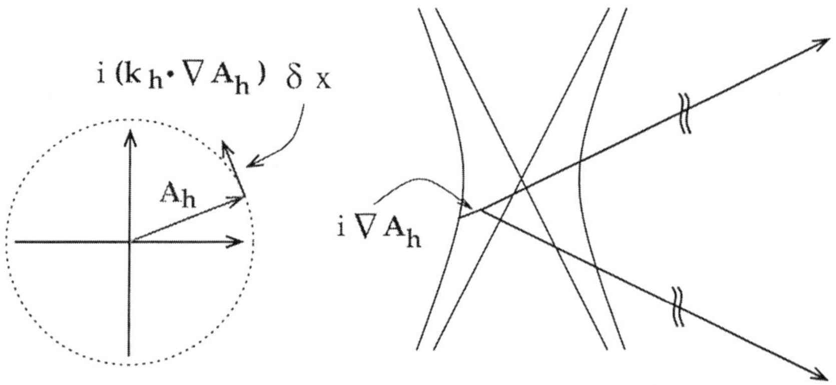

Figure 3.10: Phasor diagram of the field amplitudes. Right: A mismatch in the wave vectors relative to a tie point is taken up by i times the directional derivative of the field amplitudes, making them rotate in the phasor diagram on the left-hand side, the angle of rotation being proportional to the propagation length δx.

time derivatives $(\partial/\partial t)\tilde{\mathbf{A}}_{\mathrm{h}}$ and $\dot{\varphi}$. This discussion is carried further in sec. 3.6.

When solving eq. (3.29) numerically, one would like to keep the derivatives as small as possible, which can be arranged by a proper choice of $\mathbf{k}_0$ and $\varphi(\mathbf{r}, t)$. At the same time, this will minimize the second-order derivatives of $\tilde{\mathbf{A}}_{\mathrm{h}}$ that were neglected in eq. (3.29) and which are a source of error, independent of the numerical integration algorithm. Because the minimization removes the effect of perfect-crystal diffraction on the field amplitudes (i.e., the dispersion splitting), only the disturbance itself is left for the differential equations. Qualitative information on the effect of a disturbance on the diffracting x-rays and order-of-magnitude estimates can then be obtained by equating the disturbance terms with the derivatives of the field amplitudes (see sec. 3.8 and also sec. 3.3.2). This can be of great help in the design of x-ray optics, such as in ch. 4.

With $\mathbf{k_0}$ chosen close to what it would be in the dynamical theory of an undistorted crystal, only a small correction $\nabla\varphi$ is required to match a local dispersion surface or approximate it with a global choice of $\varphi(\mathbf{r}, t)$. If only the leading order is to be kept, $\nabla\varphi$ and $\dot\varphi$ can be neglected on the left-hand side of eq. (3.29). On the right-hand side, we approximate $(k - (\dot\varphi/c) + (\dot{\mathbf{u}}\cdot\mathbf{h})/c)^2 \approx (k - (\dot\varphi/c))^2 + 2(k/c)\dot{\mathbf{u}}\cdot\mathbf{h}$ and $(\mathbf{k_{h^\circ}} + \nabla\varphi)^2 \approx (\mathbf{k_h} + \nabla\varphi)^2 - 2\mathbf{k_h}\cdot\nabla\mathbf{u}\cdot\mathbf{h}$. This turns eq. (3.29) into

$$
\begin{aligned}
2i\frac{k}{c}\frac{\partial\tilde{\mathbf{A}}_\mathbf{h}}{\partial t} - 2i\mathbf{k_h}\cdot\nabla\tilde{\mathbf{A}}_\mathbf{h} = \\
\left(\left(k - \frac{\dot\varphi}{c}\right)^2 - (\mathbf{k_h} + \nabla\varphi)^2\right)\tilde{\mathbf{A}}_\mathbf{h} - 4\pi r_e \sum_{m,\mathbf{h}'} \tilde\eta_{m,\mathbf{h}-\mathbf{h}'}\left[\tilde{\mathbf{A}}_{\mathbf{h}'}\right]_\mathbf{h} + \\
2\left[\frac{k}{c}\dot{\mathbf{u}}\cdot\mathbf{h} + \mathbf{k_h}\cdot\nabla\mathbf{u}\cdot\mathbf{h}\right]\tilde{\mathbf{A}}_\mathbf{h} - 4\pi r_e \sum_{m,\mathbf{h}'}\left(e^{i\mathfrak{s}_m} - 1\right)\tilde\eta_{m,\mathbf{h}-\mathbf{h}'}\left[\tilde{\mathbf{A}}_{\mathbf{h}'}\right]_\mathbf{h}.
\end{aligned}
\tag{3.30}
$$

This is a very general expression which can describe both, displacive disturbances (through $\mathbf{u}$) and structure factor modulations (optical phonons, etc., through $\tilde\eta_{m,\mathbf{h}}$). Only rarely do both appear in combination, and in most cases, the equation can be simplified considerably. Examples are given in sec. 3.8. The separation into a sum with $\tilde\eta_{m,\mathbf{h}-\mathbf{h}'}$ and one with $(\exp(i\mathfrak{s}_m) - 1)\tilde\eta_{m,\mathbf{h}-\mathbf{h}'}$ was made to emphasize the role of the latter as a disturbance. If $|\mathfrak{k}_m|$ is much less than the reciprocal pendellösung period and ϖ_m is much less than the reciprocal time to traverse it, then a dipole approximation may applied and the exponential term $\exp(i\mathfrak{s}_m) - 1$ in the second sum be replaced with $i(\mathfrak{k}_m\cdot\mathbf{r} - \varpi_m t)$. There is however some subtlety to that, see sec. 3.8.1. Also, when running a computer simulation by the method of characteristics (see sec. 3.5), then there is no advantage to splitting the exponential in the way of eq. (3.30), and one might rather keep it all in one sum.

At this point, it might be helpful to discuss the different types of disturbances, and how they affect the diffracting x-rays. The vector field $\mathbf{u}(\mathbf{r}, t)$ describes a deformation of the lattice in space and time. This might, for example, be due to bending of an otherwise perfect crystal (see fig. 3.11), or it might be a periodic lattice spacing modulation, similar to a frequency modulation (FM) of a radio carrier wave. In contrast to that, $\tilde\eta_{m,\mathbf{h}}$ refers to a disturbance of a lattice whose periodicity is unaltered, similar to an amplitude modulation (AM) of a radio carrier wave.

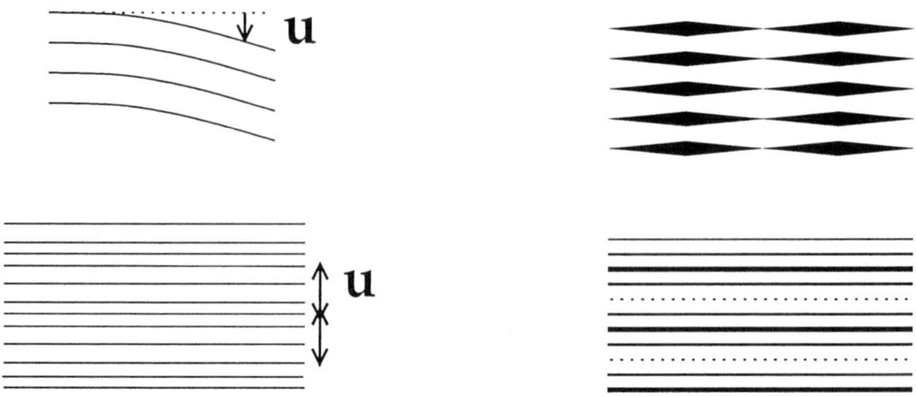

Figure 3.11: Different types of crystal disturbances. Left: displacive, right: structure factor modulation type. Upper left graph: a bent crystal (see also fig. 3.14), lower left graph: a periodically modulated lattice spacing, as produced by a sound wave. Right: The scattering power of the lattice is modulated but its spacing is the same as in an undisturbed crystal.

The two types of disturbance ($\mathbf{u}$ and $\tilde{\eta}_{m,\mathbf{h}}$) have different local effects on the derivatives of the field amplitudes. Whereas a displacement $\mathbf{u}$ couples the derivative of $\tilde{\mathbf{A}}_{\mathrm{h}}$ only to that same amplitude, does a structure factor modulation affect $\nabla\tilde{\mathbf{A}}_{\mathrm{h}}$ through the amplitudes of all waves. After some propagation, a displacive disturbance will of course also couple all field amplitudes with with each other, but only indirectly so through the scattering.

If the disturbance is periodic with a wave vector $\mathfrak{k}$ and frequency ϖ (which $\tilde{\eta}_{m,\mathbf{h}}$ is by definition and $\mathbf{u}$ may or may not be), then one might introduce more Fourier components of the field amplitudes with wave vectors $\mathbf{k}_0 + \mathbf{h} \pm \mathfrak{k}$ and frequencies $\omega \pm \varpi$ and convert the differential equation (3.30) into an algebraic equation which gives the amplitudes of the main reflections and satellites (associated with the reciprocal lattice vectors $\mathbf{h} \pm \mathfrak{k}$). This approach, being limited to periodic problems, will not be pursued here but it helps to form an expectation of the behavior of solutions of the differential equation (3.30).

Just as L_{H} is a typical length scale of the diffraction in an undisturbed crystal, so is it possible to define a length scale $L_{\mathrm{H}}^{(d)}$ and a time scale $\tau_{\mathrm{H}}^{(d)}$ of the effect of a disturbance on the diffracting x-rays. This can be done by taking L_{H}, defined in eq. (3.18), multiplied with the ratio of the disturbance terms versus the perfect-crystal diffraction term $4\pi r_e \tilde{\eta}_{\mathrm{H}}$. For a displacive disturbance, this gives

$$L_{\mathrm{H}}^{(d)} = \frac{k\pi\cos\Theta}{\mathbf{k}_{\mathrm{H}} \cdot \nabla\mathbf{u} \cdot \mathbf{H}} \qquad \tau_{\mathrm{H}}^{(d)} = \frac{\pi\cos\Theta}{\dot{\mathbf{u}} \cdot \mathbf{H}}. \tag{3.31}$$

For a structure factor modulation, we approximate $\exp(i\mathfrak{s}_m) - 1 \approx i\mathfrak{s}_m$. The phase $\mathfrak{s}_m = \mathfrak{k}_m \cdot \mathbf{r} - \varpi_m t$ is obtained by inserting L_{H} from eq. (3.18) for $\mathbf{r}$

and $\tau_{\mathbf{H}}$ from eq. (3.19) for t. In the simplest case, m ranges over 0 and ± 1 (see p. 54), with $\tilde{\eta}_{0,\mathbf{H}}$ for the Fourier component of the undisturbed crystal and $\tilde{\eta}_{\pm 1,\mathbf{H}}$ for the disturbance.

$$L_{\mathbf{H}}^{(d)} = \frac{C}{\mathfrak{k}_1} \left| \frac{\tilde{\eta}_{0,\mathbf{H}}}{\tilde{\eta}_{1,\mathbf{H}}} \right| \qquad \tau_{\mathbf{H}}^{(d)} = \frac{C \cos\Theta}{\varpi_1} \left| \frac{\tilde{\eta}_{0,\mathbf{H}}}{\tilde{\eta}_{1,\mathbf{H}}} \right| \tag{3.32}$$

3.3 Comparison with the Takagi-Taupin Theory

The difference between the Takagi-Taupin theory and the eikonal Takagi theory of this chapter can be illustrated by writing the dielectric function of the disturbed crystal as $\epsilon = 1 + \chi_p(\mathbf{r}) + \chi_d(\mathbf{r}, t)$, where χ_p is the spatially periodic susceptibility of the undisturbed crystal and $\chi_d(\mathbf{r}, t)$ is due to the disturbance. This is for illustrative purposes only, and has no bearing upon the derivations of this chapter which are based on microscopic electromagnetism.

The Takagi-Taupin theory treats the field amplitudes $\mathcal{E}_{\mathrm{h}}(\mathbf{r}, t)$ in the form $\mathbf{E}_{\mathrm{h}}^{\mathrm{T}}(\mathbf{r}) \exp(i(\mathbf{k}_{\mathrm{h}} \cdot \mathbf{r} - \omega t))$, with $\mathbf{k}_{\mathrm{h}}$ corresponding to the "1" in ϵ and the remainder $\mathbf{E}^{\mathrm{T}}$ of the spatial dependence of $\mathcal{E}_{\mathrm{h}}$, corresponding to $\chi_p + \chi_d$ in ϵ, left to be determined in a differential equation.

In the approach of Balibar and Authier [101, 102], the field amplitudes are written as $\mathbf{E}_{\mathrm{h},n}^{\mathrm{B}}(\mathbf{r}) \exp(i(\mathbf{k}_{\mathrm{h}n} \cdot \mathbf{r} - \omega t))$ (although n is omitted there), where $\mathbf{k}_{\mathrm{h},n}$ now originates from a tie point, and a full description of the fields in the crystal requires as many $\mathbf{k}_{\mathrm{h},n}$, as there are branches of the dispersion surface. The differential equations now must be solved for $\mathbf{E}_{n,\mathrm{h}}^{\mathrm{B}}$, which corresponds to only χ_d in ϵ, whereas the part corresponding to $1 + \chi_p$ needs not be solved for.

In the present theory, $\mathcal{E}_{\mathrm{h}}$ are written as $\mathbf{E}_{\mathrm{h},n}^{\mathrm{E}}(\mathbf{r}) \exp(i(\mathbf{k}_{\mathrm{h}} \cdot \mathbf{r} - \omega t + \varphi_n))$, with the contribution to $\mathcal{E}$, to be determined through a differential equation, corresponding to only part of χ_d. The terms "1", "χ_p" and part of "χ_d" are removed from the solution to be sought.

The approach of Balibar and Authier might be called a perturbation theory, with coefficients $\mathbf{E}^{\mathrm{B}}$ for the solutions $\exp(i(\mathbf{k}_{\mathrm{h}} \cdot \mathbf{r} - \omega t))$ of the exactly solvable problem. In contrast, the approach of this chapter might be called an adaptive perturbation theory. Furthermore, the eikonal Takagi theory of this chapter explicitly accomodates time-dependent structure factor changes beyond the adiabatic approximation.

3.3.1 Differential Equations. Takagi published a brief note [82] on his theory in 1962. The main work [83] appeared in 1969. Rather similar to it, except for the inclusion of a second order term, is the work of D. Taupin [84], published in 1964. Both authors base the derivations on macroscopic electromagnetism, i.e., on the $\mathbf{D}$ field and the electric susceptibility. Equation

[83].(39) in the notation of the present text reads:

$$-\frac{\mathbf{k_h}}{|\mathbf{k_h}|} \cdot \nabla(k\tilde{\mathbf{A}}_\mathrm{h}) = ik\frac{{\mathbf{k_h}_\circ}^2 - {k_r}^2}{2k^2}(k\tilde{\mathbf{A}}_\mathrm{h}) + i\frac{k}{2}\sum_{\mathrm{h'}\neq\mathrm{h}} 4\pi r_e \tilde{\eta}_{\mathrm{h}-\mathrm{h'}} k\left[\tilde{\mathbf{A}}_{\mathrm{h'}}\right]_\mathrm{h} \tag{3.33}$$

with the correspondences:

eq.(3.33)	k	k_r	$\mathbf{k_h}$	$\frac{{\mathbf{k_h}_\circ}^2-{k_r}^2}{2k^2}$	$k\tilde{\mathbf{A}}_\mathrm{h}$	$4\pi r_e k^2 \tilde{\eta}_\mathrm{h}$
[83].(39)	$2\pi K$	$2\pi k$	$-2\pi\mathbf{k_g}$	β'_h	$\mathbf{D}'_h$	$-\chi_h$

(3.34)

The system (3.33) of partial first-order differential equations is equivalent to a system of coupled ordinary differential equations [103].[ch1.5] and is usually written in that way [83].(40). For the two-wave case, Takagi converts this system into a single second order partial differential equation of hyperbolic type whose solutions can be expressed as an integral over the inital conditions by the method of Riemann [103].[ch5.4]. The characteristics of that second order hyperbolic differential equation are of course the same (along $\mathbf{k_0}$ and $\mathbf{k_h}$) as those of the first order system from which it originated.

Whereas eqs. (3.29) and (3.30) offer the possibility of adapting the wave vectors $\mathbf{k_h}_\circ + \nabla\varphi$ and frequencies $\omega + (\dot{\varphi}/c)$ to the changing conditions of propagation through the disturbed crystal, the same is not possible within the Takagi theory. One may, of course, choose $\mathbf{k_0}$ to match a tie point at a given location in the disturbed crystal but as the amplitudes will generally change in propagation, the tie point will shift and $\mathbf{k_0}$ cannot follow it (or rather, it can if $\nabla \times \mathbf{k_0} = \mathbf{0}$ is maintained but then we are back to the eikonal theory of this chapter). Because the Takagi equation lacks the means to approximate tie points, it makes the derivatives of the field amplitudes become much larger than with eq. (3.30) with eikonal functions determined appropriately for all modes. This may cause problems in a numerical simulation, and the neglected second-order derivatives may also induce an undetermined error.

3.3.2 Generalized Wave Fields. Nevertheless, if the shift of a tie point T is much less than its reciprocal distance from the symmetry center Q of the disersion surface, it is advantageous to make $\mathbf{k_0}$ match it. Doing so, Authier and Balibar [101, 102, 104] were able to introduce a local dispersion surface and generalized wave fields that correspond to the modes of standard dynamical diffraction. With that tool in hand, they could obtain qualitative information like the curvature of x-ray trajectories. The theory does, however, not permit the wave vectors to follow the shifting tie points and is therefore limited to small disturbances.

3.4 Comparison with Kato's Eikonal Theory

In his publications [77, 78, 79], N. Kato demonstrates the applicability of Fermat's principle to the diffraction of x-rays and derives the Euler-Lagrange equation that describes the beam trajectories. Starting from a modulated Bloch wave (in the notation of this text)

$$\mathbf{D}(\mathbf{r}, t) = e^{ikS_0} \sum_{\mathbf{h}} \mathbf{D}_{\mathbf{h}} e^{i\mathbf{h}\cdot\mathbf{r}}, \tag{3.35}$$

and using the properties of the perfect-crystal dispersion surface, he derives a variational principle and a differential equation that describes beam trajectories. The variational principle for a beam trajectory from A to B is obtained by use of the convex/convcave shapes of the branches of the dispersion surface [77].(44):

$$\delta \int_A^B (\mathbf{k_0} \cdot d\mathbf{r}) = 0, \tag{3.36}$$

which leads to the Euler-lagrange equation [77].(45). Using the property of the dispersion surface that the Poynting vector is perpendicular to it, this is converted to [77].(47)

$$\frac{d}{d\ell} \mathbf{k}_{0,\ell} = \frac{\partial \mathbf{k_0}}{\partial \ell} \cdot \mathbf{n} \tag{3.37}$$

where ℓ parametrizes the trajectory, $\mathbf{k}_{0,\ell}$ is the component of $\mathbf{k_0}$ along the trajectory, and $\mathbf{n}$ is a normal vector to the dispersion surface. Furthermore, making use of energy conservation (and thus excluding interbranch scattering, see sec. 3.7), the beam trajectory is related to the energy flow [77].(50).

As stated in the definition of φ on p. 54, the eikonal in the present theory is actually $(\mathbf{k_0} \cdot \mathbf{r} + \varphi)/k$, and φ itself is called the "eikonal function". Strictly speaking, even $(\mathbf{k_0} \cdot \mathbf{r} + \varphi)/k$ is not an eikonal because it does not necessarily take up *all* of the phase of the Bloch wave. In a strong disturbance, the phases of the individual field amplitudes do not evolve by the same amount in the course of propagation and no single eikonal could subsume them. It is exactly this admission of some amount of individual phase evolution of the field amplitudes that makes the present theory applicable to strong disturbances. As a consequence of the departure from a strict eikonal, the present theory cannot make use of the properties of the dispersion surface in an exact sense. Instead, properties of the solutions of the differential equation 3.30 after the choice of an optimal eikonal function will be discussed in the following sections and it will be shown that they behave approximately as one would expect from standard dynamical diffraction theory.

3.5 Numerical Solution of the Differential Equations

The most common method of solving the system (3.29) or (3.33) of differential equations is the method of characteristics [105], which relies on taking steps of fixed magnitude along the directions of information propagation. With eq. (3.33), these are simply given by $\mathbf{k}_0$ and $\mathbf{k}_\mathrm{h}$ because the left-hand side contains directional derivatives along these wave vectors. Figure 3.12 illustrates the procedure. The nodes of the computational grid on the entrance surface are assigned values of the field amplitudes, which are determined from the incident wave, i.e., in the two-wave case with a single plane wave $\tilde{\mathbf{A}}_0 \exp(i\mathbf{k}\cdot\mathbf{r})$ incident on the surface, a node with indices (m, n) and a real-space location $\mathbf{r}_{m,n}$ gets $(\mathbf{A}_0)_{m,n} = \mathbf{A}_0 \exp(i\mathbf{k}_0 \cdot \mathbf{r}_{m,n})$ and $(\mathbf{A}_\mathrm{h})_{m,n} = \mathbf{0}$. In a time-dependent problem, the computational grid must be extended to include a time grid, and boundary values would be assigned analogously. A large body of literature deals with applications of the theory in problems like x-ray section topography, and a random pick turns up [106, 107].

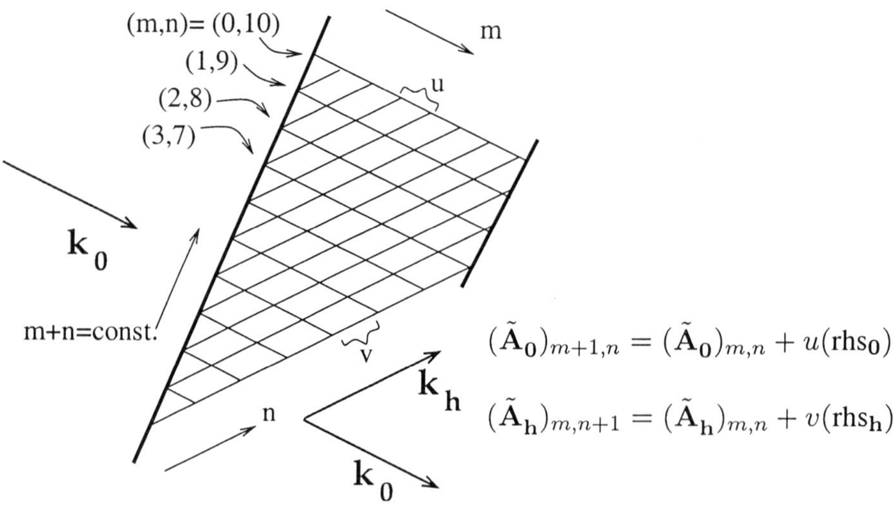

Figure 3.12: The computational grid of the method of characteristics for two waves $\mathbf{k}_0$ and $\mathbf{k}_\mathrm{h}$. Both entrance and exit surfaces are asymmetric relative to the reciprocal lattice vector $\mathbf{h}$ of the reflection. The nodes have indices m, n. The algorithm for stepping along the nodes is indicated to the right, with $(\mathrm{rhs}_\mathrm{h})$ for the right-hand side of eq. (3.30) for the wave vector $\mathbf{k}_\mathrm{h}$.

The same approach can be applied to eq. (3.29) for an N-wave case with any given set $\{\varphi_n, n = 1 \ldots N\}$ of eikonal functions. The amplitudes of the N waves are determined by the boundary conditions, just as one would do with a perfect crystal. The numerical integration, as described above, is then carried out separately for each of the N branches of the dispersion surface. Different choices of φ_n will give different magnitudes for the directional derivatives on the left-hand side of eq. (3.29). One may now apply a numerical optimization procedure for φ_n, varying them between runs of numerical solutions of eq. (3.29), to minimize the derivatives *globally*. This requires a metric for the derivatives whose choice is at the discretion of the person running the simulation, but might, for example, be

$$\int \left| \sum_{\mathbf{h}} \left(\mathbf{k}_{\mathbf{h}} \cdot \nabla \tilde{\mathbf{A}}_{\mathbf{h}}^{(+)} - \frac{k}{c} \frac{\partial \tilde{\mathbf{A}}_{\mathbf{h}}^{(+)}}{\partial t} \right) \cdot \tilde{\mathbf{A}}_{-\mathbf{h}}^{(-)} \right| , \tag{3.38}$$

where the integration covers the region in space for which the simulation is being run. The factor $\tilde{\mathbf{A}}_{-\mathbf{h}}^{(-)}$ makes sure that the gradients of the stronger field amplitudes have more weight in the minimization than those of the weaker ones, in keeping with their higher overall importance. The metric in eq. (3.38) also has a direct physical interpretation in eliminating artifical scattering between the modes used in the description of a diffraction problem ("interbranch scattering," see sec. 3.7).

3.6 The Dispersion Surface

With a given number N of waves deemed relevant to a particular diffraction problem, the dispersion surface within the tangent space to a particular point $\mathbf{r}$ is defined by the right-hand side of the first line of eq. (3.30). It is therefore equivalent in character (topology, relative phases of the field amplitudes, etc.) and has the same number of branches as the dispersion surface of the equivalent diffraction problem for an undisturbed crystal. The multitude of local dispersion surfaces will henceforth be termed a global dispersion surface, the branches of which are equivalence classes of branches of the local dispersion surfaces. To be quite clear about this: The dispersion surface and thus the tie points on it are defined locally, using only the terms in eq. (3.30) that correspond to standard dynamical diffraction, i.e., without the perturbation terms in the third line of eq. (3.30). The local dispersion surfaces, and thus the gloabl dispersion surface, are properties of the disturbed crystal. The eikonal functions for the branches of the global dispersion surface, although not solutions of a differential equation themselves, are determined by the solutions of a particular problem. Generally, $\nabla\varphi_n + \mathbf{k}_0$ does not originate from a local dispersion surface but only from a point close to it. The field amplitudes, determined as solutions of the differential equation (3.30), are only approximate solutions of the local algebraic dynamical diffraction problem and the tie point on a local dispersion surface that best approximates them need not coincide with the point on that surface that is closest to the origin of $\nabla\varphi_n + \mathbf{k}_0$. As will be demonstrated in sec. 3.8.2, solutions of the global diffraction problem can be stable or unstable with respect to disturbances of the perfect crystal lattice, and the quality of the approximations, quantified with the metric (3.38), depends on that stability.

Other than in the original eikonal theory, the deviations from the local dispersion surfaces do not limit the validity of the theory because they are picked up by the derivatives of the field amplitudes in the Takagi-type system of differential equations.

3.6.1 Propagation of the Field Amplitudes. To illustrate the dispersion surface and the role of the eikonal function, let us discuss the following two examples. First, take a perfect crystal and choose a Bloch base vector $\mathbf{k_0}$. Figure 3.13 shows the two-wave case with $\mathbf{k_0}$ originating in the symmetry center Q. An eikonal function φ_0 with a constant gradient can now be found, such

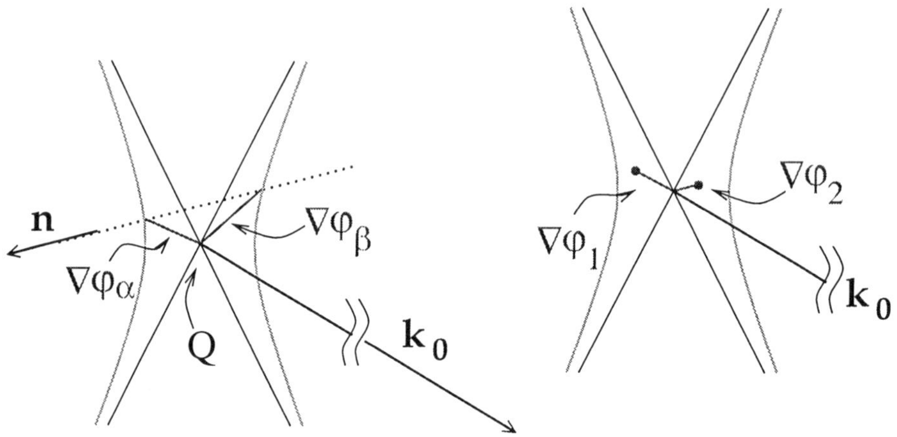

Figure 3.13: Matching tie points with eikonal functions. Left: eikonal functions φ_α, φ_β ideally matched to the dispersion surface on the α- and the β-branch and matching a particular boundary condition with a surface normal vector $\mathbf{n}$. Right: Any other choice φ_1, φ_2 of the eikonal functions is possible, too, with the derivatives of the field amplitudes picking up the balance, as shown in fig. 3.10.

that $\mathbf{k_0} + \nabla\varphi_0$ matches a tie point. Now, take another function φ. The resulting mismatched wave vector in $(\mathbf{k_0} + \nabla\varphi)^2$ on the right-hand side of eq. (3.30) is then the cause of gradients $\nabla\tilde{\mathbf{A}}_0$ and $\nabla\tilde{\mathbf{A}}_\mathrm{H}$, and (in the simple case of a perfect crystal only)

$$i\nabla\tilde{\mathbf{A}}_\mathrm{h} = (\nabla\varphi - \nabla\varphi_0)\tilde{\mathbf{A}}_\mathrm{h} \tag{3.39}$$

i.e., the field gradients pick up the wave vector mismatch and have the same direction.

The second example deals with a static lattice distortion, i.e., $\nabla^2\mathbf{u} \neq 0$. The discussion is based solely on the local dispersion surfaces. Whether this approach is valid depends on the following criteria.

1 The solution of eq. (3.30) must be stable with respect to disturbances (see discussion in sec. 3.8.2).

2 By assumption, the field amplitudes at the beginning of the propagation problem correspond to a local tie point (not any set of field amplitudes does so, see p. 45). As the amplitudes change, they must still correspond to some tie point. This is the case if the diffracting mode is coupled weakly to other modes, so that the interbranch scattering is weak (see sec. 3.7).

3 $\nabla\varphi$ is the gradient of a scalar function, which restricts it to some degree. It can therefore generally not exactly match the local tie points. However, if the above two conditions are fulfilled, then a tie point does exist globally to good approximation, and, therefore, there is a scalar function whose gradient matches it—the phase of any one of the Bloch component waves participating in the mode, and $\nabla\varphi$ is just the gradient of that phase minus some constant $\mathbf{k_0}$.

Now, we turn to the discussion of the evolution of the field amplitudes in the presence of a lattice distortion. In the absence of $\nabla\mathbf{u}$, choose an eikonal function φ_0, so that $\nabla\varphi_0+\mathbf{k_0}$ originates from that tie point. Then "turn on" $\nabla\mathbf{u}$. Because $\nabla\mathbf{u}\cdot\mathbf{h}$ are different gradients for different reciprocal lattice vectors $\mathbf{h}$, no eikonal function φ can be found, so that $\nabla\varphi$ compensates for all of them. Although φ cannot be made to fully counter the effect of $\nabla\mathbf{u}$, it can be chosen to make the sum in the integrand of the metric (3.38) vanish *locally*: By multiplying each line in eq. (3.30) with the respective complex conjugate field amplitude and making use of $\mathbf{k_0} + \nabla\varphi_0$ originating from a tie point, the following two equations are obtained (superscript $(+)$ dropped where possible):

$$2i\mathbf{k_0}\cdot\nabla\tilde{\mathbf{A}}_{\mathbf{h}}^{(+)}\cdot\tilde{\mathbf{A}}_{-\mathbf{h}}^{(-)} = |\tilde{\mathbf{A}}_{\mathbf{h}}|^2\mathbf{k_h}\cdot(\nabla(\varphi-\varphi_0)-\nabla\mathbf{u}\cdot\mathbf{h}) \tag{3.40}$$

Taking the sum over $\mathbf{h}$ and demanding the left-hand side (i.e., the integrand of eq. (3.38)) vanish, gives an equation for $\nabla(\varphi-\varphi_0)$:

$$\left(\sum_{\mathbf{h}}\mathbf{k_h}\left|\tilde{\mathbf{A}}_{\mathbf{h}}\right|^2\right)\cdot\nabla(\varphi-\varphi_0) = \sum_{\mathbf{h}}\mathbf{k_h}\left|\tilde{\mathbf{A}}_{\mathbf{h}}\right|^2\cdot\nabla\mathbf{u}\cdot\mathbf{h} \tag{3.41}$$

which means that $\nabla(\varphi-\varphi_0)$ interpolates among the gradients $\nabla(\mathbf{u}\cdot\mathbf{h})$. Locally, this interpolation can be expressed in terms of a single gradient $\nabla(\varphi-\varphi_0)$ but globally, this is not possible because the field amplitudes $\tilde{\mathbf{A}}_{\mathbf{h}}$ are variable. Equation (3.41) is discussed in more depth in sec. 3.7.

Because the gradients $\nabla\mathbf{u}\cdot\mathbf{h}$ differ from each other for different $\mathbf{h}$, there is a differential contribution of $\mathbf{u}$ to the real part of the right-hand side of eq. (3.30). Because of the factor i on the left-hand side, this induces a differential phase shift of the field amplitudes $\tilde{\mathbf{A}}_{\mathbf{h}}$ versus $\tilde{\mathbf{A}}_0$ (as seen from the perspective of the disturbed lattice, i.e., within local coordinates). After propagating some distance, this contributes to an imaginary part on the right-hand side of eq. (3.30), and thus to the real part of the field amplitude gradients. The response of the solutions of eq. (3.30) to a disturbance occurs therefore in a two-step process: First a phase shift, then a change of the modulus.

The response of the field amplitudes to the disturbance can be understood in terms of a feedback loop (discussed in more detail in sec. 3.8.2).

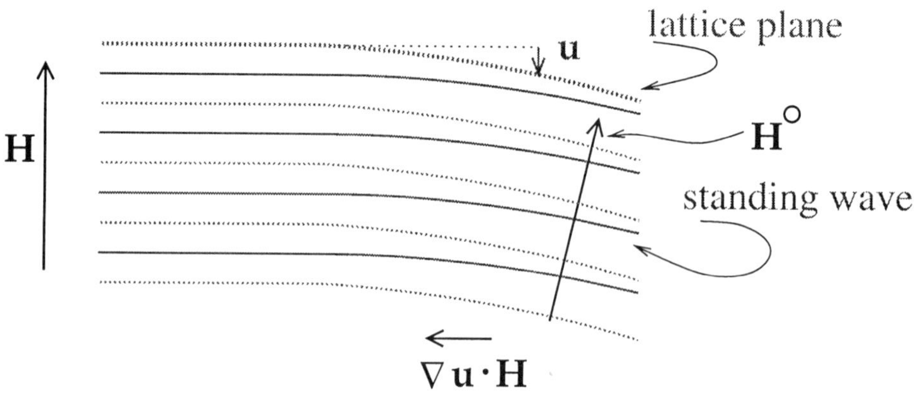

Figure 3.14: Distorted lattice and standing waves: As the lattice planes curve, they "walk off" the standing waves (see fig. 3.3) of dynamical diffraction, here shown for the α branch. From the perspective of the local lattice, a driving term $\nabla\mathbf{u}$ induces a phase shift between the amplitudes, which curves the standing wave maxima.

3.7 Beams, Adiabatic Change and Interbranch Scattering

Instead of considering the individual field amplitudes, one may also look at the evolution of a diffracting mode as an entity: If for each $\mathbf{h}$, eq. (3.30) is multiplied with $\tilde{\mathbf{A}}^{(-)}_{-\mathbf{h}}$, the sum over $\mathbf{h}$ is taken, and only the lowest order of $\exp(i\mathfrak{s}_m) - 1$ is kept, then the left-hand side reads

$$-2i\sum_{\mathbf{h}}\left(\mathbf{k}_{\mathbf{h}}\cdot\nabla\tilde{\mathbf{A}}^{(+)}_{\mathbf{h}}\right)\cdot\tilde{\mathbf{A}}^{(-)}_{-\mathbf{h}} + 2i\frac{k}{c}\sum_{\mathbf{h}}\frac{\partial\tilde{\mathbf{A}}^{(+)}_{\mathbf{h}}}{\partial t}\tilde{\mathbf{A}}^{(-)}_{-\mathbf{h}} =$$
$$\sum_{\mathbf{h}}\left[\left(k-\frac{\dot{\varphi}}{c}\right)^2 - (\mathbf{k}_{\mathbf{h}}+\nabla\varphi)^2\right]\left|\tilde{\mathbf{A}}_{\mathbf{h}}\right|^2 - 4\pi r_e\sum_{\substack{m,\mathbf{h}'\\ \mathbf{h}}}\tilde{\eta}_{m,\mathbf{h}-\mathbf{h}'}\left[\tilde{\mathbf{A}}_{\mathbf{h}'}\right]_{\mathbf{h}}\tilde{\mathbf{A}}^{(-)}_{-\mathbf{h}} +$$
$$2\sum_{\mathbf{h}}\left[\frac{k}{c}\dot{\mathbf{u}} + \mathbf{k}_{\mathbf{h}}\cdot\nabla\mathbf{u}\right]\cdot\mathbf{h}\left|\tilde{\mathbf{A}}_{\mathbf{h}}\right|^2 - 4i\pi r_e\sum_{\substack{m,\mathbf{h}'\\ \mathbf{h}}}\mathfrak{s}_m\tilde{\eta}_{m,\mathbf{h}-\mathbf{h}'}\left[\tilde{\mathbf{A}}_{\mathbf{h}'}\right]_{\mathbf{h}}\tilde{\mathbf{A}}^{(-)}_{-\mathbf{h}}. \tag{3.42}$$

The left-hand side (first line) is just the integrand of the metric (3.38). It is also the continuity equation of the electromagnetic energy density and flow:

$$-i\frac{4\pi}{kc}\nabla\cdot\mathbf{S}_n + i\frac{4\pi}{kc}\frac{\partial\mathcal{U}}{\partial t} = \text{(rhs.)}, \qquad \begin{aligned}\mathbf{S} &= \frac{kc}{4\pi}\sum_{\mathbf{h}}\mathbf{k}_{\mathbf{h}}\left|\tilde{\mathbf{A}}_{\mathbf{h}}\right|^2\\ \mathcal{U} &= \frac{k^2}{4\pi}\sum_{\mathbf{h}}\left|\tilde{\mathbf{A}}_{\mathbf{h}}\right|^2,\end{aligned} \tag{3.43}$$

where $\mathbf{S}_n$ is the Poynting vector [1].(6.109) and $\mathcal{U}_n$ is the energy density [1].(6.106) of the electromagnetic waves represented by the field amplitudes $(\tilde{\mathbf{A}}_n)_\mathbf{h}$. The abbreviation (r.h.s.) means the right-hand side of eq. (3.42), which contains contributions from the disturbance terms in the third line of eq. (3.42), and also from a possible mismatch of $\mathbf{k_0} + \nabla\varphi$ from the tie point for the amplitudes $(\tilde{\mathbf{A}}_n)_\mathbf{h}$ (supposing the amplitudes permit a tie point at all, see p. 45).

A nonvanishing divergence means that $\mathbf{S}$ changes its length, which may be due to absorption, or also due to the so-called interbranch scattering [108, 109], i.e., a reassignment of photons among the modes of dynamical diffraction (branches of the dispersion surface). Recall that an index n for the modes (α, β, etc., branch) was suppressed in an index antiproliferation effort. There is actually a Poynting vector $\mathbf{S}_n$ for each mode, and it is only the sum $\sum_n \nabla \cdot \mathbf{S}_n$ that describes the absorption losses. Redistribution of lengths among the $\mathbf{S}_n$ may be interpreted as the interbranch scattering. It should be emphasized here that these modes and the scattering between them are not anything intrinsic to the nature of the problem, because (r.h.s) in eq. (3.43) depends on the choice of φ. There is, however, a minimal scattering rate between the branches, which is obtained by minimimizing the metric in eq. (3.38), and thus $\sum_n \nabla \cdot \mathbf{S}_n$, and that rate is intrinsic to the diffraction problem, together with a given finite number of parameters to describe the eikonal functions. If the eikonal function φ makes $\mathbf{k_0} + \nabla\varphi$ originate from a local tie point, then the second line in eq. (3.42) vanishes, and the disturbances in the third line of eq. (3.42) are directly the driving terms of the continuity equation for $\mathbf{S}$. A change in $\mathbf{S}_n$ that does not contribute to $\nabla \cdot \mathbf{S}_n$ changes the direction of $\mathbf{S}_n$, and correspondingly the location of the tie point on the dispersion surface.

In the simple case of an undisturbed crystal, it is possible to relate $\nabla\mathbf{S}$ directly to a mismatch from a tie point T. Suppose that $\mathbf{k_0} + \nabla\varphi_0$ originates from T but $\mathbf{k_0} + \nabla\varphi$ does not. Neglecting $\dot{\varphi}^2$ and $(\nabla\varphi)^2$, eq. (3.42) can be written as

$$i\frac{\partial\mathcal{U}}{\partial t} - i\nabla\cdot\mathbf{S}_n = 2\,(\dot{\varphi} - \dot{\varphi}_0)\mathcal{U} - 2\mathbf{S}\cdot\nabla(\varphi - \varphi_0) \tag{3.44}$$

In the presence of a disturbance, there are additional driving terms to the derivatives of $\mathcal{U}$ and $\mathbf{S}$ from the third line of eq. (3.42).

Consider now a disturbed crystal, and assume that the field amplitudes at a given point $\mathbf{r}, t$ admit a tie point T and $\mathbf{k_0} + \nabla\varphi_0$ originates from it. Equation (3.42) reads then:

$$\begin{aligned} i\frac{\partial\mathcal{U}}{\partial t} - i\nabla\cdot\mathbf{S}_n &= \frac{k^2}{4\pi}\sum_{\mathbf{h}}\left[\dot{\mathbf{u}} + \frac{c}{k}\mathbf{k_h}\cdot\nabla\mathbf{u}\right]\cdot\mathbf{h}\left|\tilde{\mathbf{A}}_\mathbf{h}\right|^2 \\ &\quad - ikcr_e\sum_{\substack{m,\mathbf{h}'\\ \mathbf{h}}}\mathfrak{s}_m\tilde{\eta}_{m,\mathbf{h}-\mathbf{h}'}\left[\tilde{\mathbf{A}}_{\mathbf{h}'}\right]_\mathbf{h}\tilde{\mathbf{A}}^{(-)}_{-\mathbf{h}}. \end{aligned} \tag{3.45}$$

The lattice disturbance is thus the cause of a sink or source to the continuity equation of the electromagnetic energy flow.

Please note the different ways of reasoning used here and for eq. (3.41) which are complementary (see fig. 3.10). Here, the eikonal function φ_0 that matches the tie point for the amplitudes is kept, and the effect of a disturbance on the field amplitudes is studied. For eq. (3.41) another eikonal function φ was chosen to make the sum of the derivatives vanish and eq. (3.41) then gives the projection of $\nabla(\varphi - \varphi_0)$ onto $\mathbf{S}$. Equation (3.41), supplemented with the time dependent and the $\tilde{\eta}_{m,\mathbf{h}}$ terms, is:

$$\mathcal{U}(\dot{\varphi} - \dot{\varphi}_0) + \mathbf{S}\cdot\nabla(\varphi - \varphi_0) = \frac{k^2}{4\pi}\sum_{\mathbf{h}} \left|\tilde{\mathbf{A}}_{\mathbf{h}}\right|^2 \dot{\mathbf{u}}\cdot\mathbf{h} + \frac{kc}{4\pi}\sum_{\mathbf{h}} \mathbf{k}_{\mathbf{h}} \left|\tilde{\mathbf{A}}_{\mathbf{h}}\right|^2 \cdot\nabla\mathbf{u}\cdot\mathbf{h}$$
$$- i\frac{kc}{2} r_e \sum_{\substack{m,\mathbf{h}'\\ \mathbf{h}}} \mathfrak{s}_m \tilde{\eta}_{m,\mathbf{h}-\mathbf{h}'} \left[\tilde{\mathbf{A}}_{\mathbf{h}'}\right]_{\mathbf{h}} \tilde{\mathbf{A}}^{(-)}_{-\mathbf{h}}. \tag{3.46}$$

This expression will be useful in ch. 4.

3.8 Obtaining Qualitative Information

Having the dispersion surface of a disturbed crystal is of advantage for order-of-magnitude estimates, eliminating in some cases the need to run a computer simulation. The concept of the eikonal functions makes it possible to approximate throughout the sample under investigation the tie points that one would determine locally. Assuming that eikonal functions for all branches of the dispersion surface have been found, we want to know more about the local behavior of solutions to eq. (3.30) in a given point $\mathbf{r}$. This discussion is useful only if a solution is stable with respect to a disturbance of the crystal, as demonstrated for a displacive disturbance in sec. 3.8.2. In a way, the argument is circular: eikonal functions to represent tie points on the branches of the global dispersion surface of the disturbed crystal (as defined on p. 63) and field amplitudes to go with them are determined in an iterative procedure. Under the assumption that this procedure has converged well in the sense that the metric (3.38) is minimized, one may study the local behavior of the solutions, their stability, etc.

To determine the eikonal functions, the Bloch base wave vector $\mathbf{k}'_0$ is best chosen to originate from the symmetry center of the dispersion surface (point Q in fig. 3.13). However, after an eikonal function $\varphi_n^{(0)}$ for a given branch has been determined, the discussion of local behavior is facilitated by the definition of another Bloch base vector $\mathbf{k}_0 = \mathbf{k}'_0 + \nabla\varphi_n^{(0)}|_{\mathbf{r}}$, where $\nabla\varphi_n^{(0)}|_{\mathbf{r}}$ is the gradient of $\varphi_n^{(0)}$ at point $\mathbf{r}$. One may think of $\mathbf{k}_0$ simply as an abbreviation, introduced to reduce the formula clutter. Obviously, that wave vector $\mathbf{k}_0$ is only useful for local considerations. It does generally not originate exactly from the local dispersion surface and the field amplitudes do not match a tie point exactly but both are good approximations if the metric (3.38) is small which may be expected for a stable solution.

With $\mathbf{k}_0$ thus defined, the first line of the right-hand side in eq. (3.30) becomes small and of the same order of magnitude as the perturbation terms in the second and third lines of the equation. The derivatives of the field amplitudes on the left-hand side are thus equated to a right-hand side that is of the order of the perturbation, with all spatial and temporal dependence that is common to all of the field amplitudes comprising one mode being taken up by the eikonal function.

3.8.1 Example: Optical Phonons, Frequency Shifts. Here, a spatially perfect crystal is subjected to a sudden perturbation that induces a rapid change in the electron density, without a lattice distortion, i.e., $\nabla\mathbf{u} = \mathbf{0}$. An example would be a laser-pulse induced, strongly excited optical phonon. A diffracting mode shall have an eikonal function $\varphi_0(\mathbf{r})$ which matches the appropriate tie point perfectly in the absence of the disturbance. Because the unperturbed crystal is static, φ_0 is independent of time. This makes $(k^2 - (\mathbf{k_h} + \nabla\varphi_0)^2)\tilde{\mathbf{A}}_\mathbf{h} - 4\pi r_e \sum_{m,\mathbf{h}'} \tilde{\eta}_{m,\mathbf{h}-\mathbf{h}'} \left[\tilde{\mathbf{A}}_{\mathbf{h}'}\right]_\mathbf{h}$ (the second line of eq. (3.30)) vanish. No terms with $\mathbf{u}$ are present by definition of the problem. Suppose first, that the eikonal function is not adapted to the disturbance, i.e., $\varphi = \varphi_0$ is kept. Equation (3.30) is then reduced to

$$2i\left[\frac{k}{c}\frac{\partial}{\partial t} - \mathbf{k_h}\cdot\nabla\right]\tilde{\mathbf{A}}_\mathbf{h} = -4\pi r_e \sum_{m,\mathbf{h}'} \tilde{\eta}_{m,\mathbf{h}-\mathbf{h}'}\left(e^{i\mathfrak{s}_m} - 1\right)\left[\tilde{\mathbf{A}}_{\mathbf{h}'}\right]_\mathbf{h}, \quad (3.47)$$

directly relating the field amplitude derivatives to the perturbation terms. The characteristics of this differential equation are the 4-vectors (but written in (1+3) notation), $(-(k/c), \mathbf{k_h})$, and for given boundary conditions, the solutions are fully determined. When dealing separately with temporal and spatial coordinates, as in eq. (3.47), the question may however arise in which way the value of the right-hand side shall be distributed among the spatial and the temporal derivatives on the left-hand side. The following discussion shall aid in answering that question: Each term $\tilde{\eta}_{m,\mathbf{h}-\mathbf{h}'}(\exp(i\mathfrak{s}_m) - 1)$ in the sum gives a contribution to the change in a given point $(t, \mathbf{r})$ of the phase of $\tilde{\mathbf{A}}_\mathbf{h}$, expressed by the directional derivative along $(-(k/c), \mathbf{k_h})$. The contribution from that term $\tilde{\eta}_{m,\mathbf{h}-\mathbf{h}'} \exp(i\mathfrak{s}_m)$ would vanish in a reference frame that is moving along the world line of $\mathfrak{s}_m$, i.e., moving along $\mathfrak{k}_m$ with the phase velocity[8] $v_p = \varpi/|\mathfrak{k}|$ of that particular Fourier component of the disturbance. Since we are staying,

[8]Never mind that an optical phonon can have a superluminal phase velocity (at the Brillouin zone center) - this motion at velocity v_p is virtual and meant only for the sake of visualization.

instead, in the point $(t, \mathbf{r})$, the phase of the disturbance evolves there. The contribution of $\tilde{\eta}_{m,\mathbf{h}-\mathbf{h}'} \exp(i\mathfrak{s}_m)$ is thus split among the partial derivatives according to the ratio:

$$\left(\frac{k}{c}\frac{\partial \tilde{\mathbf{A}}_{\mathbf{h}}^{(+)}}{\partial t} \cdot \tilde{\mathbf{A}}_{-\mathbf{h}}^{(-)}\right) : \left(\left(\mathbf{k}_{\mathbf{h}} \cdot \nabla \tilde{\mathbf{A}}_{\mathbf{h}}^{(+)}\right) \cdot \tilde{\mathbf{A}}_{-\mathbf{h}}^{(-)}\right) = (\varpi\tau) : \left(\mathfrak{k} \cdot \frac{\mathbf{k}_{\mathbf{h}}}{|\mathbf{k}_{\mathbf{h}}|} c\tau\right) \tag{3.48}$$

where $\mathbf{k}_{\mathbf{h}}/|\mathbf{k}_{\mathbf{h}}|$ is a unit vector along $\mathbf{k}_{\mathbf{h}}$ and $c\tau$ is the distance covered in that direction at the speed of light (of the x-rays) within a time τ. Both parts of the ratio on the left-hand side in eq. (3.48) were multiplied with the same vector $\tilde{\mathbf{A}}_{-\mathbf{h}}^{(-)}$ to yield scalars.

In particular, if the disturbance has high frequencies ϖ at low wave numbers $|\mathfrak{k}|$, then the time derivative of $\tilde{\mathbf{A}}_{\mathbf{h}}$ is dominant and a frequency shift may result (heterodyne mixing) - if it is not cancelled by another Fourier component of the disturbance, see below. To further discuss the influence of a disturbance on the field amplitudes, the origin of the Fourier series of the disturbance shall be moved to the point $(t, \mathbf{r})$ under consideration. This is, of course, something one should not do when actually calculating a solution of the propagation problem, but it is valid in a discussion of local behavior. With the new origin, the dipole approximation can be applied and $\exp(i\mathfrak{s}_m) - 1$ be replaced with $i\mathfrak{s}_m$ in a small region about $(t, \mathbf{r})$. By definition (see eq. (3.22) and fig. 3.9) $\tilde{\eta}_{m,\mathbf{h}}$ and $\tilde{\eta}^*_{-m,-\mathbf{h}}$ and $\mathfrak{s}_m = -\mathfrak{s}_{-m}$, and therefore the sum $i\sum_{m,\mathbf{h}'} \tilde{\eta}_{m,\mathbf{h}-\mathbf{h}'}\mathfrak{s}_m$ is real-valued. It follows that only the imaginary part of $\tilde{\eta}_{m,\mathbf{h}-\mathbf{h}'}$ actually contributes to the derivatives of $\tilde{\mathbf{A}}_{\mathbf{h}}$, whereas the real parts of $\tilde{\eta}_{\pm m,\mathbf{h},\mathbf{h}'}$ cancel each other out. Therefore, a contribution to the derivatives of the field amplitudes comes only from the inversion antisymmetric part (relative to the origin $(t, \mathbf{r})$) of a disturbance, which may be due to a slope in space or time of the envelope of the disturbance, or due to the disturbance itself being of noncentrosymmetric character, such as an optical phonon in a noncentrosymmetric crystal (GaAs for example, see also p. 107 for an interpretation).

Instead of considering only the contribution of a single mode of the disturbance with a particular frequency ϖ and wave vector $\mathfrak{k}$, one might also consider wavepackets of optical phonons. If the group velocity of such a wavepacket matches the group velocity of the diffracting x-rays, then the action of the disturbance on the x-rays will be enhanced [110], see also ch. 4, sec. 4.4.

One final point: The eikonal function was kept unchanged from φ_0 in the above discussion. If φ is now adapted to minimize the derivatives in the presence of the disturbance, then the derivatives of $\tilde{\mathbf{A}}_{\mathbf{h}}$ become smaller and the phase shifts (frequency shifts and wave vector changes) represented by them, are taken up by $\dot{\varphi}$ and $\nabla(\varphi - \varphi_0)$. Nothing (except perhaps a higher accuracy of the solution) changes in the actual behavior of the waves (frequency shifts, etc.) by doing so.

3.8.2 Example: Static Distortion, Guided Waves.

Equation (3.30) is now reduced to:

$$\mathbf{k_h}\cdot\nabla\tilde{\mathbf{A}}_\mathbf{h} = \frac{i}{2}\left(k^2 - (\mathbf{k_h}+\nabla\varphi)^2\right)\tilde{\mathbf{A}}_\mathbf{h} - 2\pi i r_e \sum_{\mathbf{h}'} \tilde{\eta}_{\mathbf{h}-\mathbf{h}'}\left[\tilde{\mathbf{A}}_{\mathbf{h}'}\right]_\mathbf{h} + i\mathbf{k_h}\cdot\nabla\mathbf{u}\cdot\mathbf{h}\tilde{\mathbf{A}}_\mathbf{h} \tag{3.49}$$

where $\mathbf{h}$ and $\mathbf{h}'$ may take the values $\mathbf{0}$ and $\mathbf{H}$. At a given real-space point $\mathbf{r}$, the local tie point, as defined by eq. (3.30) without the perturbation term $\nabla\mathbf{u}\cdot\mathbf{h}$, shall be at a location $-\mathbf{k}_t$ in reciprocal space. The eikonal function approximates the local tie point, i.e., the wave vector $\mathbf{k}_d = \mathbf{k}_t - (\mathbf{k_0} + \nabla\varphi)$ of the deviation from the tie point is small.

We now consider the stability of a mode. To that end, we follow the evolution of the field amplitudes in propagation. The stages of the amplitude evolution in the following discussion are given letters (a), (b), (c), ... which correspond to the phasors in fig. 3.15. For the sake of this discussion, assume that the phase of $\tilde{\mathbf{A}}_0$ at $\mathbf{r}$ equals zero, which can always be arranged by multiplying eq. (3.49) with an appropriate phase factor.[9] With a perfectly matched tie point, the phase of $\tilde{\mathbf{A}}_\mathbf{H}$ would then be equal to 0 or π on the β and α branches, respectively. Consider now the α branch. At the entrance surface, where the splitting into modes occurs, the amplitudes are equal with opposite signs ((a) in fig. 3.15). Due to the lattice distortion, shown in the figure, the term $\mathbf{k_H}\cdot\nabla\mathbf{u}\cdot\mathbf{H}$ is negative, whereas $\mathbf{k_0}\cdot\nabla\mathbf{u}\cdot\mathbf{0}$ is of course zero. At a point $\mathbf{r}$ in the crystal, the amplitudes are therefore as in (b) - no change in $\tilde{\mathbf{A}}_0$, and $\tilde{\mathbf{A}}_\mathbf{H}$ has picked up a negative phase ϕ.

This negative phase of $\tilde{\mathbf{A}}_\mathbf{H}$ makes the sum in the line for $\mathbf{h} = \mathbf{0}$ in eq. (3.49) equal to a real number times a factor $\exp(i(\pi - \phi)) \approx -1 + i\phi$ (with π for the α branch). With the factor $-i$ before the sum and $\tilde{\eta}_\mathbf{H} > 0$ (electron density, not charge density!), the directional derivative of $\tilde{\mathbf{A}}_0$ on the left-hand side acquires a positive real part. Therefore, in the course of propagation, the modulus of $\tilde{\mathbf{A}}_0$ grows, and with a similar argument, the modulus of $\tilde{\mathbf{A}}_\mathbf{H}$ shrinks ((c) in fig. 3.15). Again, in the line of eq. (3.49) for $\mathbf{h} = \mathbf{0}$, the sum is now larger (less negative, due to the factor $\exp(i\pi)$ between the field amplitudes on the α branch) than necessary to match the tie point, and because of $-i$ before the sum, the right-hand side acquires a negative imaginary part ((d) in the figure). Likewise, the right-hand side of eq. (3.49) for $\mathbf{h} = \mathbf{H}$ - after multiplying both sides with the phase factor $\exp(-i\pi)$ to cancel the phase $\exp(i\pi)$ of $\tilde{\mathbf{A}}_\mathbf{H}$ - acquires a positive imaginary part (d). This tends to counter the negative phase ϕ that $\tilde{\mathbf{A}}_\mathbf{H}$ started out with. *In conclusion: A mode on the α branch is stable with respect to small perturbations* and the maxima of the standing waves (see figs. 3.3 and 3.15) tend to stay between the lattice planes. This is to be expected

[9] Fixing the phase of one amplitude at one point in space does not jeopardize the generality of this discussion.

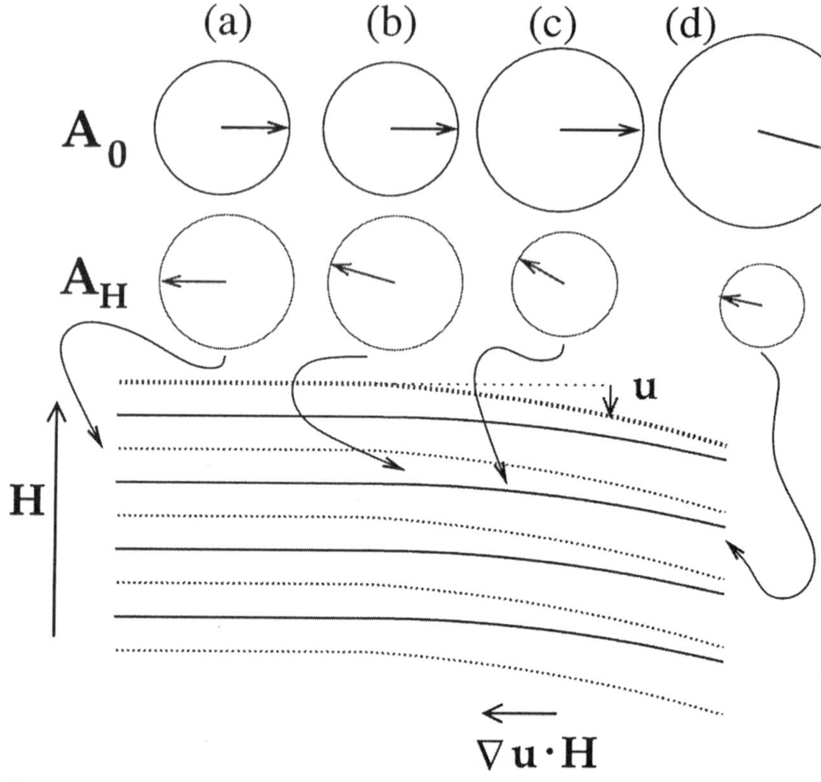

Figure 3.15: Amplitude evolution, guided waves on the α branch in a crystal with a distortion as shown in fig. 3.14. The little phasors for $\tilde{\mathbf{A}}_0$ and $\tilde{\mathbf{A}}_H$ have letters (a), (b), (c), ... which denote stages in the amplitude evolution explained in the text.

because a mode on the α branch has longer wave vectors than one on the β branch and therefore tends to be guided by a refractive index profile (just like like light in a glass fiber). Similarly, the evolution of the amplitudes can be followed through for the β branch, with the result of a β mode being *unstable* towards small perturbations.

There is of course a limit to the stability of a mode on the α branch. If the disturbance is so strong that the diffractive properties change significantly over one characteristic scattering length, i.e., the pendellösung length, then guidance of the standing waves breaks down. A quantitative criterion is obtained from eq. (3.49). For stability,

$$\mathbf{k_h} \cdot \nabla \mathbf{u} \cdot \mathbf{h} \tilde{\mathbf{A}}_\mathbf{h} \ll \sum_{\mathbf{h'}} \tilde{\eta}_{\mathbf{h}-\mathbf{h'}} \left[\tilde{\mathbf{A}}_{\mathbf{h'}} \right]_\mathbf{h} \tag{3.50}$$

must be valid. This expression depends on the magnitude of $\mathbf{u}$ vs. η, but also on the amplitude ratios.

The above discussion was about the phases of the field amplitudes and the positions of the standing waves relative to the lattice planes. What above the direction of energy flow, i.e., the beams? According to the above discussion for the α branch, the modulus of $\tilde{\mathbf{A}}_0$ grows at the expense of that of $\tilde{\mathbf{A}}_{\mathbf{H}}$, as the radiation propagates in a crystal with a curvature as shown in fig. 3.15. Therefore, the tie point slides down on the dispersion surface, and the beam direction bends downward, i.e., with the same tendency as the curvature of the lattice planes. This result was also obtained by Penning and Polder [76] on the basis of geometrical optics (see fig. 4 there and also [74], fig. 13.10) and has been shown computationally on the basis of the Takagi-Taupin theory [104].

3.9 From Boundary to Transition Conditions

In analogy to the boundary conditions for the coupling of waves across a crystal boundary (sec. 2.5), the transport of waves within a disturbed crystal is also subject to the requirement of continuity. This continuity is expressed in terms of gradients instead of the vectorial algebraic relations at a discontinuous boundary. The simplest case is a gradient of the average refractive index for a single wave due to an inhomogenenous average electron density. Upon insertion of Taylor expansions for $\tilde{\eta}_0$ and $(\mathbf{k}_0+\nabla\varphi)^2$ about a given point $\mathbf{r}$ and imposition of the condition that the left-hand side vanish, equation (3.30) becomes

$$0 = \left(k^2 - (\mathbf{k}_0 + \nabla\varphi)^2 - 2\mathbf{x}\cdot\nabla\nabla\varphi\cdot\mathbf{k}_0\right)\tilde{\mathbf{A}}_0 - 4\pi r_e(\tilde{\eta}_0 + \mathbf{x}\cdot\nabla\tilde{\eta}_0)\tilde{\mathbf{A}}_0. \tag{3.51}$$

With $\mathbf{k}_0 + \nabla\varphi$ matching a tie point at the real-space location $\mathbf{r}$ and with the requirement for this to be independent of $\mathbf{x}$, this becomes the expected generalization of Snell's law:

$$0 = \nabla\left(\nabla\varphi\cdot\mathbf{k}_0\right) + 2\pi r_e\nabla\tilde{\eta}_0. \tag{3.52}$$

For diffractive cases, i.e., those with more than a single wave in eq. (3.30), two equivalent paths of reasoning shall be pursued here. Both describe the same physics and are complementary to each other (see remarks in sec. 3.7 and fig. fig. 3.10). The first one (i) is analogous to that applied above for the single wave case. We assume that the field amplitudes permit a tie point T (see p. 45) and that $\mathbf{k}_0 + \varphi_0$ matches it. Note that the tie point is defined through the field amplitudes in eq. (3.30) *without* the disturbance in the third line. With the disturbance present, φ_0 will not make the left-hand side of eq. (3.30) vanish. However, as demonstrated with eqs. (3.41) and (3.46), another eikonal function φ can be found to make the left-hand side in a sum over $\mathbf{h}$ vanish *locally*. Equation (3.41) then relates the directions of $\nabla(\varphi - \varphi_0)$ and of the gradient $\nabla\mathbf{u}$ through their projections onto the Poynting vector. It does, however, not determine the components of $\nabla(\varphi-\varphi_0)$ perpendicular to $\mathbf{S}$, and these are given by the boundary conditions. With an eikonal function that minimizes the metric globally, there may of course be an additional contribution to $\nabla(\varphi-\varphi_0)$, whose size depends on the degree of minimization. With a disturbance that is stable towards disturbances (see sec. 3.8.2), that size can be expected to remain small. When dealing with a structure factor modulation disturbance, eq. (3.46) must be used instead of (3.41). This is demonstrated in ch. 4, sec. 4.1.

The other line (ii) of reasoning is that φ_0 is kept. The disturbance will then make the left-hand side of eq. (3.30) non-zero, and the directions of the field amplitude gradients onto their respective wave vectors are related to $\nabla\mathbf{u}$. This approach is taken in most of this chapter, particularly in sec. 3.8.

3.10 Summary and Discussion

Geometrical optics and wave optics each have merits of their own. The eikonal Takagi theory presented in this chapter combines the two approaches and (hopefully) gets the best from both worlds. The eikonal functions allow the introduction of the concept of dispersion surfaces into the Takagi theory, and with them comes the rich visual terminology of the standard dynamical diffraction. Nevertheless, the theory is still wave optical in character, and the geometrical optics that comes with the eikonal fucntions is not an approximation, but only a tool to take some load off the wave optical treatment. Coming with the minimization of the first order derivatives of the field amplitudes, one may expect the error induced by the neglect of the second order derivatives to be reduced, as well. The theory can therefore be expected to give more precise results that the bare Takagi theory in principle, quite apart from the reduction of numerical errors in practice.

It is shown in sec. 3.8.2 that solutions of the differential equations can be stable or unstable towards a disturbance. In analogy to the guidance of light in a glass fiber with a higher refractive index in the core, modes on the α branch tend to be stable. Far away (in space and time) from disturbance, a stable mode can be described in terms of geometrical optics. As it encounters a disturbance, the more complicated wave optical behavior is accomodated by the differential equations, and eventually, the solutions settle into other modes which again admit a geometrical optical description.

Time dependence is also included explicitly in the eikonal Takagi theory. It is argued that macroscopic electromagnetism on the basis of the $\mathbf{D}$ field and the electric susceptibility cannot account for phenomena where the motion of charges due to external influences becomes appreciably fast. In order to describe phenomena that occur on time scales comparable to the reciprocal energy bandwidth (or equivalently the time it takes to traverse on pendellösung length) of the x-ray reflections under study, the theory is formulated on the basis of microscopic electromagnetism. Luckily, this also fits in quite well with the nonlinear dynamical diffraction theory developed in sec. 4.

4. Nonlinear Dynamical Diffraction from Free Electrons

We now return to a perfect lattice with an electron density given by eq. (3.4), and include nonlinear optical effects due to quasi-free electrons (see ch. 2, sec. 1). Because the electrons do not reach relativistic velocities in any conceivable x-ray beam (ch. 2, sec. 1), Lorentz invariance is unnecessary. It is even obstructive because the crystal lattice defines a preferred frame. Therefore, the following derivations will be carried out in the lab frame (where the crystal is), using the transverse gauge $\nabla \cdot \mathbf{A} = 0$, [1].(6.43).

Several effects can be identified as contributing to the nonlinearity: Considering the Lorentz equation

$$\dot{\mathbf{v}} = \frac{q}{mc}\left(-c\nabla\Phi - \dot{\mathbf{A}} + \mathbf{v} \times (\nabla \times \mathbf{A})\right), \tag{3.53}$$

optical nonlinearities can arise from an anharmonic potential Φ, but much stronger (relatively speaking) at x-ray frequencies (see ch. 2, sec. 1) is the nonlinear effect of the Lorentz force: once an electron is oscillating under the influence of the electric field $\dot{\mathbf{A}}/c$, its first-order velocity re-enters the Lorentz equation through $\mathbf{v} \times (\nabla \times \mathbf{A})$, giving rise to a second-order acceleration.

Yet another nonlinear effect is based on the spatial modulation of the electron density: The current generated at $\mathbf{r}$ is due to the oscillation of electrons whose density is given by $\eta(\mathbf{r} - \mathbf{x})$. This effect is quite different from the ones due to the spatial inhomogeneity of the electromagnetic field discussed on p. 22.

Instead of calculating the second-order current by integrating eq. (3.53), a more direct approach shall be taken here, based upon the results of ch. 2, sec. 2.2. Anharmonic potentials are neglected in that approach.

4.1 Multiple Bloch Waves

In the sections on linear dynamical diffraction, it was sufficient to consider only one Bloch sum because a more complicated electromagnetic field could be expressed as a superposition of simple Bloch sums. This is no longer possible if nonlinear coupling between modes is admitted, and strictly speaking, a solution can generally not be expressed in the form of a Bloch wave at all. However, for a weak nonlinearity, a multiple Bloch sum can be used, containing a finite number of Bloch sums for the strong incident waves, and a finite number of Bloch sums, representing relatively weak waves at sum and difference frequencies and wave vectors. Each of the individual Bloch sums has its own index n, with frequency (ω_n), base vector $(\mathbf{k}_n)_0$ and set of component amplitudes $(\tilde{\mathbf{A}}_n)_\mathbf{h}$.

The vector potential is real-valued and therefore, the definitions of the Bloch waves in eqs. (3.5) and (3.24) explicitly contain terms $\tilde{\mathbf{A}}_\mathbf{h} \exp(i(\mathbf{k}_\mathbf{h} \cdot \mathbf{r} - \omega t))$ and $\tilde{\mathbf{A}}_\mathbf{h}^* \exp(-i(\mathbf{k}_\mathbf{h} \cdot \mathbf{r} - \omega t))$. Instead of doing the same here, we make use of the multiple Bloch waves that has to be used anyway, and define that for each index n, an index $-n$ is also present and $(\mathbf{k}_n)_0 = -(\mathbf{k}_{-n})_0$ and $(\omega_n) = -(\omega_{-n})$ and

$(\tilde{\mathbf{A}}_n)_\mathbf{h} = (\tilde{\mathbf{A}}^*_{-n})_{-\mathbf{h}}$. This makes no difference to the physics and considerably reduces the formula clutter. The set of Bloch base vectors $(\mathbf{k}_n)_\mathbf{0}$ is taken to be closed in the sense that $(\mathbf{k}_n)_\mathbf{0} = (\mathbf{k}_{n'})_\mathbf{0}+(\mathbf{k}_{n''})_\mathbf{0}$ is a member of the set if $(\mathbf{k}_{n'})_\mathbf{0}$ and $(\mathbf{k}_{n''})_\mathbf{0}$ are. This generally implies an infinite number of base wave vectors $(\mathbf{k}_n)_\mathbf{0}$ but usually, only a few of these have appreciable amplitudes associated with themselves or any $(\mathbf{k}_n)_\mathbf{0} + \mathbf{h}$. The vector potential is now:

$$\mathbf{A} = \sum_{n,\mathbf{h}} (\mathbf{A}_n)_\mathbf{h} = \sum_{n,\mathbf{h}} (\tilde{\mathbf{A}}_n)_\mathbf{h} e^{i((\mathbf{k}_n)_\mathbf{h}\cdot\mathbf{r}-\omega_n t)} \tag{3.54}$$

with $(\mathbf{k}_n)_\mathbf{h} = (\mathbf{k}_n)_\mathbf{0} + \mathbf{h}$. All Bloch component waves are taken to be in the transverse gauge [1].(6.43), i.e., $(\mathbf{k}_n)_\mathbf{h} \cdot (\tilde{\mathbf{A}}_n)_\mathbf{h} = 0$, with no condition on Φ.

4.2 The System of Nonlinear Equations

If the electron does not reach relativistic velocities, then its mass, as seen in the lab frame, stays constant and its velocity is proportional to the kinematic momentum $\mathbf{p}_k$, given by eq. (2.19)[10].

$$\begin{aligned}\mathbf{p}_k = \frac{e}{c}\mathbf{A} + \mathbf{p}_\mathbf{0} - \frac{e}{c}\sum_{n,\mathbf{h}} (\mathbf{k}_n)_\mathbf{h} \frac{\left[\mathbf{p}_\mathbf{0}\cdot(\tilde{\mathbf{A}}_n)_\mathbf{h} - mc\,(\tilde{\Phi}_n)_\mathbf{h}\right] e^{i((\mathbf{k}_n)_\mathbf{h}\cdot\mathbf{r}-\omega_n t)}}{m\omega_n - \mathbf{p}_\mathbf{0}\cdot\mathbf{k}_{n,\mathbf{h}}} \\ -\frac{e^2}{2c^2}\sum_{\substack{n,\mathbf{h}\\ n',\mathbf{h}'}} \left[(\mathbf{k}_n)_\mathbf{h} + (\mathbf{k}_{n'})_{\mathbf{h}'}\right] \frac{\left[(\tilde{\mathbf{A}}_n)_\mathbf{h}\cdot(\tilde{\mathbf{A}}_{n'})_{\mathbf{h}'} - (\tilde{\Phi}_n)_\mathbf{h}(\tilde{\Phi}_{n'})_{\mathbf{h}'}\right]}{m(\omega_n+\omega_{n'}) - \mathbf{p}_\mathbf{0}\cdot((\mathbf{k}_n)_\mathbf{h} + (\mathbf{k}_{n'})_{\mathbf{h}'})} \\ e^{i(((\mathbf{k}_n)_\mathbf{h}+(\mathbf{k}_{n'})_{\mathbf{h}'})\cdot\mathbf{r}-(\omega_n+\omega_{n'})t}\end{aligned} \tag{3.55}$$

where $\mathbf{p}_\mathbf{0}$ is the momentum before the fields were present (in which case the kinematical and the canonical momentum are the same). Because the electron is part of a crystal that is resting in the lab frame, $\mathbf{p}_\mathbf{0} = \mathbf{0}$ here. The term in eq. (2.19) that contains $|\mathbf{A}^0_j|^2$ is omitted in eq. (3.55) because it is constant in space and in time and does therefore not contribute to the scattering of x-rays.

In the jellium model introduced in sec. 2 of this chapter, the scalar potential Φ is related linearly to the vector potential $\mathbf{A}$ and smaller than $\mathbf{A}$ by a factor of about 10^{-6}. That makes the contribution of $(\tilde{\Phi}_n)_\mathbf{h}$ in the first sum of eq. (3.55) very much smaller than other linear terms, but presumably larger than the non-linear terms in the second sum in eq. (3.55). However, it is clearly separated from the nonlinear terms in frequency and wave vectors. The contributions

[10] Here, the tilde on $(\tilde{\mathbf{A}}_n)_\mathbf{h}$ takes the function of the superscript 0 in eq. (2.19) to denote the amplitude without the phase in the exponential

of $(\tilde{\Phi}_n)_{\mathbf{h}}(\tilde{\Phi}_{n'})_{\mathbf{h}'}$ in the second sum in eq. (3.55) are of second order, but very much smaller (ca. 10^{-12}) than those from $(\tilde{\mathbf{A}}_n)_{\mathbf{h}} \cdot (\tilde{\mathbf{A}}_{n'})_{\mathbf{h}'}$. Therefore, the induced scalar potential plays only a minor role in both, linear and nonlinear contributions and it will be neglected in the following considerations for the sake of conciseness. Likewise, the scattering of x-rays due to the small mobility of the nuclei is stronger than nonlinear scattering from x-rays, but is also clearly separated from it in frequency and momentum and may therfore be ignored.

With this, it is now possible to determine the first- and second-order velocities $\mathbf{v}^{(1)}$, $\mathbf{v}^{(2)}$ of the electrons, and thence the current density that is to be inserted into eq. (3.2). With $\mathbf{J}(\mathbf{r},t) = -e\eta(\mathbf{r},t)\mathbf{v}(\mathbf{r},t)$, the first-order current is simply $\mathbf{J}^{(1)}(\mathbf{r},t) = -e\eta(\mathbf{r},0)\mathbf{v}^{(1)}(\mathbf{r},t)$. The second-order current density has two contributions $\mathbf{J}^{(2)}(\mathbf{r},t) = \mathbf{J}^{(2a)}(\mathbf{r},t) + \mathbf{J}^{(2b)}(\mathbf{r},t)$ with $\mathbf{J}^{(2a)}(\mathbf{r},t) = -e\eta(\mathbf{r},0)\mathbf{v}^{(2)}(\mathbf{r},t)$ and $\mathbf{J}^{(2b)}(\mathbf{r},t) = -e\eta(\mathbf{r},t)\mathbf{v}^{(1)}(\mathbf{r},t)$, where $\eta(\mathbf{r},t) = \eta(\mathbf{r},0) + \mathbf{x}^{(1)} \cdot \nabla\eta|_{\mathbf{r},0}$ is a Taylor expansion up to the first order.

Equation (3.55) gives the first- and second-order velocities $\mathbf{v}^{(1)}$ and $\mathbf{v}^{(2)}$

$$\begin{aligned}
\mathbf{v}^{(1)} &= \frac{e}{mc}\sum_{n,\mathbf{h}}(\tilde{\mathbf{A}}_n)_{\mathbf{h}} e^{i((\mathbf{k}_n)_{\mathbf{h}}\cdot\mathbf{r}-\omega_n t)} \qquad (3.56)\\
\mathbf{v}^{(2)} &= -\frac{e^2}{2m^2c^3}\sum_{\substack{n',\mathbf{h}'\\ n'',\mathbf{h}''}}\left[\frac{(\mathbf{k}_{n'})_{\mathbf{h}'}+(\mathbf{k}_{n''})_{\mathbf{h}''}}{k_{n'}+k_{n''}}\right](\tilde{\mathbf{A}}_{n'})_{\mathbf{h}'}\cdot(\tilde{\mathbf{A}}_{n''})_{\mathbf{h}''}\\
&\qquad e^{i(((\mathbf{k}_{n'})_{\mathbf{h}'}+(\mathbf{k}_{n''})_{\mathbf{h}''})\cdot\mathbf{r}-(\omega_{n'}+\omega_{n''})t}
\end{aligned}$$

with $k_n = \omega_n/c$. The scalar potential terms were neglected as explained above. A variable reassignment with $\omega_n = \omega_{n'} + \omega_{n''}$ and $(\mathbf{k}_n)_{\mathbf{O}} = (\mathbf{k}_{n'})_{\mathbf{O}} + (\mathbf{k}_{n''})_{\mathbf{O}}$, $\mathbf{h} = \mathbf{h}' + \mathbf{h}''$, and thus $(\mathbf{k}_n)_{\mathbf{h}} = (\mathbf{k}_{n'})_{\mathbf{h}'} + (\mathbf{k}_{n''})_{\mathbf{h}''}$, makes the exponents comparable. It has the same intent, but a somewhat different realization, than that used for eqs. (3.8) and (3.26). There is no obvious arithmetic relationship relating the indices to the wave vectors and frequencies, and therefore index retrieval functions must be introduced for n: $n'' = N_m(n,n')$, where the index m runs over all cases for which $(\mathbf{k}_n)_{\mathbf{O}} = (\mathbf{k}_{n'})_{\mathbf{O}} + (\mathbf{k}_{n''})_{\mathbf{O}}$. In the particular cases considered below, there will be only one. With the index retrieval functions, $(\tilde{\mathbf{A}}_{n''})_{\mathbf{h}''}$ is replaced with $(\tilde{\mathbf{A}}_{N_m(n,n')})_{\mathbf{h}-\mathbf{h}'}$

$$\begin{aligned}
\mathbf{v}^{(1)} &= \frac{e}{mc}\sum_{n,\mathbf{h}}(\tilde{\mathbf{A}}_n)_{\mathbf{h}} e^{i((\mathbf{k}_n)_{\mathbf{h}}\cdot\mathbf{r}-\omega_n t)} \qquad (3.57)\\
\mathbf{v}^{(2)} &= -\frac{e^2}{2m^2c^3}\sum_{\substack{n,n',m\\ \mathbf{h},\mathbf{h}'}}\frac{(\mathbf{k}_n)_{\mathbf{h}}}{k_n}(\tilde{\mathbf{A}}_{N_m(n,n')})_{\mathbf{h}-\mathbf{h}'}\cdot(\tilde{\mathbf{A}}_{n'})_{\mathbf{h}'} e^{i((\mathbf{k}_n)_{\mathbf{h}}\cdot\mathbf{r}-\omega_n t}
\end{aligned}$$

After multiplying with $-e\eta(\mathbf{r},0)$, the first-order current density $\mathbf{J}^{(1)}$ and the contribution $\mathbf{J}^{(2a)} = -e\eta(\mathbf{r},t)\mathbf{v}^{(2)}$ to the second-order current density are obtained. The former is after the usual index transformation (compare with eq. (3.8)):

$$\mathbf{J}^{(1)} = -\frac{e^2}{mc}\sum_{n,\mathbf{h},\mathbf{h}'} \tilde{\eta}_{\mathbf{h}-\mathbf{h}'}(\tilde{\mathbf{A}}_n)_{\mathbf{h}'} e^{i((\mathbf{k}_n)_\mathbf{h}\cdot\mathbf{r}-\omega_n t)} \tag{3.58}$$

For the second-order current, a similar index transformation can be found, except that it involves three, not just two reciprocal lattice vectors. If $\mathbf{h}$ is used in the Fourier sum for η and $\mathbf{h}, \mathbf{h}'$ in the double sum for $\mathbf{v}^{(2)}$ are replaced with $\mathbf{h}', \mathbf{h}''$, then the index transformation is: $\mathbf{h}+\mathbf{h}' \to \mathbf{h}$, $\mathbf{h}' \to \mathbf{h}'$, $\mathbf{h}'' \to \mathbf{h}''$, and thus $\mathbf{h} \to \mathbf{h}-\mathbf{h}'$. This makes $\mathbf{J}^{(2a)}$:

$$\mathbf{J}^{(2a)} = \frac{e^3}{2m^2c^3}\sum_{\substack{n,n',m\\ \mathbf{h},\mathbf{h}',\mathbf{h}''}} \tilde{\eta}_{\mathbf{h}-\mathbf{h}'}\frac{(\mathbf{k}_n)_{\mathbf{h}'}}{k_n}(\tilde{\mathbf{A}}_{N_m(n,n')})_{\mathbf{h}'-\mathbf{h}''}\cdot(\tilde{\mathbf{A}}_{n'})_{\mathbf{h}''} e^{i((\mathbf{k}_n)_\mathbf{h}\cdot\mathbf{r}-\omega_n t} \tag{3.59}$$

For the other contribution $\mathbf{J}^{(2b)}$ to the second-order current density, we integrate $\mathbf{v}^{(1)}$ to obtain

$$\mathbf{x}^{(1)} = i\frac{e}{mc}\sum_{n,\mathbf{h}'}\frac{(\tilde{\mathbf{A}}_n)_{\mathbf{h}'}}{\omega_n} e^{i((\mathbf{k}_n)_{\mathbf{h}'}\cdot\mathbf{r}-\omega_n t)} \tag{3.60}$$

After multiplying with $\nabla\eta = i\sum_\mathbf{h} \mathbf{h}\tilde{\eta}_\mathbf{h}\exp(i\mathbf{h}\cdot\mathbf{r})$, using $\omega_n = ck_n$ and applying the usual index transformation, we get

$$\mathbf{x}\cdot\nabla\eta = -\frac{e}{mc^2}\sum_{n\mathbf{h}\mathbf{h}'} \tilde{\eta}_{\mathbf{h}-\mathbf{h}'}\frac{(\tilde{\mathbf{A}}_n)_{\mathbf{h}'}\cdot(\mathbf{h}-\mathbf{h}')}{k_n} e^{i((\mathbf{k}_n)_\mathbf{h}\cdot\mathbf{r}-\omega_n t)} \tag{3.61}$$

This is now multiplied with $-e\mathbf{v}^{(1)}$ (using $\mathbf{h}''$ in its sum) to obtain $\mathbf{J}^{(2b)}$. An index transformation $\mathbf{h}+\mathbf{h}'' \to \mathbf{h}$:

$$\mathbf{J}^{(2b)} = \frac{e^3}{m^2c^3}\sum_{\substack{n,n',m\\ \mathbf{h},\mathbf{h}',\mathbf{h}''}} (\tilde{\mathbf{A}}_{n'})_{\mathbf{h}''}\tilde{\eta}_{\mathbf{h}-\mathbf{h}'-\mathbf{h}''}\frac{(\tilde{\mathbf{A}}_{N_m(n,n')})_{\mathbf{h}'}\cdot(\mathbf{h}-\mathbf{h}'-\mathbf{h}'')}{k_n} e^{i((\mathbf{k}_n)_\mathbf{h}\cdot\mathbf{r}-\omega_n t)} \tag{3.62}$$

Everything is now ready for the final assembly. After passing the current densities $\mathbf{J}^{(1)}$, $\mathbf{J}^{(2a)}$ and $\mathbf{J}^{(2b)}$ through eq. (3.3) to obtain the transverse current and inserting the Bloch wave, eq. (3.54) into the wave equation eq. (3.2), the following fundamental system of equations of second-order nonlinear dynamical diffraction is obtained for each n and $\mathbf{h}$ (meaning, each Fourier-Bloch component with exponent[11] $\exp(i((\mathbf{k}_n)_\mathbf{h}\cdot\mathbf{r}-\omega_n t))$ may be considered separately and

[11] NB, there is no $\pm$ in this exponent because other than in sec. 2, positive- and negative-running waves are taken separately, with indices $\pm n$.

the wave equation is recovered if each of the following equations is multiplied with $\exp(i((\mathbf{k}_n)_\mathbf{h} \cdot \mathbf{r} - \omega_n t))$ and the sum over n and $\mathbf{h}$ is taken):

$$\begin{aligned} 0 &= \left(k_n^2 - (\mathbf{k}_n)_\mathbf{h}^2\right) (\tilde{\mathbf{A}}_n)_\mathbf{h} - 4\pi r_e \sum_{\mathbf{h}'} \left[(\tilde{\mathbf{A}}_n)_{\mathbf{h}'}\right]_\mathbf{h} \tilde{\eta}_{\mathbf{h}-\mathbf{h}'} \\ &+ 4\pi \frac{r_e^2}{e} \sum_{\substack{n',m \\ \mathbf{h}',\mathbf{h}''}} [(\mathbf{k}_n)_{\mathbf{h}'}]_\mathbf{h} \frac{\tilde{\eta}_{\mathbf{h}-\mathbf{h}'} (\tilde{\mathbf{A}}_{N_m(n,n')})_{\mathbf{h}'-\mathbf{h}''} \cdot (\tilde{\mathbf{A}}_{n'})_{\mathbf{h}''}}{2k_n} \\ &+ 4\pi \frac{r_e^2}{e} \sum_{\substack{n',m \\ \mathbf{h}',\mathbf{h}''}} \left[(\tilde{\mathbf{A}}_{n'})_{\mathbf{h}''}\right]_\mathbf{h} \frac{\tilde{\eta}_{\mathbf{h}-\mathbf{h}'-\mathbf{h}''} (\tilde{\mathbf{A}}_{N_m(n,n')})_{\mathbf{h}'} \cdot (\mathbf{h} - \mathbf{h}' - \mathbf{h}'')}{k_n}, \end{aligned} \tag{3.63}$$

where transverse parts $[(\mathbf{k}_n)_{\mathbf{h}'}]_\mathbf{h} = (\mathbf{k}_n)_\mathbf{h} \times (\mathbf{k}_n)_{\mathbf{h}'} \times (\mathbf{k}_n)_\mathbf{h} / |(\mathbf{k}_n)_\mathbf{h}|^2$ and $\left[(\tilde{\mathbf{A}}_{n'})_{\mathbf{h}''}\right]_\mathbf{h} = (\mathbf{k}_n)_\mathbf{h} \times (\tilde{\mathbf{A}}_{n'})_{\mathbf{h}''} \times (\mathbf{k}_n)_\mathbf{h} / |(\mathbf{k}_n)_\mathbf{h}|^2$ are defined. The first line represents standard linear dynamical diffraction and corresponds to eq. (3.11). There are two types of nonlinear contributions in the other two lines, being due to $\mathbf{A} \cdot \mathbf{A}$ and the nonlinear displacement current, respectively. The $\mathbf{A} \cdot \mathbf{A}$ contribution has a vectorial direction, given by the part transverse to $(\mathbf{k}_n)_\mathbf{h}$ of the longitudinal component $(\mathbf{k}_n)_{\mathbf{h}'}$, see fig. 2.2. Instead of seeking an algebraic solution of eq. (3.63), one may also take a Takagi-style approach (compare to eq. (3.33)), under the assumption that in the absence of nonlinear scattering (in the second and third lines of eq. (3.63)), the first line of eq. (3.63) is exactly zero for a given field amplitude $(\tilde{\mathbf{A}}_n)_\mathbf{h}$ and its wave vector $(\mathbf{k}_n)_\mathbf{h}$. The nonlinear scattering contribution is therefore taken as a small perturbation to a mode of standard dynamical diffraction. The growth of the amplitude $(\tilde{\mathbf{A}}_n)_\mathbf{h}$ during the propagation of x-rays due to this perturbation is then

$$-2i(\mathbf{k}_n)_\mathbf{h} \cdot \nabla (\tilde{\mathbf{A}}_n)_\mathbf{h} = (\text{n.l.t.}) \tag{3.64}$$

where n.l.t. means the nonlinear terms in the second and third lines of eq. (3.63).

Equation (3.63) is rather unwieldy because it contains sums over multiple Bloch waves and their component waves. Luckily, only very few indices and reciprocal lattice vectors remain in practically relevant cases, such as the one given in the next section.

4.3 An Example: Parametric Down Conversion

All successful experiments to demonstrate parametric down conversion of x-rays which have been described to date [111, 112, 113, 114], see also ch. 5, were done with a wave vector matching scheme that involved four waves, namely the pump wave $(\mathbf{k}_p, \omega_p, \tilde{\mathbf{A}}_p)$, the diffracted pump wave $(\mathbf{k}_p + \mathbf{h}, \omega_p, \tilde{\mathbf{A}}_{p\mathbf{h}})$ and the so-called signal $(\mathbf{k}_s, \omega_s, \tilde{\mathbf{A}}_s)$ and idler $(\mathbf{k}_i, \omega_i, \tilde{\mathbf{A}}_i)$ waves, plus the complex conjugate waves $(-\mathbf{k}_p, -\omega_p, \tilde{\mathbf{A}}_p^*)$, $(-\mathbf{k}_p - \mathbf{h}, -\omega_p, \tilde{\mathbf{A}}_{p\mathbf{h}}^*)$, $(-\mathbf{k}_s, -\omega_s, \tilde{\mathbf{A}}_s^*)$ and $(-\mathbf{k}_i, -\omega_i, \tilde{\mathbf{A}}_i^*)$. The wave vector and frequency matching conditions are $\mathbf{k}_s + \mathbf{k}_i = \mathbf{k}_p + \mathbf{h}$ and $\omega_i + \omega_s = \omega_p$ (see fig. 3.16). This combination is the simplest

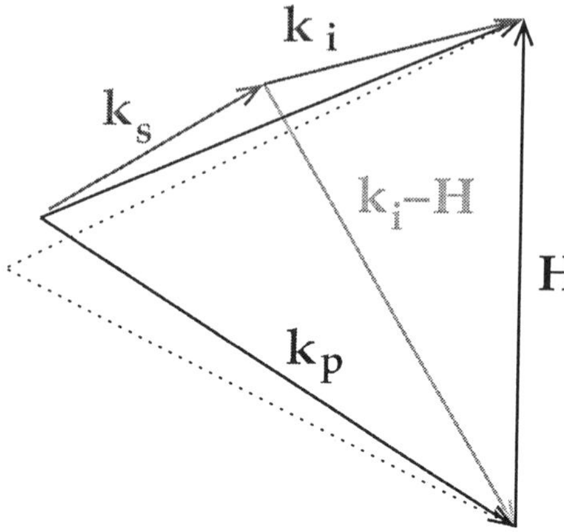

Figure 3.16: Wave vector matching in parametric down conversion of x-rays, as realized in [111, 112, 113, 114]

example of nonlinear dynamical diffraction, in which only the pump is diffracted by the crystal lattice. More complicated cases are however possible, such as the one discussed in ch. 5, sec. 2.8.

In the simple case, the system (3.63) of nonlinear equations is reduced to four&four lines for the above waves and their complex conjugates and the sums in eq. (3.63) run over the reciprocal lattice vectors $\mathbf{0}$ and $\mathbf{H}$. The set of Bloch base vectors in eq. (3.54) is *not* simply taken from the eight (four&four) above wave vectors, but rather consists of $(\mathbf{k}_P)_0 = \mathbf{k}_p$, $(\mathbf{k}_S)_0 = \mathbf{k}_s$ and $(\mathbf{k}_I)_0 = \mathbf{k}_i - \mathbf{H}$ plus the corresponding opposed vectors. This is necessary to make sure the condition $(\mathbf{k}_n)_0 = (\mathbf{k}_{n'})_0 + (\mathbf{k}_{n''})_0$ can be fulfilled for the combinations of wave vectors that are relevant here.

The lines of eq. (3.63) that refer to the pump and diffracted pump waves (i.e., have exponents $\exp(\pm i(\mathbf{k}_p \cdot \mathbf{r} - \omega_p t))$ and $\exp(\pm i((\mathbf{k}_p + \mathbf{h}) \cdot \mathbf{r} - \omega_p t))$ associated with them) can be written in disregard of the nonlinear contributions because the signal and idler waves are so much weaker. For the pump and diffracted pump wave, eq. (3.63) is then nothing but the fundamental equation of standard dynamcial diffraction.

The lines for the signal and the idler wave are more complicated because all four&four amplitudes appear in them. Tables 3.1 and 3.2 list the contributions to eq. (3.63) of all possible combinations of indices that refer to the nonvanishing amplitudes $(\tilde{\mathbf{A}}_P)_0$, $(\tilde{\mathbf{A}}_P)_\mathbf{H}$, $(\tilde{\mathbf{A}}_S)_0$ and $(\tilde{\mathbf{A}}_I)_\mathbf{H}$. When reading the tables, it is important to keep in mind that, due to the way the idler Bloch sum is defined, there are two notations for the wave vectors and amplitudes, which differ in particular for the idler wave: the idler wave vector is $\mathbf{k}_i$ and the idler amplitude

(n, n', n'')	$(\mathbf{h}, \mathbf{h}', \mathbf{h}'')$	$k_n^2\,[(\mathbf{k}_n)_{\mathbf{h}'}]_{\mathbf{h}}$	$\tilde{\eta}_{\mathbf{h}-\mathbf{h}'}$	$(\tilde{\mathbf{A}}_{n''})_{\mathbf{h}'-\mathbf{h}''}$	$(\tilde{\mathbf{A}}_{n'})_{\mathbf{h}''}$	
$(s, -i, p)$	$(\mathbf{0}, \mathbf{0}, -\mathbf{H})$	$\mathbf{k}_s \times \mathbf{k}_s \times \mathbf{k}_s = 0$	$\tilde{\eta}_{\mathbf{O}}$	$(\tilde{\mathbf{A}}_P)_{\mathbf{H}}$	$(\tilde{\mathbf{A}}_I^*)_{-\mathbf{H}}$	
$(s, -i, p)$	$(\mathbf{0}, -\mathbf{H}, -\mathbf{H})$	$\mathbf{k}_s \times (\mathbf{k}_s - \mathbf{H}) \times \mathbf{k}_s$	$\tilde{\eta}_{\mathbf{H}}$	$(\tilde{\mathbf{A}}_P)_{\mathbf{O}}$	$(\tilde{\mathbf{A}}_I^*)_{-\mathbf{H}}$	✓
$(s, p, -i)$	$(\mathbf{0}, \mathbf{0}, \mathbf{H})$	$\mathbf{k}_s \times \mathbf{k}_s \times \mathbf{k}_s = 0$	$\tilde{\eta}_{\mathbf{O}}$	$(\tilde{\mathbf{A}}_I^*)_{-\mathbf{H}}$	$(\tilde{\mathbf{A}}_P)_{\mathbf{H}}$	
$(s, p, -i)$	$(\mathbf{0}, -\mathbf{H}, \mathbf{0})$	$\mathbf{k}_s \times (\mathbf{k}_s - \mathbf{H}) \times \mathbf{k}_s$	$\tilde{\eta}_{\mathbf{H}}$	$(\tilde{\mathbf{A}}_I^*)_{-\mathbf{H}}$	$(\tilde{\mathbf{A}}_P)_{\mathbf{O}}$	✓
$(i, -s, p)$	$(\mathbf{H}, \mathbf{0}, \mathbf{0})$	$\mathbf{k}_i \times (\mathbf{k}_i - \mathbf{H}) \times \mathbf{k}_i$	$\tilde{\eta}_{\mathbf{H}}$	$(\tilde{\mathbf{A}}_P)_{\mathbf{O}}$	$(\tilde{\mathbf{A}}_S^*)_{\mathbf{O}}$	✓
$(i, -s, p)$	$(\mathbf{H}, \mathbf{H}, \mathbf{0})$	$\mathbf{k}_i \times \mathbf{k}_i \times \mathbf{k}_i = 0$	$\tilde{\eta}_{\mathbf{O}}$	$(\tilde{\mathbf{A}}_P)_{\mathbf{H}}$	$(\tilde{\mathbf{A}}_S^*)_{\mathbf{O}}$	
$(i, p, -s)$	$(\mathbf{H}, \mathbf{0}, \mathbf{0})$	$\mathbf{k}_i \times (\mathbf{k}_i - \mathbf{H}) \times \mathbf{k}_i$	$\tilde{\eta}_{\mathbf{H}}$	$(\tilde{\mathbf{A}}_S^*)_{\mathbf{O}}$	$(\tilde{\mathbf{A}}_P)_{\mathbf{O}}$	✓
$(i, p, -s)$	$(\mathbf{H}, \mathbf{H}, \mathbf{H})$	$\mathbf{k}_i \times \mathbf{k}_i \times \mathbf{k}_i = 0$	$\tilde{\eta}_{\mathbf{O}}$	$(\tilde{\mathbf{A}}_S^*)_{\mathbf{O}}$	$(\tilde{\mathbf{A}}_P)_{\mathbf{H}}$	

Table 3.1: Contributions of type $\mathbf{k}(\mathbf{A} \cdot \mathbf{A})$ to eq. (3.63), containing the nonvanishing amplitudes $(\tilde{\mathbf{A}}_P)_{\mathbf{O}}$, $(\tilde{\mathbf{A}}_P)_{\mathbf{H}}$, $(\tilde{\mathbf{A}}_S)_{\mathbf{O}}$ and $(\tilde{\mathbf{A}}_I)_{\mathbf{H}} = \tilde{\mathbf{A}}_i$ (see text). The four nonvanishing contributions, indicated by check marks (✓), all contain the same amplitudes $(\tilde{\mathbf{A}}_p)_{\mathbf{O}}$, $(\tilde{\mathbf{A}}_s)_{\mathbf{O}}$ and $(\tilde{\mathbf{A}}_i)_{\mathbf{H}}$.

is $\tilde{\mathbf{A}}_i = (\tilde{\mathbf{A}}_I)_{\mathbf{H}}$. The indices $-s$ and $-i$ in the column for (n, n', n'') mean the respective complex conjugate waves.

(n, n', n'')	$(\mathbf{h}, \mathbf{h}', \mathbf{h}'')$	$\tilde{\eta}_{\mathbf{h}-\mathbf{h}'-\mathbf{h}''}$	$(\tilde{\mathbf{A}}_{n''})_{\mathbf{h}'} \cdot (\mathbf{h}-\mathbf{h}'-\mathbf{h}'')$	$k_n^2\left[(\tilde{\mathbf{A}}_{n'})_{\mathbf{h}''}\right]_{\mathbf{h}}$	
$(s, -i, p)$	$(\mathbf{0}, \mathbf{0}, -\mathbf{H})$	$\tilde{\eta}_{\mathbf{H}}$	$(\tilde{\mathbf{A}}_P)_{\mathbf{O}} \cdot \mathbf{H}$	$\mathbf{k}_s \times (\tilde{\mathbf{A}}_I^*)_{-\mathbf{H}} \times \mathbf{k}_s$	✓
$(s, -i, p)$	$(\mathbf{0}, \mathbf{H}, -\mathbf{H})$	$\tilde{\eta}_{\mathbf{O}}$	$(\tilde{\mathbf{A}}_P)_{\mathbf{H}} \cdot \mathbf{0}$	$\mathbf{k}_s \times (\tilde{\mathbf{A}}_I^*)_{-\mathbf{H}} \times \mathbf{k}_s$	
$(s, p, -i)$	$(\mathbf{0}, -\mathbf{H}, \mathbf{0})$	$\tilde{\eta}_{\mathbf{H}}$	$(\tilde{\mathbf{A}}_I^*)_{-\mathbf{H}} \cdot \mathbf{H}$	$\mathbf{k}_s \times (\tilde{\mathbf{A}}_P)_{\mathbf{O}} \times \mathbf{k}_s$	✓
$(s, p, -i)$	$(\mathbf{0}, -\mathbf{H}, \mathbf{H})$	$\tilde{\eta}_{\mathbf{O}}$	$(\tilde{\mathbf{A}}_I^*)_{-\mathbf{H}} \cdot \mathbf{0}$	$\mathbf{k}_s \times (\tilde{\mathbf{A}}_P)_{\mathbf{H}} \times \mathbf{k}_s$	
$(i, -s, p)$	$(\mathbf{H}, \mathbf{0}, \mathbf{0})$	$\tilde{\eta}_{\mathbf{H}}$	$(\tilde{\mathbf{A}}_P)_{\mathbf{O}} \cdot \mathbf{H}$	$\mathbf{k}_i \times (\tilde{\mathbf{A}}_S^*)_{\mathbf{O}} \times \mathbf{k}_i$	✓
$(i, -s, p)$	$(\mathbf{H}, \mathbf{H}, \mathbf{0})$	$\tilde{\eta}_{\mathbf{O}}$	$(\tilde{\mathbf{A}}_P)_{\mathbf{H}} \cdot \mathbf{0}$	$\mathbf{k}_i \times (\tilde{\mathbf{A}}_S^*)_{\mathbf{O}} \times \mathbf{k}_i$	
$(i, p, -s)$	$(\mathbf{H}, \mathbf{0}, \mathbf{0})$	$\tilde{\eta}_{\mathbf{H}}$	$(\tilde{\mathbf{A}}_S^*)_{\mathbf{O}} \cdot \mathbf{H}$	$\mathbf{k}_i \times (\tilde{\mathbf{A}}_P)_{\mathbf{O}} \times \mathbf{k}_i$	✓
$(i, p, -s)$	$(\mathbf{H}, \mathbf{0}, \mathbf{H})$	$\tilde{\eta}_{\mathbf{O}}$	$(\tilde{\mathbf{A}}_S^*)_{\mathbf{O}} \cdot \mathbf{0}$	$\mathbf{k}_i \times (\tilde{\mathbf{A}}_P)_{\mathbf{H}} \times \mathbf{k}_i$	

Table 3.2: Contributions of type $\mathbf{A}(\mathbf{A} \cdot \mathbf{h})$ to eq. (3.63), containing the nonvanishing amplitudes $(\tilde{\mathbf{A}}_P)_{\mathbf{O}}$, $(\tilde{\mathbf{A}}_P)_{\mathbf{H}}$, $(\tilde{\mathbf{A}}_S)_{\mathbf{O}}$ and $(\tilde{\mathbf{A}}_I)_{\mathbf{H}} = \tilde{\mathbf{A}}_i$ (see text). The four nonvanishing contributions, indicated by check marks (✓), all contain the same amplitudes $(\tilde{\mathbf{A}}_p)_{\mathbf{O}}$, $(\tilde{\mathbf{A}}_s)_{\mathbf{O}}$ and $(\tilde{\mathbf{A}}_i)_{\mathbf{H}}$.

The two tables show that the diffracted pump wave $(\mathbf{k}_p + \mathbf{h}, \omega_p, \tilde{\mathbf{A}}_{p\mathbf{h}})$ plays no role. Taking the nonzero contributions for the signal amplitude and inserting them into eq. (3.63) gives the equation:

$$0 = (k_s^2 - \mathbf{k}_s^2)(\tilde{\mathbf{A}}_s) - 4\pi r_e \tilde{\eta}_0 (\tilde{\mathbf{A}}_s) + 4\pi \frac{r_e^2}{e} \tilde{\eta}_{\mathbf{H}} \frac{(\mathbf{k}_s \times \mathbf{H} \times \mathbf{k}_s)(\mathbf{A}_i^* \cdot \mathbf{A}_p)}{k_s |\mathbf{k}_s|^2}$$
$$+4\pi \frac{r_e^2}{e} \tilde{\eta}_{\mathbf{H}} \left[\frac{\left(\mathbf{k}_s \times \tilde{\mathbf{A}}_i^* \times \mathbf{k}_s\right)\left(\tilde{\mathbf{A}}_p \cdot \mathbf{H}\right) + \left(\mathbf{k}_s \times \tilde{\mathbf{A}}_p \times \mathbf{k}_s\right)\left(\tilde{\mathbf{A}}_i^* \cdot \mathbf{H}\right)}{k_s |\mathbf{k}_s|^2} \right] \quad (3.65)$$

and a similar line for the complex-conjugate signal wave.

Given a pump amplitude, and the idler amplitude being due to vacuum fluctuations, this expression can be solved for the signal amplitude. Signal and idler are interchangeable in parametric down conversion, and a similar expression for the idler amplitude due to the pump and the signal is also found from eq. (3.63).

Although no-one has to date measured the polarizations of down-converted x-ray photons, this property is of particular interest to potential applications in tests of the quantum theory, using entangled photons. Equation (3.65) can be used to make predictions. Photons in the signal mode are generated through three nonlinear terms in eq. (3.65), called I (end of the first line in eq. (3.65) and IIa, IIb (in the second line of the equation). In each of them, the polarization of the signal is determined by a double cross product $\mathbf{k}_s \times \mathbf{X} \times \mathbf{k}_s$, where $\mathbf{X}$ is either $\mathbf{H}$ (term I) or the idler or pump amlitude (terms IIa, IIb). This ensures that the amplitude of the signal wave will be transverse to $\mathbf{k}_s$. Table 3.3 gives the possible polarization combinations of signal, idler and pump due to these cross products and the projection factors $\mathbf{A}_i^* \cdot \mathbf{A}_p$ (term I) and $\mathbf{A}_p \cdot \mathbf{H}$ and $\mathbf{A}_i^* \cdot \mathbf{H}$ (terms IIa, IIb). The terminology for polarizations in the table is that of crystallography, 'π' meaning a polarization vector within the scattering plane defined by $\mathbf{k}_p$ and $\mathbf{H}$, and 'σ' a polarization perpendicular to that plane.

term	polarization			angular
no.	pump	idler	signal	dependence
I	π	π	π	$\cos(\Theta + \alpha_s)\cos(2\Theta)$
I	σ	σ	π	$\cos(\Theta + \alpha_s)\cos(2\Theta)$
IIa	π	π	π	$\cos(\alpha_i + \alpha_s)\cos(\Theta)$
IIa	π	σ	σ	$\cos(\Theta)$
IIb	π	π	π	$\cos(2\Theta + \alpha_s)\cos(\Theta + \alpha_i)$
IIb	σ	π	σ	$\cos(\Theta + \alpha_i)$

Table 3.3: Polarizations of signal and idler for a given pump polarization and the terms responsible for them.

5. Appendix

5.1 Dynamical Diffraction in Macroscopic Form

Usually[12] [73], dynamical diffraction is derived on the basis of Maxwell's equations in macroscopic form [1].(6.28):

$$\begin{aligned} &(a)\quad \nabla\cdot\mathbf{B}=0 \qquad &(b)\quad \nabla\times\mathbf{E}+\tfrac{1}{c}\tfrac{\partial\mathbf{B}}{\partial t}=0 \\ &(c)\quad \nabla\cdot\mathbf{D}=4\pi\rho \qquad &(d)\quad \nabla\times\mathbf{H}=\tfrac{4\pi}{c}\mathbf{j}+\tfrac{1}{c}\tfrac{\partial\mathbf{D}}{\partial t} \end{aligned} \tag{3.66}$$

with $\mathbf{D}=\mathbf{E}+4\pi\mathbf{P}$ ([1].(6.71)). The polarization $\mathbf{P}$ can, in turn, be expressed in terms of the linear susceptibility tensor χ, and $\mathbf{E}$ as $\mathbf{P}=\chi\mathbf{E}$ ([1].(4.36)). With x-rays, χ is of the order of 10^{-5} or less, and one may also write $\mathbf{P}=\chi\mathbf{D}$.

In the absence of free charges, $\rho=0$ and $\mathbf{j}=\mathbf{0}$, and with zero magnetic susceptibility, we have $\mathbf{H}=\mathbf{B}$. We now apply the curl operator to equation (3.66.b) and insert (3.66.d). Then, $\mathbf{E}=\mathbf{D}-4\pi\mathbf{P}=\mathbf{D}-4\pi\chi\mathbf{D}$ is applied to obtain the wave equation

$$\nabla\times\nabla\times(\mathbf{D}-4\pi\chi\mathbf{D})+\frac{1}{c^2}\frac{\partial^2\mathbf{D}}{\partial t^2}=0 \tag{3.67}$$

With $\nabla\times\nabla\times\mathbf{D}=\nabla(\nabla\cdot\mathbf{D})-\nabla^2\mathbf{D}$, and $\rho=0$, making $\nabla\cdot\mathbf{D}=0$

$$\nabla^2\mathbf{D}-\frac{1}{c^2}\frac{\partial^2\mathbf{D}}{\partial t^2}+4\pi\nabla\times\nabla\times(\chi\mathbf{D})=0 \tag{3.68}$$

This is formally the same as eq. (3.2), with $\mathbf{D}$ instead of $\mathbf{A}$ and $\chi\mathbf{D}$ instead of the integral in eq. (3.3). Harmonically bound electrons with a resonance at ω_0 oscillate with an amplitude $x=(e/m)|\mathbf{D}|/(\omega_0^2-\omega^2)$, making the electric polarization $\mathbf{P}=(e\eta)(e/m)\mathbf{D}/(\omega_0^2-\omega^2)$ and the susceptibility

$$\chi=\frac{r_e c^2}{\omega_0^2-\omega^2}\eta. \tag{3.69}$$

Table 3.4 compares the notation used here and in other works [73, 74].

A Fourier expansion for χ is

$$\chi(\mathbf{r})=\sum_{\mathbf{h}}\tilde{\chi}_{\mathbf{h}}e^{i\mathbf{h}\cdot\mathbf{r}} \tag{3.70}$$

and the Bloch wave is

$$\mathbf{D}=e^{-i\omega t}\sum_{\mathbf{h}}\tilde{\mathbf{D}}_{\mathbf{h}}e^{i\mathbf{k}_{\mathbf{h}}\cdot\mathbf{h}} \qquad \mathbf{k}_{\mathbf{h}}=\mathbf{k}_0+\mathbf{h}. \tag{3.71}$$

[12] Please see also the remarks on the notation of wave vectors, etc. on p. xxiii.

Using the index transformation $\mathbf{h}+\mathbf{h}' \to \mathbf{h}$ and $\mathbf{h}' \to \mathbf{h}'$, as before, we get

$$0 = (k^2 - \mathbf{k_h}^2)\tilde{\mathbf{D}}_\mathbf{h} + 4\pi \sum_{\mathbf{h}'} \mathbf{k_h}^2 \chi_{\mathbf{h}-\mathbf{h}'} \left[\tilde{\mathbf{D}}_{\mathbf{h}'}\right]_\mathbf{h}, \quad \left[\tilde{\mathbf{D}}_{\mathbf{h}'}\right]_\mathbf{h} = \frac{\mathbf{k_h} \times \tilde{\mathbf{D}}_{\mathbf{h}'} \times \mathbf{k_h}}{\mathbf{k_h}^2} \tag{3.72}$$

The average refractive index is obtained by keeping only the term with $\mathbf{h}' = \mathbf{h}$ in the sum of eq. (3.72):

$$n = \sqrt{1 + 4\pi\chi_0} \approx 1 + 2\pi\chi_0, \tag{3.73}$$

which, for $\omega \gg \omega_0$, is the same as eq. (3.12).

Likewise, the equation for the dispersion hyperbola reads

$$\xi_0\xi_h = 4\pi^2 k^2 C^2 \chi_\mathbf{H}\chi_{-\mathbf{H}}. \tag{3.74}$$

which is the same as eq. (3.14).

The rocking width, eq. (3.17), becomes in the terminology of macroscopic electromagnetism:

$$\Delta\Theta = \frac{8\pi C|\chi_\mathbf{H}|}{\sin 2\Theta} \tag{3.75}$$

In another commonly used notation [73], the symbol $\Gamma = r_e\lambda^2/(\pi V) = 4\pi r_e/(k^2 V)$ is defined (where V is the unit cell volume of the crystal). For $\omega_0 \ll \omega$, this makes $\chi_0 = -\Gamma F_0/4\pi$, where the real part of F_0 is the number of electrons in one unit cell of volume V. The average refractive index is then (compare [73].[p. 687])

$$n = \sqrt{1 + 4\pi\chi_0} \approx 1 - \frac{\Gamma F_0}{2} \tag{3.76}$$

and the equation for the dispersion hyperbola reads [73].(21)

$$\xi_0\xi_h = \frac{1}{4} k^2 C^2 \Gamma^2 F_\mathbf{h} F_{-\mathbf{h}}, \tag{3.77}$$

where $F_\mathbf{h}$ is the unit cell structure factor (see p. 39).

In the notation of [73], eq. (3.18) for the maximum pendellösung period is

$$L_\mathbf{H} = \frac{\cos\Theta}{\Gamma|F_H|}\lambda \tag{3.78}$$

5.2 The Longitudinal Current

The first-order velocity of an electron of charge $-e$ in an electromagnetic wave is, according to eq. (2.19), $\mathbf{v} = (e/mc)\mathbf{A}$, from which the first-order displacement can be determined as $\mathbf{x} = (ie/m\omega c)\mathbf{A}$. Due to this displacement,

micr. here	macr. here	[74]	[73]	meaning
k	$k = \frac{2\pi}{\lambda}$	$2\pi k$	$2\pi k$	wave number
$-\frac{r_e \eta}{k^2}$	χ	$\frac{\chi}{4\pi}$	$\frac{\Gamma F}{4\pi}$	susceptibility
$n \approx 1 - \frac{2\pi r_e \eta}{k^2}$	$n \approx 1 + 2\pi\chi$	$n \approx 1 + \frac{\chi}{2}$	$n \approx 1 - \frac{\Gamma F}{2}$	refractive index
$\Delta\Theta$	$\Delta\Theta$	2δ	$\Delta\Theta$	rocking width

Table 3.4: Corresponding quantities in different notations of dynamical diffraction theory: micr. here and macr. here mean the microscopic and macroscopic formulation of this chapter and the citations [74] and [73] refer to Authier's book and the Batterman-Cole article.

the field-free electron density $\eta(\mathbf{r})$ is replaced with $\eta(\mathbf{r} - \mathbf{x})$, while the background charge, tied to the nuclei, can be taken to be immobile. The net charge density is therefore $\rho(\mathbf{r}, t) = (-e)(\eta(\mathbf{r} - \mathbf{x}) - \eta(\mathbf{r}))$. A Taylor expansion of η in eq. (3.4) to first order gives

$$\rho(\mathbf{r}, t) = -ie \sum_{\mathbf{h}} \mathbf{x}(\mathbf{r}, t) \cdot \mathbf{h}\ \tilde{\eta}_{\mathbf{h}} e^{i\mathbf{h}\cdot\mathbf{r}} \tag{3.79}$$

and upon insertion of eq. (3.5) into $\mathbf{x} = (ie/m\omega c)\mathbf{A}$, application of the usual index transformation (as for eq. (3.8)), and using also $k = \omega/c$ and the classical electron radius $r_e = e^2/mc^2$:

$$\rho^{(\pm)}(\mathbf{r}, t) = \frac{r_e}{k} \sum_{\mathbf{h},\mathbf{h}'} \tilde{\eta}_{\mathbf{h}\mp\mathbf{h}'} \tilde{\mathbf{A}}^{(\pm)}_{\pm\mathbf{h}'} \cdot (\mathbf{h} \mp \mathbf{h}')\ e^{\pm i(\mathbf{k}_{\mathbf{h}}\cdot\mathbf{r} - \omega t)}, \tag{3.80}$$

where $\rho^{(\pm)}$ is the charge density due to $\mathbf{A}^{\pm}$, respectively.

We can now apply $\nabla^2\Phi = 4\pi\rho$ and integrate twice to get

$$\Phi^{(\pm)} = -4\pi\frac{r_e}{k} \sum_{\mathbf{h},\mathbf{h}'} \frac{\tilde{\eta}_{\mathbf{h}\mp\mathbf{h}'} \tilde{\mathbf{A}}^{(\pm)}_{\pm\mathbf{h}'} \cdot (\mathbf{h} \mp \mathbf{h}')}{{\mathbf{k}_{\mathbf{h}}}^2} e^{\pm i(\mathbf{k}_{\mathbf{h}}\cdot\mathbf{r} - \omega t)}, \tag{3.81}$$

$$\tag{3.82}$$

With this expression, we can now estimate the magnitude of Φ. The reciprocal lattice vectors $\mathbf{h}$ are of the order of 10^{10} m^{-1}, the wave number k is typically 10^{11} m^{-1}, the classical electron radius is $r_e = 2.82 \cdot 10^{-15}$m and the electron density η is of the order of 10^{30} m^{-3}. That makes $|\Phi| \approx (10^{-7} \ldots 10^{-6})|\mathbf{A}|$.

We now insert eqs. (3.4) for η, (3.5) for $\mathbf{A}$ and (3.81) for Φ into eq. (2.19):

$$\mathbf{J}^{(\pm)} = -r_e\, c \sum_{\mathbf{h},\mathbf{h}'} \tilde{\eta}_{\mathbf{h}\mp\mathbf{h}'} \left[\tilde{\mathbf{A}}^{(\pm)}_{\pm\mathbf{h}'} \mp \frac{\mathbf{k}_{\mathbf{h}'}}{k} \tilde{\Phi}^{(\pm)}_{\pm\mathbf{h}'} \right] e^{\pm i(\mathbf{k}_{\mathbf{h}'}\cdot\mathbf{r} - \omega t)} \tag{3.83}$$

With the introduction of the abbreviations ${}^{\times}\tilde{\mathbf{A}}_{\mathbf{h}}^{(+)}$ and ${}^{\times}\tilde{\mathbf{A}}_{-\mathbf{h}}^{(-)}$ (with '×' meaning 'neither transverse, nor longitudinal') for the contents of the square brackets, this becomes:

$$\mathbf{J}^{(\pm)} = -r_e c \sum_{\mathbf{h}\mathbf{h}'} \tilde{\eta}_{\mathbf{h}\mp\mathbf{h}'} \, {}^{\times}\tilde{\mathbf{A}}_{\pm\mathbf{h}'}^{(\pm)} e^{\pm i(\mathbf{k_h}\cdot\mathbf{r}-\omega t)} \tag{3.84}$$

The transverse current, corresponding to eq. (3.9) is then

$$\mathbf{J}_t^{(\pm)} = -r_e c \sum_{\mathbf{h},\mathbf{h}'} \tilde{\eta}_{\mathbf{h}\mp\mathbf{h}'} \left[{}^{\times}\tilde{\mathbf{A}}_{\pm\mathbf{h}'}^{(\pm)}\right]_{\mathbf{h}} e^{\pm i(\mathbf{k_h}\cdot\mathbf{r}-\omega t)} \tag{3.85}$$

with

$$\left[{}^{\times}\tilde{\mathbf{A}}_{\pm\mathbf{h}'}^{(\pm)}\right]_{\mathbf{h}} = \frac{\mathbf{k_h} \times {}^{\times}\tilde{\mathbf{A}}_{\pm\mathbf{h}'}^{(\pm)} \times \mathbf{k_h}}{|\mathbf{k_h}|^2} \tag{3.86}$$

5.3 Applicability of Macroscopic Electromagnetism

Two conditions must be met for macroscopic electromagnetism to be applicable. First, the displacement of charges that produce the polarization $\mathbf{P}$ (and, depending on the type of problem also the magnetization $\mathbb{M}$) must be much smaller than the relevant length scales of the problem. What the length scales are depends on the nature of the problem. In linear optics, it is usually the wavelength. With x-rays, the wavelength is of the order of interatomic distances and the oscillatory motion of the electrons is much smaller than an atom diameter if the x-ray photon energy is far from an absorption edge, or of the order of an atom diameter right at an absorption edge. In either case, the macroscopic approach can be justified for linear x-ray optics in perfect crystals because the electron density is periodic over distances much larger than the wavelength and can be expanded into a Fourier series. This is not true for nonlinear x-ray optics because the anharmonic electronic motion cannot be expressed in terms of a nonlinear susceptibility (see discussion on p. 23), and therefore, no such Fourier expansion is possible.

The other requirement that must be fulfilled for the macroscopic description of electromagnetism is that the motion of charges (and perhaps spins) that is the cause of a polarization, or a magnetization, must be due to the fields only - or can at least be approximated in that way. If the charges move under the influence of, for example, the rapid coherent atomic oscillations in an optical phonon, then it may not be possible to employ a quasistatic approximation. Only if the electromagnetic waves encounter a quasi-static lattice throughout their traversal of a typical length scale of interaction with the matter, is the quasi-static approximation valid. With x-rays, such am interaction length is the pendellösung period, and the traversal time is roughly the reciprocal energy width of a plane-wave x-ray reflection. With strong reflections, such as Si (111), that time is a few femtoseconds and thus still below optical phonon frequencies. There are, however, weaker reflections, such as GaP (666), chosen as an example because it might be a good candidate for inducing noticable frequency shifts in x-rays (see ch. 4), due to its noncentrosymmetric structure and readily excitable optical phonons. At an energy of 20.0 keV, the energy width of the reflection is ca. 0.025 eV [115], which is roughly half the optical phonon energy in the same material [116]. The x-rays will therefore traverse roughly two pendellösung periods within one optical phonon oscillation period, providing for a strong coupling.

5.4 The Position of a Tie Point in Reciprocal Space

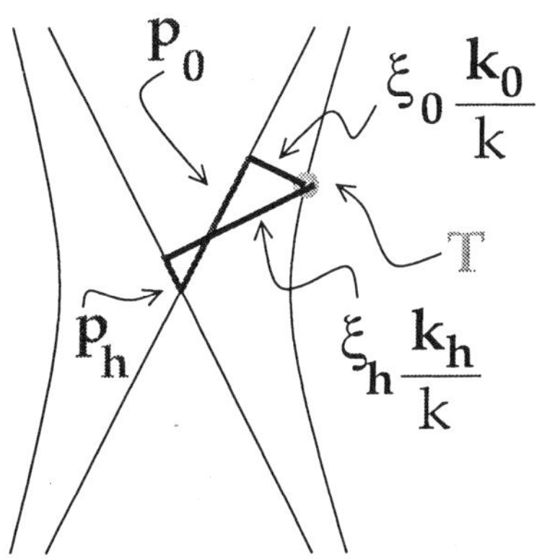

$$\mathbf{T} = \mathbf{p}_0 + \xi_0 \frac{\mathbf{k}_0}{k_r} = \mathbf{p}_h + \xi_h \frac{\mathbf{k}_h}{k_r}$$

Figure 3.17: The position of a tie point $\mathbf{T}$ with coordinates ξ_0 and ξ_h in reciprocal space.

The vectors $\mathbf{p}_0$ and $\mathbf{p}_h$ are oriented along the asymptotes of the two-wave dispersion hyperbola. The position of a tie point $\mathbf{T}$ relative to the symmetry point of the hyperbola can be written as

$$\mathbf{T} = \mathbf{p}_h + \xi_h \frac{\mathbf{k}_h}{k_r} = \mathbf{p}_0 + \xi_0 \frac{\mathbf{k}_0}{k_r} \tag{3.87}$$

We now need expressions for $\mathbf{p}_0$ and $\mathbf{p}_h$. A vector in the direction of $\mathbf{p}_0$ (not a unit vector) is $\mathbf{k}_h - \mathbf{k}_0(\mathbf{k}_h \cdot \mathbf{k}_0)/{k_r}^2 = \mathbf{k}_h - \mathbf{k}_0 \cos 2\Theta$ and likewise for $\mathbf{p}_h$. We can thus define vectors along the asymptotes

$$\mathfrak{k}_0 = \mathbf{k}_h - \mathbf{k}_0 \cos 2\Theta \qquad \mathfrak{k}_h = \mathbf{k}_0 - \mathbf{k}_h \cos 2\Theta \tag{3.88}$$

By definition, $\mathfrak{k}_0 \cdot \mathbf{k}_0 = \mathfrak{k}_h \cdot \mathbf{k}_h = 0$, and

$$\mathbf{k}_0 \cdot \mathfrak{k}_h = \mathbf{k}_h \cdot \mathfrak{k}_0 = {k_r}^2(1 - \cos^2 2\Theta) = {k_r}^2 \sin^2 2\Theta \tag{3.89}$$

We can now multiply eq. (3.87) with either $\mathbf{k_0}/k_r$, or $\mathbf{k_h}/k_r$ to obtain two equations which determine $\mathbf{p}_0$ and $\mathbf{p}_h$.

$$\mathbf{k_0} \cdot \mathbf{p_h} = k_r\,(\xi_0 - \xi_h \cos 2\Theta) \qquad \mathbf{k_h} \cdot \mathbf{p_0} = k_r\,(\xi_h - \xi_0 \cos 2\Theta) \quad (3.90)$$

comparing eq. (3.90) with eq. (3.89) and (3.88) yields

$$\begin{aligned} \mathbf{p_h} &= (\mathbf{k_0} - \mathbf{k_h}\cos 2\Theta)\,\frac{\xi_0 - \xi_h \cos 2\Theta}{k_r \sin^2 2\Theta} \\ \mathbf{p_0} &= (\mathbf{k_h} - \mathbf{k_0}\cos 2\Theta)\,\frac{\xi_h - \xi_0 \cos 2\Theta}{k_r \sin^2 2\Theta} \end{aligned} \quad (3.91)$$

Inserting both lines into eq. (3.87), and adding gives

$$\mathbf{T} = \frac{1}{k_r \sin^2 2\Theta}\left(\underbrace{(\mathbf{k_h} - \mathbf{k_0}\cos 2\Theta)}_{\mathfrak{k}_0}\xi_h + \underbrace{(\mathbf{k_0} - \mathbf{k_h}\cos 2\Theta)}_{\mathfrak{k}_h}\xi_0\right) \quad (3.92)$$

5.5 The Direction of the Poynting Vector

It is well known that the Poynting vector is perpendicular to the real part of the dispersion surface, as shown in fig. 3.2. A proof is demonstrated here for the two-wave case. The real part of the time-averaged Poynting vector is $\mathbf{S} = (c/8\pi)\mathbf{E}(\mathbf{r}) \times \mathbf{H}^*(\mathbf{r})$ (see [1]).(6.132) and the equation before [1].(7.13). Its average over one lattice plane spacing (corresponding to the reciprocal lattice vector $\mathbf{h}$) can be written in terms of the indvidual waves, with the cross terms $\mathbf{E_0} \times \mathbf{H_h^*} \exp(i(\mathbf{k_0} - \mathbf{k_h}) \cdot \mathbf{r})$ and $\mathbf{E_h} \times \mathbf{H_0^*} \exp(i(\mathbf{k_h} - \mathbf{k_0}) \cdot \mathbf{r})$ averaging out, giving

$$\mathbf{S} = \mathbf{S_0} + \mathbf{S}_h = \omega \mathbf{k_0} \left|\tilde{\mathbf{A}}_\mathbf{0}\right|^2 + \omega \mathbf{k_h} \left|\tilde{\mathbf{A}}_\mathbf{h}\right|^2 \quad (3.93)$$

With eq. (3.15), this becomes

$$\mathbf{S} = \omega \mathbf{k_0} \left|\tilde{\mathbf{A}}_\mathbf{0}\right|^2 + \omega \mathbf{k_h} \frac{\xi_0}{\xi_h} \left|\tilde{\mathbf{A}}_\mathbf{0}\right|^2 \quad (3.94)$$

Moving to an infinitesimally neighboring point $\mathbf{T} + d\mathbf{T}$ on the dispersion surface with coordinates $\xi_0 + d\xi_0$ and $\xi_h + d\xi_h$ is subject to eq. (3.14), and thus $d\xi_h = -(\xi_h/\xi_0)d\xi_0$. This gives

$$d\mathbf{T} = \frac{1}{k_r \sin^2 2\Theta}\left(\underbrace{\mathbf{k_0} - \mathbf{k_h}\cos 2\Theta}_{\perp \mathbf{k_h}} - \frac{\xi_h}{\xi_0}\underbrace{\mathbf{k_h} - \mathbf{k_0}\cos 2\Theta}_{\perp \mathbf{k_0}}\right) d\xi_0 \quad (3.95)$$

after replacing (ξ_h/ξ_0) with the amplitude ratio, according to eq. (3.15), we have $d\mathbf{T} \cdot \mathbf{S} = \mathbf{0}$. Because $d\mathbf{T}$ is tangential to the dispersion surface, it follows that $\mathbf{S}$ is perpendicular to it.

5.6 Details of Derivations

5.6.1 Amplitude Ratio, Equation (3.15). Rewrite eq. (3.11) for the two-wave case as

$$\begin{aligned} ({k_r}^2 - \mathbf{k}_0^2)\tilde{\mathbf{A}}_0^{(+)} + 4\pi r_e \tilde{\eta}_{-\mathbf{h}}\left[\tilde{\mathbf{A}}_\mathbf{h}\right]_0 &= 0 \\ ({k_r}^2 - {\mathbf{k}_\mathbf{h}}^2)\tilde{\mathbf{A}}_\mathbf{h}^{(+)} + 4\pi r_e \tilde{\eta}_\mathbf{h}\left[\tilde{\mathbf{A}}_0\right]_\mathbf{h} &= 0 \end{aligned} \tag{3.96}$$

and approximate ${k_r}^2 - \mathbf{k}_0^2 = (k_r + |\mathbf{k}_0|)(k_r - |\mathbf{k}_0|) \approx 2k_r\xi_0$, and likewise ${k_r}^2 - {\mathbf{k}_\mathbf{h}}^2 \approx 2k_r\xi_h$. This makes eq. (3.96), each line multiplied with the complex conjugate of 'its' field amplitude

$$\begin{aligned} 2k_r\xi_0\left|\tilde{\mathbf{A}}_0^{(+)}\right|^2 &= -4\pi r_e \tilde{\eta}_{-\mathbf{h}}\left[\tilde{\mathbf{A}}_\mathbf{h}\right]_0 \cdot \tilde{\mathbf{A}}_{-0}^{(-)} \\ 2k_r\xi_h\left|\tilde{\mathbf{A}}_\mathbf{h}^{(+)}\right|^2 &= -4\pi r_e \tilde{\eta}_\mathbf{h}\left[\tilde{\mathbf{A}}_0\right]_\mathbf{h} \cdot \tilde{\mathbf{A}}_{-\mathbf{h}}^{(-)} \end{aligned} \tag{3.97}$$

In the absence of absorption, the wave vectors, and thus ξ_0 and ξ_h are real-valued. Although $\tilde{\eta}_\mathbf{h}$ and $\tilde{\eta}_{-\mathbf{h}}^*$ may be complex, depending on the choice of coordinate origin, they are complex conjugates of each other $\tilde{\eta}_\mathbf{h} = \tilde{\eta}_{-\mathbf{h}}^*$ (see page 39). Therefore, if the phase of $\tilde{\eta}_{-\mathbf{h}}$ is ϕ, then the phase of $\left[\tilde{\mathbf{A}}_\mathbf{h}\right]_0 \cdot \tilde{\mathbf{A}}_0^{(+)}$ must be $-\phi$ or $\pi - \phi$ to make the right-hand sides in eq. (3.97) real-valued, and the same argument holds for $\tilde{\eta}_\mathbf{h}$ vs. $\left[\tilde{\mathbf{A}}_0\right]_\mathbf{h} \cdot \tilde{\mathbf{A}}_\mathbf{h}^{(+)}$. Furthermore, $\left|\left[\tilde{\mathbf{A}}_\mathbf{h}\right]_0 \cdot \tilde{\mathbf{A}}_0^{(+)}\right| \approx \left|\left[\tilde{\mathbf{A}}_0\right]_\mathbf{h} \cdot \tilde{\mathbf{A}}_\mathbf{h}^{(+)}\right|$ for σ and π polarization because $|\mathbf{k}_0| \approx |\mathbf{k}_\mathbf{h}|$. Dividing the first line by the second yields the result for vanishing absorption

$$\frac{\xi_0}{\xi_h} = \frac{\left|\tilde{\mathbf{A}}_\mathbf{h}^{(+)}\right|^2}{\left|\tilde{\mathbf{A}}_0^{(+)}\right|^2} \tag{3.98}$$

5.6.2 Equation (3.27). To compute the transverse current (3.3) from (3.26), the spatial dependences of $\tilde{\mathbf{A}}_\mathbf{h}^{(+)}$, $S_{n,\mathbf{h}}$ (through φ_n and $\mathbf{u}$) and $\mathfrak{s}_m$ (through $\mathbf{u}$) can be expressed in Taylor expansions

$$\mathbf{h}\cdot\mathbf{u}(\mathbf{r}',t) = \mathbf{h}\cdot\mathbf{u}(\mathbf{r},t) + (\mathbf{r}'-\mathbf{r})\cdot\nabla\mathbf{u}|_\mathbf{r}\cdot\mathbf{h} + \dots \tag{3.99}$$

$$\varphi_n(\mathbf{r}',t) = \varphi_n(\mathbf{r},t) + (\mathbf{r}'-\mathbf{r})\cdot\nabla\varphi_n|_\mathbf{r} + \dots \tag{3.100}$$

$$\tilde{\mathbf{A}}_{\mathbf{h}'}(\mathbf{r}',t) = \tilde{\mathbf{A}}_{\mathbf{h}'}(\mathbf{r},t) + (\mathbf{r}'-\mathbf{r})\cdot\nabla\tilde{\mathbf{A}}_{\mathbf{h}'}|_\mathbf{r} + \dots \tag{3.101}$$

which give us (using the variable substitution $\mathbf{r}' - \mathbf{r} = \mathbf{r}''$):

$$\mathbf{J}_t(\mathbf{r}, t) = -\frac{r_e c}{4\pi} \sum_{\substack{mn \\ \mathbf{hh}'}} \tilde{\eta}_{m,\mathbf{h}-\mathbf{h}'} e^{-i(\omega+\varpi_m)t} \nabla \times \nabla \times \Bigg[e^{i\mathbf{r}\cdot(\mathbf{k_h}+\mathfrak{k}_m)+\varphi_n-\mathbf{h}\cdot\mathbf{u}} \int d\mathbf{r}'' \frac{\tilde{\mathbf{A}}_{\mathbf{h}'}(\mathbf{r}) + \mathbf{r}'' \cdot \nabla \tilde{\mathbf{A}}_{\mathbf{h}'}\Big|_{\mathbf{r}}}{|\mathbf{r}''|} e^{i(\mathbf{r}''\cdot(\mathbf{k_h}-\nabla\mathbf{u}\cdot\mathbf{h}+\mathfrak{k}_m+\nabla\varphi_n))} \Bigg] \quad (3.102)$$

The $\tilde{\mathbf{A}}_{\mathbf{h}'}(\mathbf{r})$ term in the integral gives $(4\pi/|\mathbf{k_h} - \nabla\mathbf{u}\cdot\mathbf{h} + \mathfrak{k}_m + \nabla\varphi_n|^2)$, as in the derivation of eq. (3.9). More troublesome is the next term $\mathbf{r}'' \cdot \nabla\tilde{\mathbf{A}}_{\mathbf{h}'}$ because, in the first-order Taylor expansion, $\tilde{\mathbf{A}}$ grows without limits for $|\mathbf{r}| \to \infty$. That term is, however, found to vanish by splitting $\nabla\tilde{\mathbf{A}}_{\mathbf{h}'} = \nabla\tilde{\mathbf{A}}^{\perp}_{\mathbf{h}'} + \nabla\tilde{\mathbf{A}}^{\parallel}_{\mathbf{h}'}$, with

$$\nabla\tilde{\mathbf{A}}^{\parallel}_{\mathbf{h}'} = \frac{\left[\nabla\tilde{\mathbf{A}}_{\mathbf{h}'} \cdot (\mathbf{k}_{\mathbf{h}^\circ} + \mathfrak{k}_m + \nabla\varphi_n)\right] \otimes (\mathbf{k}_{\mathbf{h}^\circ} + \mathfrak{k}_m + \nabla\varphi_n)}{|\mathbf{k}_{\mathbf{h}^\circ} + \mathfrak{k}_m + \nabla\varphi_n|^2}, \quad (3.103)$$

where the abbreviation $\mathbf{k}_{\mathbf{h}^\circ}$ (see p. 55) was used. The vector $\mathbf{r}'' \cdot \nabla\tilde{\mathbf{A}}^{\perp}_{\mathbf{h}'}$ in eq. (3.101) is perpendicular to $(\mathbf{k}_{\mathbf{h}^\circ} + \mathfrak{k}_m + \nabla\varphi_n)$ in the exponent, and symmetry makes the integral over the term with $\mathbf{r}'' \cdot \nabla\tilde{\mathbf{A}}^{\perp}_{\mathbf{h}'}$ vanish. That leaves the term with the vector $\mathbf{r}'' \cdot \nabla\tilde{\mathbf{A}}^{\parallel}_{\mathbf{h}'}$, which is parallel to $\mathbf{k}_{\mathbf{h}^\circ} + \mathfrak{k}_m + \nabla\varphi_n$ by definition. Application of the operator $\nabla\times$ to the exponential before the integral yields $(\mathbf{r}'' \cdot \nabla\tilde{\mathbf{A}}^{\parallel}_{\mathbf{h}'}) \times (\mathbf{k}_{\mathbf{h}^\circ} + \mathfrak{k}_m + \nabla\varphi_n) = \mathbf{0}$. That leaves:

$$\mathbf{J}_t(\mathbf{r}, t) = -r_e c \sum_{nm} e^{i(\mathbf{r}\cdot(\mathbf{k_o}+\mathfrak{k}_m)+\varphi_n)-(\omega+\varpi_m)t} \sum_{\mathbf{hh}'} \tilde{\eta}_{m,\mathbf{h}-\mathbf{h}'} \left[\tilde{\mathbf{A}}_{\mathbf{h}'}\right]_{\mathbf{h}} e^{i\mathbf{h}\cdot(\mathbf{r}-\mathbf{u})}, \quad (3.104)$$

$$\left[\tilde{\mathbf{A}}_{\mathbf{h}'}\right]_{\mathbf{h}} = \frac{(\mathbf{k}_{\mathbf{h}^\circ} + \mathfrak{k}_m + \nabla\varphi_n) \times \tilde{\mathbf{A}}_{\mathbf{h}'} \times (\mathbf{k}_{\mathbf{h}^\circ} + \mathfrak{k}_m + \nabla\varphi_n)}{|\mathbf{k}_{\mathbf{h}^\circ} + \mathfrak{k}_m + \nabla\varphi_n|^2} \quad (3.105)$$

To simplify the formulas, $\mathbf{r}$ and t are omitted as arguments to φ_n, $\tilde{\mathbf{A}}_{\mathbf{h}'}$, etc. The above evaluation of the integral (3.3) is equivalent to the averaging procedure of macroscopic electromagnetism. Taking spatio-temporal changes in the domain of integration into account will, however, lead to frequency shifts in $\mathbf{J}_t$.

5.6.3 An Integral.

$$\int \frac{e^{i\mathbf{k}\cdot\mathbf{r}'}}{|\mathbf{r}' - \mathbf{r}|} d\mathbf{r}' = e^{i\mathbf{k}\cdot\mathbf{r}} \int \frac{e^{i\mathbf{k}\cdot\mathbf{r}''}}{|\mathbf{r}''|} d\mathbf{r}'' = e^{i\mathbf{k}\cdot\mathbf{r}} \int_0^\infty dr(2\pi r) \int_{-\infty}^{\infty} dz \frac{e^{i|\mathbf{k}|z}}{\sqrt{r^2 + z^2}} \quad (3.106)$$

With [117].(3.754) for the z-integral, and then [117].(6.561.16):

$$\int \frac{e^{i\mathbf{k}\cdot\mathbf{r}'}}{|\mathbf{r}' - \mathbf{r}|} d\mathbf{r}' = e^{i\mathbf{k}\cdot\mathbf{r}} 4\pi \int_0^\infty r K_0(|\mathbf{k}|r) dr = \frac{4\pi e^{i\mathbf{k}\cdot\mathbf{r}}}{|\mathbf{k}|^2} \quad (3.107)$$

Chapter 4

ULTRAFAST DIFFRACTIVE X-RAY OPTICS

Research is an organized method for keeping you reasonably dissatisfied with what you have – Charles Kettering

B.W. Adams

Advanced Photon Source, Argonne National Laboratory, Argonne, IL 60439, USA

adams@mail.desy.de

Abstract A class of x-ray optical elements is proposed that use effects of dynamical x-ray diffraction under conditions of rapidly changing structure factors to manipulate x-rays.

1. Introduction

The first experiments with ultrafast lasers and coherent x-rays will certainly be done using simply the raw or monochromatized output of a free electron laser in a laser pump, x-ray probe configuration. Going from tens of picoseconds in x-ray pulse duration to ca. 100 fs is a tremendous advance, and it will take some time to cover that territory. Eventually, though, the few-femtosecond time domain will become interesting because the most elementary electronic processes in chemical reaction dynamics and semiconductor carrier dynamics occur on that time scale. There are a few accelerator-based ideas for creating x-ray pulses of tens of femtoseconds duration from an x-ray free-electron laser (XFEL), but none yet for the few-femtosecond domain. The x-ray optical elements described here should be capable of reaching that time scale, and they could also be used to complement accelerator-based techniques. No attempt is made here to optimize the parameters of the proposed devices to achieve the best performance and it is quite possible that some of the ideas presented here will suffocate under the details of technical problems. However, the intention of this chapter is not to present ready-made designs but rather to circulate a few ideas and to discuss some of the issues that arise.

The key idea is to change the diffractive properties of a crystal with an ultrafast laser and to use effects of dynamical diffraction to achieve a coherent effect on x-rays diffracting in it. The extended Takagi theory developed in ch. 3 is used to describe the effects of rapid structure factor changes on the diffraction of x-rays. It is of great value for the qualitative and order-of-magnitude discussion of these effects that, in contrast to the standard Takagi theory, the extended theory makes available the concept of dispersion surfaces.

Typical time scales for changes in crystal properties are tens of picoseconds for impulsively created acoustic phonons, tens to hundreds of femtoseconds for laser-generated optical phonons, and a few femtoseconds for electronic excitations. The respective effect on x-rays diffracting in such a crystal can be even faster if use is made of subcycle techniques, similar to the principle of a streak camera. Somewhere around the hundred-femtosecond to picosecond time scale, it becomes necessary to tailor the spatial/temporal profile of the disturbance to the group velocity of the x-rays. This will be a recurring theme in the following sections.

Even if the crystal diffractive properties are constant in time, interesting effects arise with short-pulse x-rays [118, 119, 120, 121, 122]. In a real-space description, these effects occur because x-rays scattered from different parts of the crystal have different optical paths, and thus different arrival times at a detector. Alternatively, one calculate the dynamical diffraction of each spectral component of an x-ray wave packet separately, and then Fourier-synthesize a diffracted wave packet. This chapter is, however, concerned only with x-ray diffraction under the conditions of a rapidly changing crystal.

2. Laser-Induced Changes in Crystal Diffractive Properties

Three ways of manipulating the diffractive properties of a crystal will be considered here, a cartoon of which is shown in fig. 4.1. In the order of decreasing time constants, they are: Generation of acoustic phonon wave packets (i.e., sound waves), generation of optical phonon wave packets, and direct electronic modification of the electron density. Use of impulsively generated acoustic phonons as an x-ray diffractive switch has been demonstrated [123, 124] (see also ch. 9 of this book). The shortest switching time yet observed is of the order of one picosecond. Optical phonons typically have oscillation periods of one-to-several hundred femtoseconds, permitting correspondingly faster response times. Coherent massive excitation of optical phonons is, however, hard to achieve, and an optical-phonon-based x-ray switch has not yet been realized. The fastest response would be possible with changes of the electron density due to massive electronic excitation: When electrons are transferred from the valence to the conduction band, corresponding to changes from bonding to antibonding orbitals, then the scattering contribution of the valence electrons changes as well. The time constants are given by the reciprocal band gap energy and can therefore be of the order of femtoseconds. Because only the valence

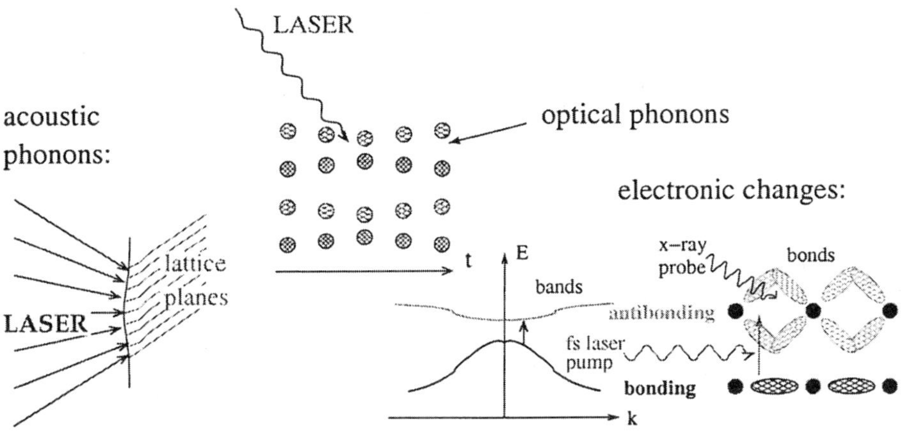

Figure 4.1: How to manipulate crystal diffraction with a laser. Left: Rapid heating of a crystal surface launches a sound wave(for details see ch. 9), which distorts the lattice planes and thus changes the diffractive properties. Middle: Electrostriction due to the strong electric field of a short intense laser pulse induces optical phonons. The Fourier components of the electron density are thus modulated in space and time. Right: In the band picture, electrons are transferred from the valence to the conduction band. This corresponds to a change from bonding to antibonding orbitals in the bond picture, and thus a change in the Fourier components of the electron density.

electrons participate, the effect is not as strong as the phonon-based ones, but it can be enhanced in some x-ray reflections, such as (222), (666), etc., in the diamond structure.

Because of the relatively weak interaction of x-rays with matter, the diffractive properties must be controlled in extended regions of the crystal. Typical length scales are the pendellösung length L_{H} defined in eq. (3.18) and the disturbance length $L_{\mathrm{H}}^{(d)}$ defined in eq. (3.32). The former is of the order of many micrometers, and the latter is even longer, depending on the strength of the disturbance. X-rays, traveling at the speed of light, take many picoseconds to traverse these lengths. If femtosecond time resolution is to be achieved, one has to give careful attention to the proper matching of group velocities and the relation of events in space and time. When the expressions "before" and "after" are used here, they are meant as timelike separations. This may sound like relativistic terminology but the discussion will be firmly rooted in the lab frame.

3. Bragg Reflection

Perhaps the first thing that comes to mind when thinking of an x-ray switch is to manipulate a Bragg reflection. A possible scheme is shown in fig. 4.2. The device might be made out of a monolithic block of crystal material, such as for example GaAs, ensuring that there is no dispersion between the several

Bragg reflections that are necessary, i.e., the Bragg condition is fulfilled equally for all wavelengths in the incident beam. It is therefore legitimate to continue the discussion under the assumption of a strictly monochromatic and slightly divergent x-ray beam. The x-rays are first reflected in a series of asymmetric Bragg reflections to prepare a beam that is narrower in divergence than the symmetric rocking curve of the crystal would permit (explained below). The x-rays incident on the portion of the crystal to the right in fig. 4.2 therefore fill only part of its rocking width. By tuning the angle of that part of the crystal relative to the rest of it (a millidegree, or so, which can be done with elastic deformation in a weak link of the monolithic structure), the incident x-rays are brought close to the limit of the angular acceptance. A reduction in structure factor will then result in a narrowing of the rocking curve and radiation that was reflected before, will then be transmitted into the crystal.

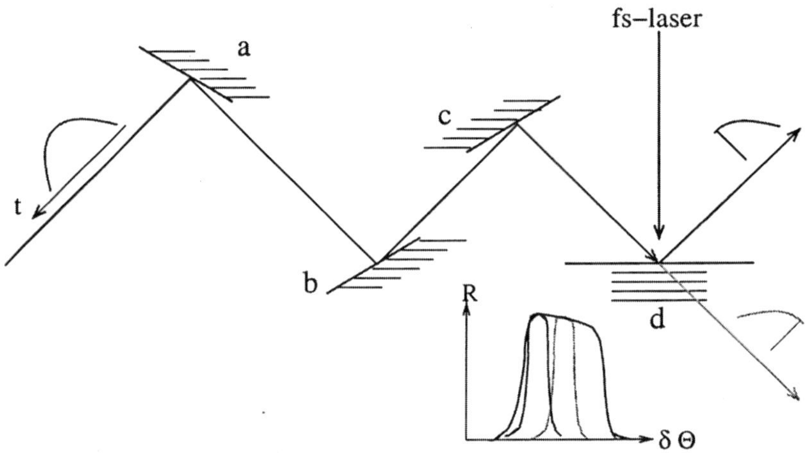

Figure 4.2: A Bragg switch for x-rays. All Bragg-reflecting surfaces are cut out of the same crystal. The first three are asymmetric and prepare a beam that fills only part of the symmetric rocking curve, which is manipulated with a fs laser.

How does the series of asymmetric reflections prepare a narrowly divergent incident beam? Figure 4.2 shows several effects in combination. Most important is the effect that an asymmetric Bragg reflection generally has a narrower rocking width than a symmetric one, because the angular range in which the surface normal in fig. 3.6 does not hit a tie point is smaller. The rocking width can thus go all the way down to zero. Second, the average refractive index of the crystal material induces a shift of the Bragg angle, which depends on the surface orientation. The rocking curves relevant at the crystal surfaces a and b in fig. 4.2 are therefore slightly shifted relative to each other, and the total reflection range of the combination is thus restricted. Finally, an asymmetric reflection changes the beam cross section versus its divergence, while keeping the phase space volume constant. The two asymmetric reflections a and b have the effect of increasing the divergence, and thus, surface b is offered a larger divergence than it can reflect. The third asymmetric reflection brings the divergence of what did pass through reflections a and b back to its initial value, making it less than the rocking width of the laser-modified reflection.

4. Laue Transmission

More promising than Bragg reflection is the use of Laue transmission because it offers considerably more flexibility and a larger interaction volume. Several effects offer themselves for use in the coherent rapid control of x-rays. One might make use of the sensitive dependence of the beam direction on the directions of the wave vectors inside a crystal, one might induce phase shifts into the interference between branches of the dispersion surface, and one might modify the conditions of anomalous transmission in Laue transmission (the Borrmann effect [90], see ch. 3, sec. 2.5). Examples of the use of these effects are given below.

4.1 Redirection of the Poynting Vector

A disturbance of the crystal lattice makes the field amplitudes of a diffracting mode change, as described in ch. 3, secs. 3.6, 3.7 and 3.8. This has then the effect of changing the tie points and the Poynting vectors of diffracting modes. By inducing a disturbance in a controlled way with a laser, diffracting modes can thus be steered through the crystal.

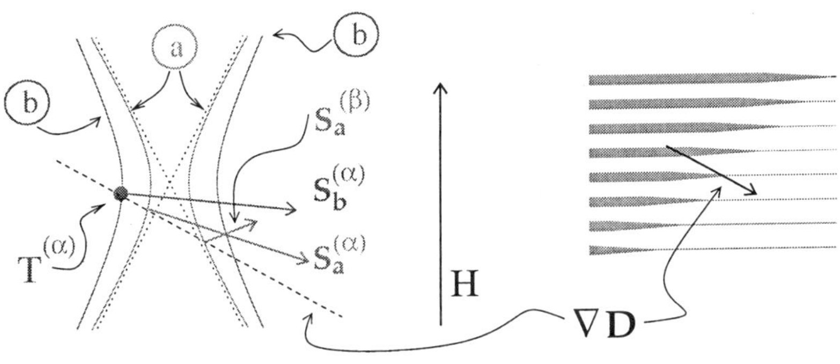

Figure 4.3: Redirecting the Poynting vector by reducing the Fourier component $\tilde{\eta}_{\mathbf{H}}$, quasi-static case (compare with fig. 4.4). Shown is the dispersion hyperbola of the two-wave case, before ("ⓑ") and after ("ⓐ") the x-rays encounter a region of reduced structure factor $\tilde{\eta}_{\mathbf{H}}$. Shown are the Poynting vector $\mathbf{S}_b^{(\alpha)}$ of a mode on the α branch before the change and $\mathbf{S}_a^{(\alpha)}$ and $\mathbf{S}_a^{(\beta)}$ are the Poynting vectors after the change. The modes before and after the change are related by the gradient ∇D of the disturbance that leads to a structure factor reduction, indicated on the right-hand side of the figure.

The following discussion refers to the two-wave case, but a generalization to more waves is straightforward. To get started, let us look at a time-independent problem: A partial disturbance is induced in a perfect crystal, as indicated on the right-hand side in fig. 4.3. The thick lines indicate the lattice planes corresponding to the reciprocal lattice vector $\mathbf{H}$ in their "pristine" form, and the thin lines mean a reduced scattering power due to the disturbance. The gradient of the scattering power is given by the gradient ∇D of the disturbance. For the sake of simplicity, assume that the disturbance has a smooth step shape, i.e., there are just two regions, each with a uniform Fourier component $\tilde{\eta}_{\mathrm{H}}$ of the electron density, and a transition region of zero or nonzero thickness. To be precise, the width of these Fourier components due to the finite sizes of the respective regions will be much less than the difference between them.

X-rays enter the crystal from the left in fig. 4.3 into the undisturbed region and are split into diffracting modes on the α and β branches of the undisturbed dispersion surface ⓑ. The tie points are matched with two eikonal functions $\varphi_b^{(\alpha)}$ and $\varphi_b^{(\beta)}$ ("b" for "before"), i.e., $\mathbf{k}_0 + \nabla\varphi_b^\alpha$ and $\mathbf{k}_0 + \nabla\varphi_b^{(\beta)}$ originate from the tie points (see ch. 3, sec. 3 for details about the eikonal function). For the sake of simplicity, only the x-rays that begin their propagation into the crystal on the α branch will be considered here. Practical applications would also be more complicated if both initial modes have to be reckoned with, but the initial β mode can be removed by making use of the anomalously strong absorption on the β branch (details below).

In the disturbed region, the dispersion surface has a shape ⓐ that differs from the undisturbed one ⓑ in fig. 4.3. If $\varphi_b^{(\alpha)}$ is kept, then the field amplitudes begin to evolve according to eq. (3.29), as the x-rays propagate into the disturbed region. Because of the factor i on the left-hand side of eq. (3.29), this evolution concerns at first only the phases of the field amplitudes. Only after propagating some length, whose order of magnitude is given by eq. (3.32), do the moduli also change. This phase evolution of the field amplitudes can largely be taken up by properly chosen eikonal functions $\varphi_a^{(\alpha)}$ and $\varphi_a^{(\beta)}$ ("a" for "after"), as described in ch. 3, sec. 3. Generally, the x-rays will have to be distributed among the α and the β branches of the disturbed dispersion surface. This so-called interbranch scattering is discussed in ch. 3, sec. 3.7. In analogy to the matching of wave vectors across interfaces, one may expect the differences $\nabla(\varphi_a^{(\alpha)} - \varphi_b^{(\alpha)})$ and $\nabla(\varphi_a^{(\beta)} - \varphi_b^{(\alpha)})$ to be directed along the gradient of the disturbance (see sec. 3.9).

As this chapter is about the manipulation of x-rays by ultrafast structure factor changes, the quasistatic approximation is actually not tenable. The above arguments can, however, be extended to the time domain, and the reassignment of x-rays to the modified dispersion surface of the disturbed crystal follows the space-time gradient of the structure factor change, as indicated in fig. 4.4.

This qualitative discussion can be substantiated by use of eq. (3.46). Because the disturbance discussed here is not of the displacive, but rather of the structure factor modulation type (see p. 52), it is not possible to relate gradients directly, such as with eq. (3.41). Instead, one takes the space-time gradient on both sides of eq. (3.46).

As stated above, changes in the moduli of the field amplitudes lag behind the actual occurence of the disturbance. Therefore, $\mathbf{S}$, $|\tilde{\mathbf{A}}_0|^2$ and $|\tilde{\mathbf{A}}_\mathbf{H}|^2$ can be approximated as constants. Furthermore, $[\tilde{\mathbf{A}}_{\mathbf{h}'}]_\mathbf{h}\tilde{\mathbf{A}}^{(-)}_{-\mathbf{h}}$ can also be taken to stay constant at first because both lines (for $\tilde{\mathbf{A}}_0$ and $\tilde{\mathbf{A}}_\mathbf{H}$) in eq. (3.30) contain the same disturbance term $\mathfrak{s}_{1,\mathbf{H}}\tilde{\eta}_{1,\mathbf{H}}[\tilde{\mathbf{A}}_{\mathbf{h}'}]_\mathbf{h}\tilde{\mathbf{A}}^{(-)}_{-\mathbf{h}}$. Within a distance from the boundary that is less than $L^{(d)}_\mathbf{H}$ of eq. (3.32), the space-time gradient $\partial/c\partial t - \nabla$ of both sides of eq. (3.46) gives therefore for the *change* of the wave vector $\mathbf{k}_0 + \nabla\varphi$ and the frequency $\omega + \dot{\varphi}$:

$$\mathcal{U}\frac{\partial^2\varphi}{c\partial t^2} - \mathbf{S}\cdot\nabla\nabla\varphi = \left(-ikcr_e\tilde{\eta}_{1,\mathbf{H}}\left[\tilde{\mathbf{A}}_{\mathbf{h}'}\right]_\mathbf{h}\tilde{\mathbf{A}}^{(-)}_{-\mathbf{h}}\right)(\mathfrak{k}_{1,\mathbf{H}} + \varpi_{1,\mathbf{H}})\,. \quad (4.1)$$

The second derivatives of φ_0 do not appear in eq. (4.1) because the crystal is perfect before the disturbance. A graphical representation of eq. (4.1) is given on the left-hand side of fig. 4.3 for the spatial derivatives and in fig. 4.4 for the spatial and temporal derivatives.

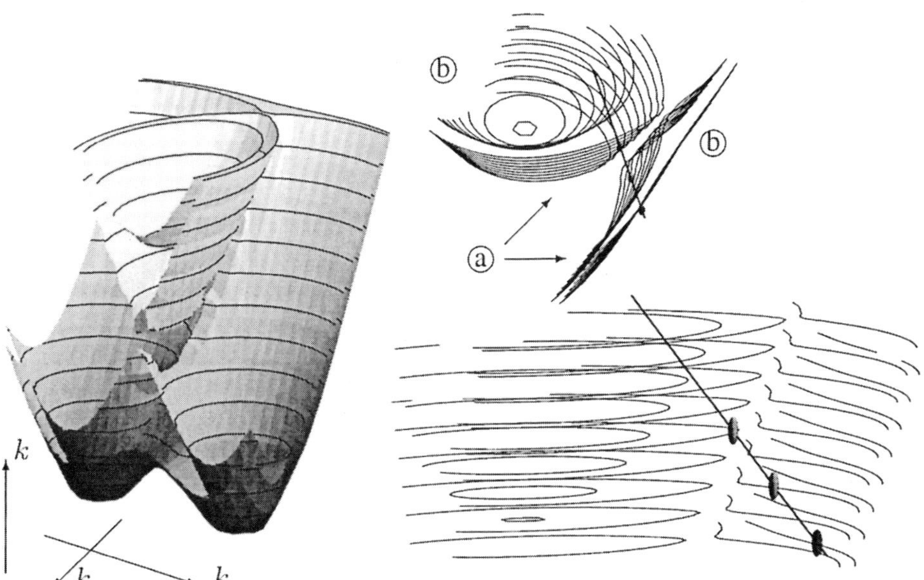

Figure 4.4: Interbranch scattering in momentum-energy space, a graphical representation of eq. (4.1). Shown are the undisturbed (ⓑ for "before") and disturbed (ⓐ for "after") dispersion surfaces in (k_x, k_y, k)-coordinates (compare fig. 3.2), a line along the space-time gradient of the disturbance, a tie point T^α_b on the α branch before the disturbance and the corresponding tie points T^α_a and T^β_a after the disturbance. See text for the meaning of "before" and "after."

4.2 An X-Ray Optical Femtosecond Streak Camera

A particular application of the laser-directed x-ray beam deflection shown in sec. 4.1 bears some resemblance to a streak camera. An ordinary streak camera is similar to an oscilloscope in that the deflection of an electron beam, passing through the gap between two electrodes, is changed by application of a ramped voltage on the electrodes. Rapid ramping is achieved by charging the pair of deflection plates to a high voltage and short-circuiting them with a laser-triggered photoconductive switch. The electron beam is generated at a photocathode, which translates an incident intensity of light, x-rays, etc., into a proportional current. This intensity is then mapped onto the streak that the deflected electron beam leaves on a phosphor screen. Other than in an ordinary oscilloscope, the time resolution is pushed to extremes, and the transit time of the electrons between the electrode plates does play a role. In order to avoid a smearing of the resulting deflection, the field on the electrodes is not applied uniformly but rather as a traveling wave whose propagation speed is matched to that of the the electron beam. One way of slowing the traveling wave down is to give the deflection electrodes a meandering shape (see fig. 4.5).[1]

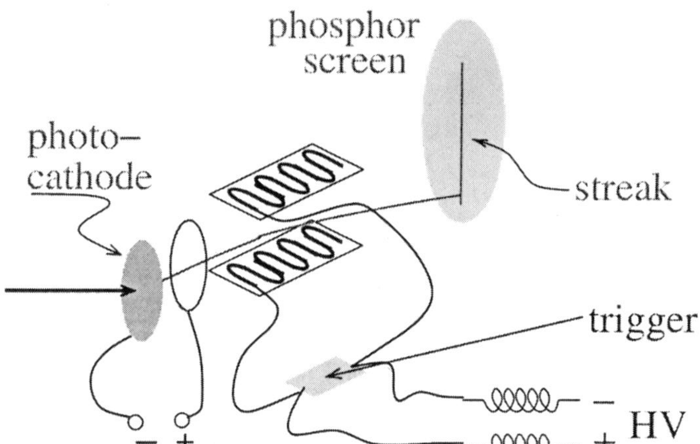

Figure 4.5: Schematic of a single-shot streak camera. Electrons are liberated from a photocathode by x-rays or light whose time structure is to be analzed. The current is roughly proportional to the intensity. After some acceleration, the electrons pass between a pair of electrodes, comprising a capacitor, which at rest is charged to a high voltage (typically a few kV). At the trigger of an ultrashort laser pulse, a photoconductive switch short-circuits the capacitor. This discharge travels along the meandering electrodes of the capacitor plates at a velocity that matches that of the electron beam along its direction. This ensures that a particular electron experiences the same voltage (fully/partially/discharged) between the plates along all of its travel between the plates.

[1]The same trick is applied in high-speed electronics, such as a state-of-the-art PC motherboard: the signal lines contain meandering sections to match signal propagation delays within a bus.

Instead of electrons, the device proposed here deflects a diffracting x-ray beam by use of laser-induced rapid structure factor changes. When based upon purely electronic structure factor changes, the device should be capable of better than 10 fs time resolution (see details below). Just as with the electrons in an ordinary streak camera, the transit time of the x-rays through the deflection region plays a significant role, and, to avoid smearing effects, the laser excitation of the disturbance must be swept across the crystal at a carefully matched velocity (details below).

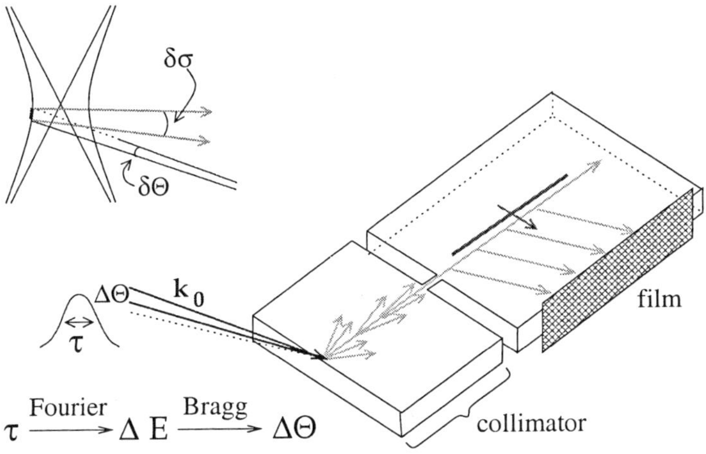

Figure 4.6: An x-ray optical streak camera with potentially down to a few femtoseconds time resolution. X-rays are focused onto a spot, at grazing incidence just above the angle of total reflection. The portion between the focal spot and the waist is a collimator that prepares a spectrally broad, femtosecond wave packet with a well-defined Poynting vector. After passing through the collimator, the x-rays propagate in a well-defined direction. A line-shaped laser excitation sweeps across the surface and changes the propagation direction, giving a projection of the x-ray time structure onto a sensor (film, CCD, ..). See text for details.

Figure 4.6 shows a schematic of the device. Its main components are a collimator and the deflection region. These (and others) shall now be discussed in detail.

4.2.1 Grazing Incidence. To have a sufficient effect on the structure factor of the diffracting material, the laser radiation must suffer strong absorption and cannot penetrate more than, say, a micrometer. A thicker interaction region would also lead to a temporal smearing (300 nm corresponds to 1 fs at the speed of light). The x-rays must therefore be matched to the thin surface layer that can be excited with the laser. This can be achieved by letting the x-rays be incident at an angle towards the surface that is just slightly above the angle of total reflection.

The Collimator. An x-ray wave packet of one femtosecond duration has a spectral width of 4.1 eV. This is beyond the plane-wave energy bandpass of most crystal reflections (the rocking width determined in ch. 3, sec. 2 can be converted to a bandpass by use of Bragg's law). Even worse, a plane wave generates a beam with a well-defined Poynting vector only if the energy bandwidth is a small fraction of the total energy bandpass. To preserve the femtosecond time structure of the x-rays, it is therefore necessary to deal with x-rays that have an angular divergence (or convergence) $\Delta\Theta$ of typically tens to hundreds of microradians and to make sure that, for each spectral component, only those plane waves within the range $\Delta\Theta$ are used that give a particular direction of beam propagation. That is the purpose of the collimator in fig. 4.6. The x-rays are focused onto a small spot of diameter f with a convergence angle $\Delta\Theta$ (numerical values of an exemplary design are given below). The radiation propagates some length ℓ to the other end of the collimator where only those rays with a well-defined direction make it through a waist of width w.

The x-rays that make it through the collimator have a narrow spread $\delta\sigma$ of propagation directions, given by the geometry of the collimator or the angular narrowing of anomalous transmission (see p. 45), whichever is smaller. The geometry of the dispersion surface relates $\delta\sigma$ to a divergence $\delta\Theta$ of the wave vectors of each spectral component (see fig. 4.6, upper left-hand side and fig. 4.7). The x-rays experience a quasi-diffraction effect at the waist (see below) which must be taken into account, and the value of w must be balanced between geometrical narrowing and diffraction widening of the radiation. A typical value of w is 0.1 mm (see below).

Deflection Region. With a transform-limited femtosecond wave packet of x-rays incident onto the crystal, the x-rays entering the deflection region have a pancake-like shape (see fig. 4.7) with a thickness of a few micrometers and a lateral extent of hundreds of micrometers. To avoid smearing by the projection onto the deflected direction, the laser excitation must be swept (see below) across the crystal surface at a velocity that is carefully matched to the x-ray propagation. Figure 4.7 explains this schematically. Suppose for the sake of simplicity that the disturbance reduces the structure factor associated with $\mathbf{H}$ to zero within a time that is comparable to, or less than, the duration of the x-ray wave packet. This time is potentially a few femtosecods (see sec. 4.2.4). Before the disturbance, the x-rays are propagating in the direction of $\mathbf{S}_b$ at a group velocity v_g. Because the velocity along the wave vectors $\mathbf{k}_0$ and $\mathbf{k}_\mathrm{H}$ is close to c, the group velocity at the symmetry point of the α branch is $c\cos\Theta$. If the intersection (point "X" in fig. 4.7) of the laser line focus with the direction of the wave vector $\mathbf{k}_0$ (line "L" in fig. 4.7) travels at the speed of light, then all the x-rays contained in a wave packet that also intersects "L" in "X" are channeled into the new direction along "L", and there is no projection smearing.

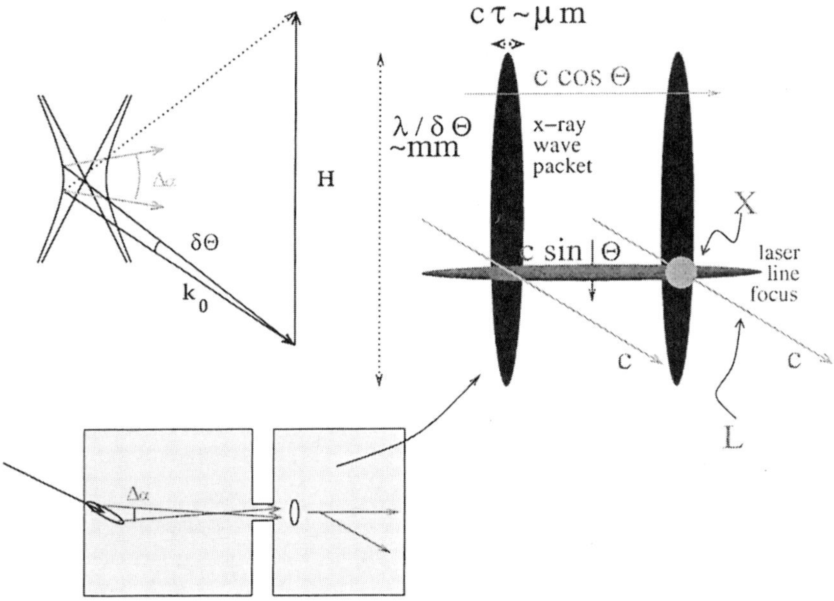

Figure 4.7: A diffracting femtosecond x-ray wave packet entering the deflection region and being deflected by a swept laser excitation. The intersection X of the line-shaped laser excitation and the x-ray wave packet travels along the line L at the speed of light. See text for details.

The sweep direction of the laser excitation need not be perpendicular to the x-ray propagation direction before the disturbance, as shown in fig. 4.7. It may in fact be simpler to make the laser beam hit the crystal surface at a small angle and along the direction of $\mathbf{k_0}$.

The x-rays passing through the waist are coupled wave modes, with wave vectors $\mathbf{k_0}$ and $\mathbf{k_H}$. At the edges of the waist, the x-rays actually do not propagate in the direction of the Poynting vector selected in the collimator but rather along the individual wave vectors. These waves are reflected back into the main beam within one pendellösung period. Therefore, the x-rays emanating from the waist spread as they propagate and are confined by the x-ray reflection. Absorption also helps in the confinement of the beam because the rays at the edge of the main beam have unbalanced amplitudes in their two component waves, and therefore anomalous transmission is diminished. Without the absorption confinement, the intensity downstream of the waist is rather similar to a diffraction pattern at a wavelength that corresponds to the pendellösung length (called quasi-diffraction above, see fig. 4.8). This quasi-diffraction pattern does, however, not distort the wave packets shown in fig. 4.7. Because all group velocities have the same component along the wave vectors

k_0 (line "L" in fig. 4.7) and k_H, the straight flat shape of a wave packet is preserved and only its lateral extent expands.

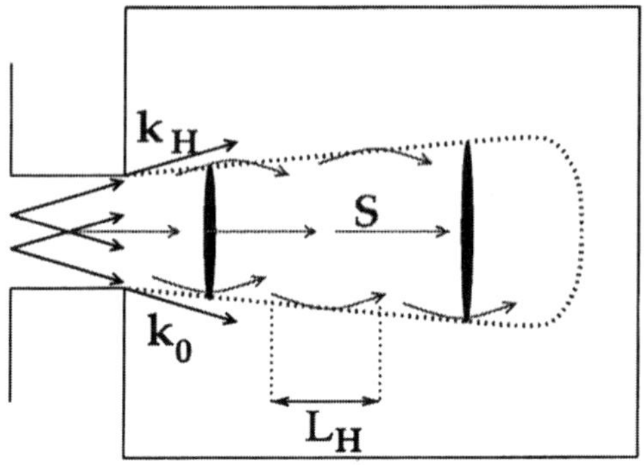

Figure 4.8: Quasi-diffraction of x-rays passing through a waist. On the periphery of the beam emanating from the waist, the x-rays propagate along the wave vectors but are reflected back into the main beam within one pendellösung period L_H. Other than in the usual diffraction of a wave, the straight shape of the x-ray wave packets is preserved.

4.2.2 Swept Laser Excitation. Although this book is about x-rays, the issue of sweeping the laser excitation in a line across the crystal surface shall be addressed briefly. Figure 4.9 shows how this might be done. It is not meant as a real design but only to demonstrate the possible use of dispersion in the control of laser focusing.

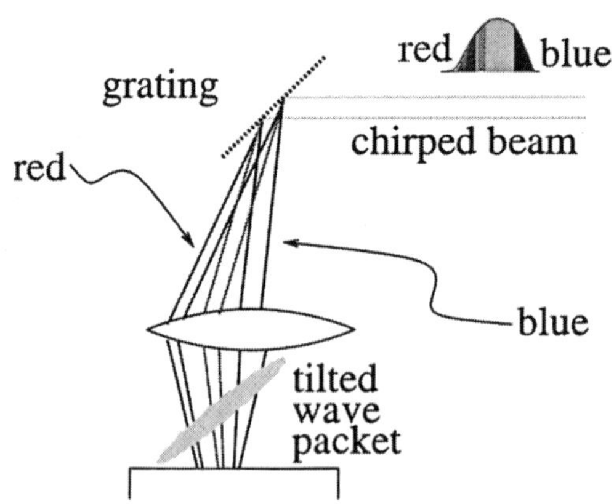

Figure 4.9: Sweeping a laser excitation across a crystal surface.

4.2.3 An Example. To illustrate the above design, consider a particular example, based on a silicon (4 0 0) reflection of 15 keV x-rays. Silicon is chosen here only to discuss the x-ray side of the device but not any particulars of changing the structure factor. For the sake of simplicity, assume that the laser pulse converts the silicon into a plasma and the structure factor is reduced to zero within, say, 25 fs. The Bragg angle is $\Theta = 17.7°$, and the attenuation length without anomalous transmission is about 450 μm. The pendellösung period, calculated with eq. (3.18) (or (3.78)), is 34 μm. For a 30-mm-long collimator, a quick calculation with the program package XOP [115] gives $\delta\Theta \approx \pm 0.5$ μrad, and thence, $\delta\sigma \approx 2.3°$ can be determined. The transform limit for 25 fs is ca. 0.17 eV, which translates into $\Delta\Theta = \pm 1.8$ μrad. With $\delta\Theta \approx \pm 0.5$ μrad and $\lambda = 0.83$ Å, the diffraction limit for the waist width is 0.16 mm. With this waist and a focal spot of the incident of 0.1 mm or less, $\delta\Theta$ is determined by the geometry to about $\pm 0.15°$, which is less than the limitation due to anomalous transmission. With a 25 fs x-ray wave packet, diffracting at $\Theta = 17.7°$ on the symmetry point of the α branch, the pancake of fig. 4.7 has a thickness of 7.1 μm, which is far less than the minimum width of the waist. The angle at which the beam spreads due to the quasi-diffraction described above is given by $L_{\rm h}/w$ and is about $\pm 11°$.

4.2.4 Discussion. Ordinary streak cameras, operating on the principle shown in fig. 4.6, can reach time resolutions of a few hundred femtoseconds. Although future technology may be able to push parameters a bit, it does not seem possible to achieve less than about 100 fs resolution:

- Streak cameras are essentially radiofrequency (RF) devices whose characteristic frequency response is determined by millimeter- to centimeter-sized components. It is actually amazing how far below the period of a typical RF oscillation the state-of-the-art cameras have come. The current limit of about 200 fs corresponds to an RF wavelength of 60 μm.

- X-rays create primary photoelectrons of several keV in the photocathode. These generate secondary electrons, which then actually leave the cathode and are accelerated towards the deflection plates. Therefore, the electrons traversing the streak camera have an energy spread that is considerably higher than in visible-light applications. For this reason, x-ray streak cameras have only reached time resolutions of not much below one picosecond.

Times below about 100 fs are therefore the domain of new concepts, such as the one proposed here. The time resolution of the x-ray optical streak camera is ultimately given by the rapidity of the structure factor change and the precision of the group velocity matching. The latter is probably limited by optics (laser focal size, etc.), and the former is yet to be determined. Leaving money matters aside, one may think of using a diamond crystal and transforming it (actually, a thin surface layer) into a plasma with an intense laser pulse of a few femtoseconds duration. Because the binding energy of the core electrons in carbon is rather low, the laser can delocalize them rather efficiently, and the structure factor is reduced almost to zero even without the nuclei leaving their lattice sites. If the technology of CVD diamond film generation makes sufficient advances to provide the crystal quality required for dynamical diffraction, then the idea of "zapping diamonds" may be less outrageous than it sounds now.

The very intense laser pulses required for the femtosecond creation of a plasma may generate an x-ray background. To eliminate this background, another crystal of the same material (i.e., diamond) may be used. If adjusted to the same orientation as the streak camera crystal, the signal x-rays (which leave the streak camera at the correct Bragg angle) are reflected, and the background is strongly suppressed.

The rapid change of the structure factor does, of course, induce a frequency shift in the diffracting x-rays, according to fig. 4.4. This is rather immaterial if the x-rays are detected directly but must be taken into consideration if an analyzer crystal is used to suppress x-rays generated by the laser.

4.3 An Ultrafast Phase Retarder

Whereas the streak camera application makes use of laser-induced changes of beam propagation directions, this device addresses the phases of the propagating field amplitudes. In particular, a change of a structure factor has an effect on the difference between the dispersion surface for σ and π polarization. It is therefore possible to manipulate the helicity pendellösung shown in fig. 3.7 on the right-hand side. This effect opens the potential of subpicosecond magnetic

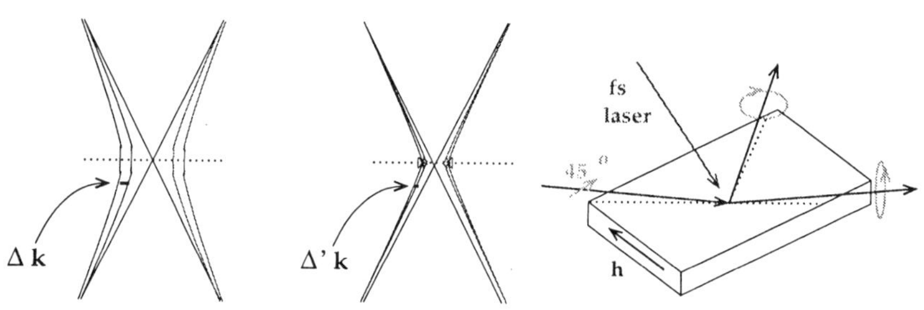

Figure 4.10: A femtosecond switchable phase retarder, making use of the dependence of the shape of the dispersion surfaces for σ and π polarization on the structure factor.

x-ray diffraction studies.

4.4 Spectral Concentration of X-Rays

Several experimental techniques, most notably Mössbauer spectroscopy, require highly monochromatized x-rays. The only way of producing x-rays of narrow energy bandwidth at a synchrotron radiation facility is to use a monochromator, i.e., an energy bandpass filter. All the x-rays outside of the pass band are discarded. It would be highly desirable to be able to make better use of the full photon flux provided by the source. The device proposed here may provide that capability. The basic idea is to extract narrow-band x-rays from a wide spectrum, shift the remaining x-rays in frequency, extract again, and so on. This is shown schematically in fig. 4.11. A theoretical description of the frequency shifts is given in sec. 3.8.1.

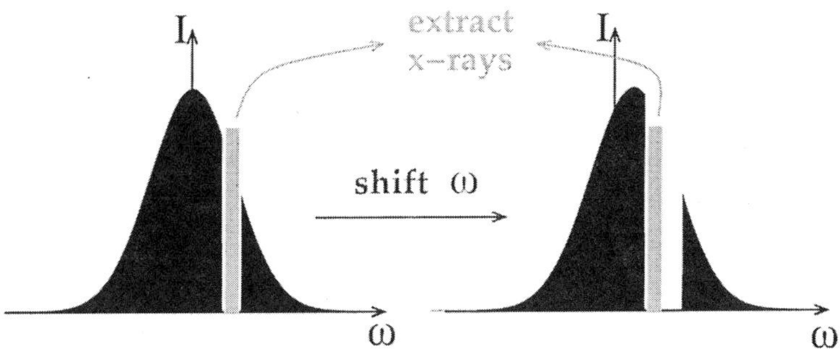

Figure 4.11: Principle of spectral concentration of x-rays by a sequence of narrow-band extractions from a wide distribution and shifts of the whole distribution.

A schematic of a device for spectral concentration is shown in fig. 4.12. Highly collimated but spectrally broadband (as compared to the requirement of the application) x-rays are incident onto a crystal, which is adjusted to reflect out a narrow energy band (the upper crystal in the figure, called "extractor"). This might be done with a diamond crystal, adjusted to the (222) reflection, which is "weakly forbidden," i.e., very narrow. Bulk diamond material is currently not available in sufficient quality to preserve the theoretically narrow width of the (222) reflection, but that may change. Also, the device can be as thin as ca. 0.1 mm, which helps in finding high-quality material. Diamond, or some other material made of light elements, is necessary to concentrate x-rays at the Fe Mössbauer resonance because the x-rays need to propagate through several millimeters of material. With higher energy photons, one might use Si instead.

After traversing one-half of the pendellösung length of the extractor, the x-rays pass into another crystal, which is cut into a comb-like shape with lamellae sticking up into grooves cut into the extractor crystal. This crystal is called the "shifter" in the figure. It is adjusted to an x-ray reflection that is well matched to that of the extractor (which is necessary because of the wavelength dispersion of Bragg's law). For example, the (440) reflection of InP has a lattice spacing that is very close to diamond (222). As the x-rays propagate through a lamella of the shifter crystal, a laser pulse modulates the structure factor and induces a frequency shift. The x-rays then enter the next section of the extractor, and so on.

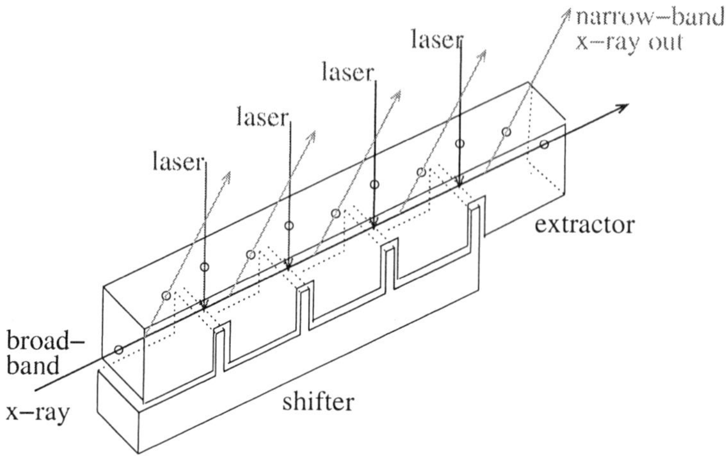

Figure 4.12: A device for spectral concentration of x-rays, consisting of two interleaved crystals. The upper one is adjusted to a narrow-band x-ray reflection and extracts radiation between laser-induced frequency shifts in the lower crystal.

The result is a bundle of parallel narrow-band x-xray beams leaving the extractor. It should be emphasized that this device does not increase the brilliance. A spectral concentration of the x-rays is obtained at the expense of a larger lateral size of the beam. Although the spectral concentrator is a nonlinear optical device (or more precisely, bilinear in the x-ray and laser intensity), Liouville's theorem of the conservation of phase space volume is applicable in modified form. The crystal-mediated interaction between laser radiation and x-rays is sufficiently weak that one may describe the spectral concentrator in terms of different slices, at different photon energies, of roughly the same portion of position-momentum space being shifted to different portions of position-momentum space at the same photon energy. Due to unavoidable losses, each of the beams from the extractor actually has a smaller spectral flux density than a simple energy bandpass filter would provide at the same source. The total flux within a spectral bandwidth is, however, larger than that from a simple filter. It depends therefore on the type of experiment as to whether the spectral concentrator is of any value. It may be necessary to monochromatize the x-rays further, below the narrow bandwidth of the reflection in the extractor. This might be done with a standard high-resolution monochromator that converts the spectrally concentrated flux at its input to a correspondingly higher flux at the output. The device shown in the figure is compatible with standard high-resolution monochromators, because all beams emanating from it are parallel to each other.

An estimate for the lengths of the extraction sections was given above to be one-half of the pertinent pendellösung length. Similarly, one may obtain an estimate for the thicknesses of the shifter sections by the following reasoning. According to the Lindemann melting criterion, the maximum lattice vibration amplitude (here in the laser-induced optical phonons) that can be tolerated by a crystal is roughly ten percent[2] of the interatomic spacing. The perturbation terms $\tilde{\eta}_{m,\mathbf{H}}$ in eqs. (3.30) and (3.47) can therefore be as large as ca. 10% of $\tilde{\eta}_{m,\mathbf{H}}$. This makes the interaction length in the lamellae, required for a frequency shift of ϖ, about ten times the usual pendellösung length. Although the latter depends very much on the type of reflection, x-ray energy, etc., a typical value is 10 μm, making the lamellae about 0.1 mm thick. To obtain a maximal effect, the group velocity of the optical phonons that induce the frequency shifts should be matched to the group velocity of the x-rays in the shifter sections. One may then interpret the process in terms of successive Doppler shifts of the x-rays from lattice planes in coherent motion.

4.5 A Fast Borrmann Shutter

Anomalous transmission (the Borrmann effect, see p. 45) takes place in a perfect crystal with x-rays occupying modes close to the symmetry center of the α branch of the dispersion surface. Under these conditions, the radiation can traverse a thickness of the crystal material in which it would otherwise be damped very strongly. This effect is very sensitive towards disturbances that lead to a shift of the tie point away from the symmetry center, or to interbranch scattering, or to a strong reduction of the structure factor. The idea of operating an x-ray shutter on the basis of the Borrmann effect is not new, see [125] and ch. 9. However, it is discussed here with emphasis on a high time resolution.

Two versions of a shutter based on the Borrmann effect are proposed here. The first one is relatively slow, that is, compared with the femtosecond devices shown above. The other one is potentially capable of femtosecond switching times and is in fact rather simple in principle.

The first version of a Borrmann shutter relies on interbranch scattering of x-rays. Photons that are transferred to the β branch by that effect are lost to the anomalous transmission because they experience an anomalously strong absorption. Only part of the x-ray photons are scattered, and some remain on the α branch. In order to achieve a high switching contrast, it is necessary to subject the x-rays to multiple disturbances, such as the stripe pattern shown in

---

[2]Lindemann says more like 20%, but this is an order-of-magnitude estimate and some allowance must be made for thermal vibrations.

fig. 4.13. This multiple disturbance necessarily has a spatial extent of hundreds

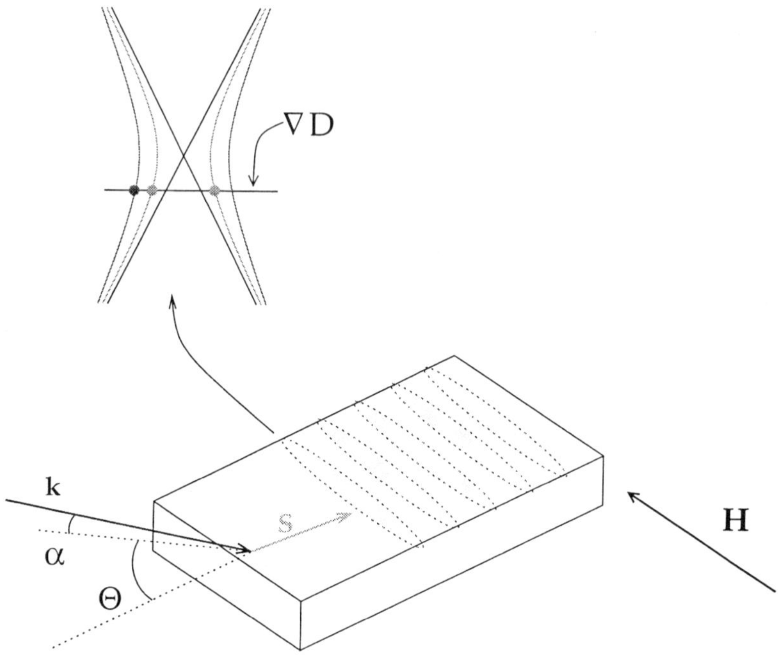

Figure 4.13: An x-ray shutter based on the Borrmann effect. A spatially periodic disturbance D leads to a structure factor modulation, and interbranch scattering occurs along ∇D. Each time the x-rays cross a transition region between disturbed and undisturbed crystal, some of the photons on the α branch are scattered onto the β branch and are lost to absorption.

of micrometers, and, to affect the same x-rays multiple times, it must travel at a velocity differing strongly from the x-ray group velocity (or not at all). Therefore, x-ray optics place a lower limit of the order of picoseconds on the switching time, even if the disturbance may be faster than that (such as optical phonons with 100 fs oscillation periods).

The second variant of the Borrmann switch does not rely on any subtleties of dynamical diffraction, like interbranch scattering: A disturbance is made to travel in parallel with the x-rays along the surface. This is not possible with a simple laser wave packet traveling across the crystal surface because the diffracting x-rays propagate significantly slower than light in free space. Velocity matching can however be achieved in the way shown in fig. 4.9. With a very strong disturbance, the structure factor is all but eliminated, and the x-rays that would travel many millimeters under anomalous transmission conditions are strongly absorbed. Because the disturbance traveling with the x-rays can be very narrow in space, the device can be very fast and still have a high switching contrast.

Chapter 5

PARAMETRIC DOWN CONVERSION

I have regularly set up shop now above the wash-house at the gate, in a garret. I have an old door set on two barrels, and two chairs, of which one is safe, and a skylight above, which will slide up and down. On the door (or table) there is a lot of bowls, jugs, plates, jam pigs (jars) etc. containing water, salt, soda, sulphuric acid, blue vitriol, plumbage ore; also broken glass, iron, and copper wire, copper and zinc plate, bee's wax, sealing wax, clay, rosin, charcoal, a lens, a Smee's Galvanic apparatus, and a countless variety of little beetles, spiders and wood lice, which fall into the different liquids and poison themselves . . .
— James Clerk Maxwell

B.W. Adams
Advanced Photon Source, Argonne National Laboratory, Argonne, IL 60439, USA
adams@mail.desy.de

Abstract Parametric down conversion, the spontaneous decay of single photons into highly correlated pairs, is the prime example of quantum optics with x-rays. The effect is very weak, with a cross section of the order of 10^{-9} of the Thomson scattering cross section. A number of experiments to demonstrate the effect and their results are described here.

1. Introduction

Spontaneous parametric down conversion (PDC) is well known in the laser physics community and has gained tremendous importance during the last decade as the prime source of entangled photons for quantum computation, quantum-encrypting communication, etc. It can be thought of as the next-order nonlinear optical analog of spontaneous emission of photons. The excited state whose decay is triggered by vacuum fluctuations is not located in an atom, but is rather a photon (commonly called the "pump"), occupying some mode of the electromagnetic field, and the final state is not a single photon, but rather a pair of highly correlated photons (called the "idler" and the "signal"). Energy is conserved, i.e., the signal and idler frequencies add up to that of the pump: $\omega_s + \omega_i = \omega_p$. A nonlinear optical medium is required to couple the participating modes. At x-ray frequencies, optical nonlinearities are very weak (see ch. 2), and down conversion is observable at all only because of the enormous virtual power density of vacuum fluctuations at x-ray frequencies (see app. 6.1).

The pioneering experiment [111] to demonstrate the effect of parametric down conversion for x-rays and the underlying proposal [126, 127] are about thirty years old. At the time, it took months to collect the data at an x-ray tube, and the results exhibited a large level of background events. Only about a quarter century later were these results corroborated by experiments performed at synchrotron radiation facilities [112, 113, 114]. Even at a modern synchrotron beamline and with the latest detectors and electronic technology, the experiment is challenging.

1.1 Nonlinear Medium

The most important ingredient for down conversion is a nonlinear optical medium with refractive properties that permit a matching of wave vectors, so that the nonlinear effect can occur coherently throughout the medium and produce a (relatively) strong effect. With visible light, nonlinear birefringent crystals are used, such as KTP, BBO, $LiNbO_3$, etc. The nonlinearity resides in the chemical bonds, i.e., the bonding orbitals have a strongly anharmonic character. This can be described phenomenologically by an expansion as in eq. (2.5) of the electrical polarization (the magnetic field plays no role) in terms of nonlinear electric susceptibilities.

As shown on p. 23, this does not work for diffracting x-rays. The nonlinearity involves both the electric and the magnetic components of electromagnetic waves, and an inversion symmetry of the medium does not exclude the even-ordered nonlinearities. For the experiments shown in this chapter, the model of free electrons is quite sufficient, and the nonlinear response can be understood by an iteration of the Lorentz equation (although this approach is not the best in a development of the theory). In the first iteration, the electron is at rest and is induced to harmonic oscillation by harmonic fields. This is the cause of Thomson scattering. Once the electron is moving, it experiences a Lorentz force in the magnetic components of the incident fields, leading to a second-order velocity containing the sums and differences of the frequencies present in the incident fields. The electron is also not at its rest position and thus samples the fields somewhere else, and it generates a current somewhere else—but these two effects cancel each other out. Obviously, the appearance of the Lorentz force invalidates a description in terms of only the electric fields, and for the same reason, the electron's motion in harmonic fields is anti-centrosymmetric relative to the rest position. A more elegant approach than that based on the Lorentz equation can be found with the results of ch. 2 and ch. 3, sec. 4.

All even orders of the nonlinear response of free electrons (the first order being the linear response, giving rise to Thomson scattering) lead to a longitudinal motion which can contribute to scattering only if several intersecting waves are coupled coherently to each other (see ch. 2). With x-rays, this implies the presence of a Bloch wave in a crystal, to be described by the nonlinear dynamical diffraction theory of ch. 3, sec. 4. Another reason for the necessity of diffraction is given in the next section.

1.2 Wave Vector and Frequency Matching

Energy is conserved in parametric down conversion. A measurement of the signal and idler photon frequencies, done within a time interval comparable to or longer than the reciprocal bandwidth of the pump, will yield values that add up to the pump frequency. However, the conversion process can take place within a much shorter time, resulting in correspondingly large bandwidths of the signal and idler frequencies, as seen individually [128, 129]. Likewise, the individual momenta of the signal and the idler can lie within loosely defined boundaries, but the total momentum must add up to that of the pump (within its usually much tighter limits), modulo a momentum transfer to the nonlinear medium. This statement implies that the conversion occurs coherently within a volume of the nonlinear medium, which is so large and massive that the transfer of momentum goes with no discernible energy loss. The coherence volume determines how restrictively the wave vectors $\mathbf{k}_i$, $\mathbf{k}_s$ and $\mathbf{k}_p$ of the idler, signal and pump must be matched.

With x-rays, the dispersion of the refractive index in the absence of resonances (see eq. (3.12)) generally makes the sum of the idler and signal wave numbers fall short of the pump wave number, i.e., $|\mathbf{k}_i|+|\mathbf{k}_s| < |\mathbf{k}_p|$. It is therefore impossible to match $\mathbf{k}_i + \mathbf{k}_s = \mathbf{k}_p$. Even if dynamical diffraction effects (see sec. 2.8 and fig. 2.2) or the anomalous dispersion at an x-ray absorption edge are used to make $|\mathbf{k}_i| + |\mathbf{k}_s| \geq |\mathbf{k}_p|$, will the angular dependence of the nonlinearity of free electrons suppress any down conversion (see ch. 2, sec. 1 and ch. 3, sec. 4.3). It is, however, possible to use diffraction from a crystal lattice, as shown in fig. 3.16, to satisfy the wave vector matching condition and to induce an angle between the pump wave and the signal and idler waves.

The wave vectors can now be matched as $\mathbf{k}_i + \mathbf{k}_s = \mathbf{k}_p + \mathbf{H}$ by detuning slightly from the Bragg angle Θ with a positive $\delta\Theta$, so that the point where $\mathbf{k}_p$ and $\mathbf{k}_i + \mathbf{k}_s$ originate, moves closer to $\mathbf{H}$. The dependence of $\delta\Theta$ on the angles α_i and α_s of the idler and the signal wave vectors $\mathbf{k}_i$ and $\mathbf{k}_s$ relative to the diffracted pump $\mathbf{k}_p + \mathbf{H}$ can be calculated by taking into account the geometry, the frequency-dependent refractive corrections to the wave vectors, and, for small angles, also dynamical diffraction effects for $\mathbf{k}_p$. The calculation is somewhat lengthy and is carried out in app. 6.4.1, but a quick answer for small $\alpha_i = \alpha_s$, and in disregard of dynamical diffraction effects is:

$$\delta\Theta = \frac{\alpha_s^2/2 + 3\delta}{\sin 2\Theta} \qquad \boxed{\alpha_i = \alpha_s}, \tag{5.1}$$

where $\delta = 2\pi r_e \tilde{\eta}_0 / k_p^2$ is the refractive index decrement (see eq. (3.12)) at the pump frequency.

Although two waves of the pump frequency are now present in the crystal, the contribution of the diffracted wave ($\mathbf{k}_p + \mathbf{H}$) is negligible, leaving essentially only the incident wave ($\mathbf{k}_p$) as the pump. This is due to the following two

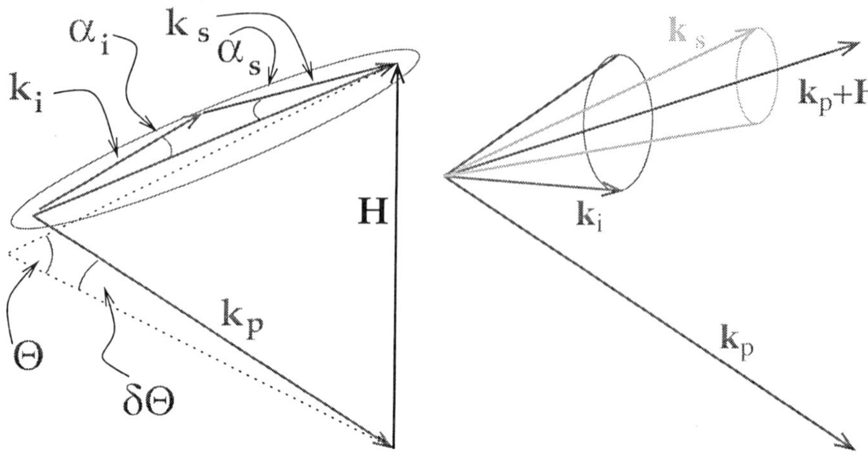

Figure 5.1: Wave vector matching in parametric down conversion of x-rays, as realized in the experiments described in [111, 112, 113, 114]. Left: The pump wave vector $\mathbf{k}_p$, a reciprocal lattice vector $\mathbf{H}$ and the idler and signal wave vectors $\mathbf{k}_i$ and $\mathbf{k}_s$ with $\mathbf{k}_i + \mathbf{k}_s = \mathbf{k}_p + \mathbf{H}$. NB: the role of the diffracted pump with wave vector $\mathbf{k}_p + \mathbf{H}$ in the down conversion is minimal (see text). Right: for given $|\mathbf{k}_i|$ and $|\mathbf{k}_s|$, the condition $\mathbf{k}_i + \mathbf{k}_s = \mathbf{k}_p + \mathbf{H}$ defines cones centered on $\mathbf{k}_p + \mathbf{H}$. The wave vectors $\mathbf{k}_i$ and $\mathbf{k}_s$ may lie anywhere on their respective cones but for a given down conversion event, they must lie on opposite sides of $\mathbf{k}_p + \mathbf{H}$.

reasons: If $\delta\Theta$ is large, i.e., greater than a few microradian, then the amplitude of the diffracted pump wave is very small (see eq. (3.15)). For small $\delta\Theta$, however, the angles α_i and α_s are also small by eq. (5.1), and therefore suppress the second-order nonlinear amplitude coupling, which depends on electronic motion along the wave vectors (see discussion in ch. 2, sec. 1). Being driven almost only by the pump wave $\mathbf{k}_p$, down conversion does not exhibit the narrowly peaked dependence on $\delta\Theta$ as a Bragg reflection does.

Another possible configuration, also using a Bragg reflection, is discussed in sec. 2.8.

1.3 Strength of the Effect

An estimate of the rate of down conversion events to expect can be obtained by use of results of ch. 3, sec. 4.2 for the nonlinear response of free electrons in a lattice, and insertion of the virtual power density of the vacuum fluctuations as a driving force. This is shown in app. 6.2. There are two possible approaches. The first is to calculate a cross section for the down conversion at one electron, then multiply it with the number of electrons in the sample, and to take into account interference enhancement. The cross section is (eq. (5.12))

$$\frac{d\sigma}{d\Omega} = \frac{r_e^4 \hbar\omega\delta}{3\pi c q^2} = \frac{137 r_e^4 k^2}{3\pi}\frac{\delta}{\omega}, \tag{5.2}$$

which is lower than the Thomson scattering cross section r_e^2 by a factor of $1.8 \cdot 10^{-8}$ for 20 keV pump photons and a 1%-bandwidth δ/ω. Geometry factors (i.e., sines and cosines of angles) are not included in this estimate. These depend, of course, very much on the details of the experiment, but they can easily take away another factor of 10 to 100. Furthermore, the down-converted photons do not go into well-defined beams but rather into the cones shown in fig. 5.1. Depending on the size of the sensitive areas in the detectors, this takes another large factor away (roughly 100 with the Si drift detectors used in the experiments described below). With Thomson scattering, all photons are reflected within one-half pendellösung period (ca. 30 μm for 25 keV in diamond). With the above estimates, the same amount of sample material should lead to an observed down-conversion event rate of one in $10^{11} \ldots 10^{12}$ incident pump photons. The rate actually observed (see for example sec. 2.4) is more like one in 10^{13} incident photons, which should be close enough, given the above rough estimates.

The other approach is to calculate the coherent generation of the down-converted amplitudes in a large sample volume, and then determine a photon flux. The former is applicable in simple scattering problems but fails when dynamical diffraction effects come into play. It then becomes necessary to apply an extension of the dynamical diffraction theory to nonlinear optics. This is done in app. 6.3 on the basis of the theory developed in ch. 3, sec. 4.3.

Leaving absorption (or stimulated emission) aside, diffraction and other interference phenomena leave the number of photons conserved ("the photons have to go somewhere"). In nonlinear optics, this kind of reasoning cannot be applied separately to the photons of any given energy - only the total electromagnetic energy is conserved. It is therefore quite possible that interference suppresses the effect of down conversion. This is discussed further in sec. 2.7.4.

2. Experiments

2.1 The Classical Experiment by Eisenberger and McCall

Results from the first successful attempt at demonstrating down conversion of x-rays were published in 1971 [111]. The authors used radiation from a 2 kW x-ray tube with Mo anode, i.e., a time-continuous source of 17 keV photons, and scintillation detectors with a time resolution of ca. 10 ns. The sample was a single crystal of beryllium, chosen for its low x-ray absorption (see sec. 4.7).

Figure 5.2 shows the setup and a time correlation spectrum of energy-discriminated events in the two detectors. The peak over the background of statistical correlations indicates coincidences due to down conversion. The background is due to random photons from Compton and diffuse elastic scattering in the sample, hitting the two detectors at the respective time difference and both being falsely identified as having the right energies (see discussion in

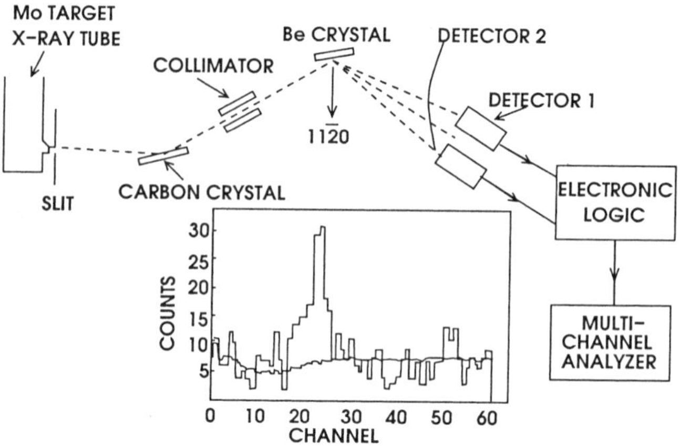

Figure 5.2: The setup of Eisenberger and McCall and a time correlation spectrum with coincidence appearing as a peak. One channel corresponds to 10 ns time difference. Figure redrawn after fig. 1 of [111].

secs. 4.3 and 4.4). The data shown in fig. 5.2 are the results of ca. 400 hours of data collection with a maximum event rate of about 1 per hour.

2.2 The First Synchrotron Results, Yoda et al.

It took a quarter century before another successful experiment was published [112]. This one was done at the Photon Factory synchrotron radiation facility,

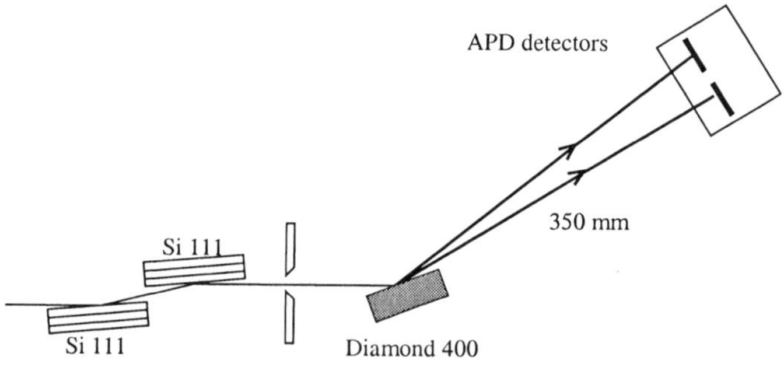

Figure 5.3: The setup of Yoda et al., figure 2 of [112] redrawn.

which is a pulsed source with a 2 ns repetition rate. The wavelength of the pump was 0.65 Å, giving a photon energy of about 19 keV. Diamond samples were used with the (400) and the (111) reflections. This material absorbs x-rays slightly more than beryllium, but the quality of available crystals is better. Instead of scintillation detectors, avalanche photodiodes (APDs) with sensitive areas of 5 mm diameter were used. With a time resolution of about 0.5 ns, these

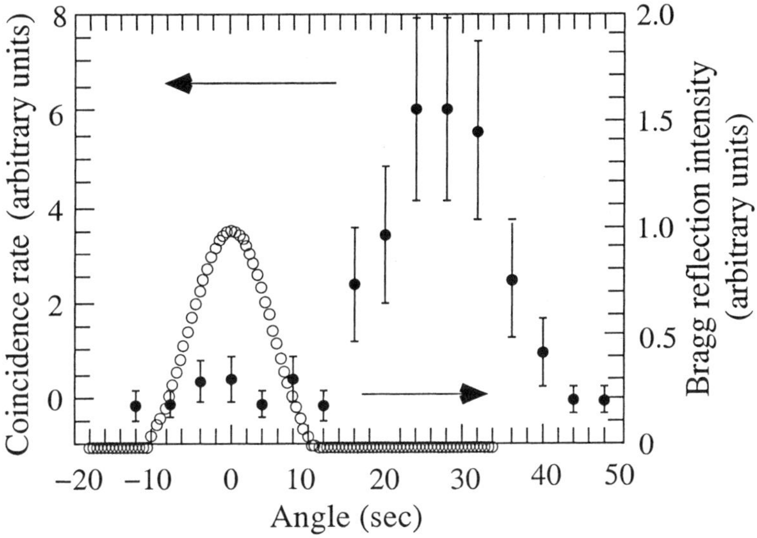

Figure 5.4: Results from Yoda et al., figure 3b of [112] redrawn. Shown are the Bragg reflected intensity (open circles) and the coincidences (solid circles with error bars). The maximum of coincidences occurs at a detuning angle of 29 seconds, i.e., 0.145 mrad, which is slightly larger than the calculated value of 24 seconds. The reflection shown here is diamond (111). The data for the (400) reflection are qualitatively similar and are shown in fig. 3a of [112]. The arrows refer to the ordinates for the two curves.

detectors were well matched to the time structure of the source. Their energy resolution is about 50% and their quantum efficiency at 9.5 keV is ca. 60%. The APDs were placed 18 mm apart and 350 mm from the sample for the (400) reflection (see fig. 5.3) and 1135 mm from the sample for the (111) reflection. Insertion of $\delta = r_e \tilde{\eta}_0 \lambda^2 / 2\pi = 2.02 \cdot 10^{-6}$ (with $r_e = 2.818 \cdot 10^{-15}$m and $\tilde{\eta}_0 = 1.065 \cdot 10^{30}\mathrm{m}^{-3}$) and $\alpha_s = 1.47°$ for (400) and $\alpha_s = 0.454°$ for (111) into eq. (5.1) yields $\delta\Theta = 4.912 \cdot 10^{-4}$ rad (= 101.3 arcsec) for the (400), and $\delta\Theta = 1.194 \cdot 10^{-4}$ (= 24.6 arcsec) for the (111) reflection. The respective values given in [112] are 96 arcsec and 24.4 arcsec, respectively. Figure 5.4 shows the data for the (111) reflection. The event rate was 6 per hour for both reflections, with a maximum background of 2.7 per hour for (400) and 0.2 per hour for (111). The authors checked the background level by introducing a delay into one of the signal chains to eliminate coincidence from the down conversion while keeping the level of accidental coincidences constant (fig. 4 in [112]).

2.3 Energy Discrimination and Time Correlation

This, and all other experiments described below, were done by me and collaborators. Our first results [113] were obtained at the European Synchrotron

Radiation Facility (ESRF) in Grenoble.[1] The sample was a synthetic type II diamond,[2] made by Sumitomo and cut to a plate of about 0.5 mm thickness in (400) orientation. It exhibited a twinned structure (fig. 3 in [113]). Figure 5.5 shows

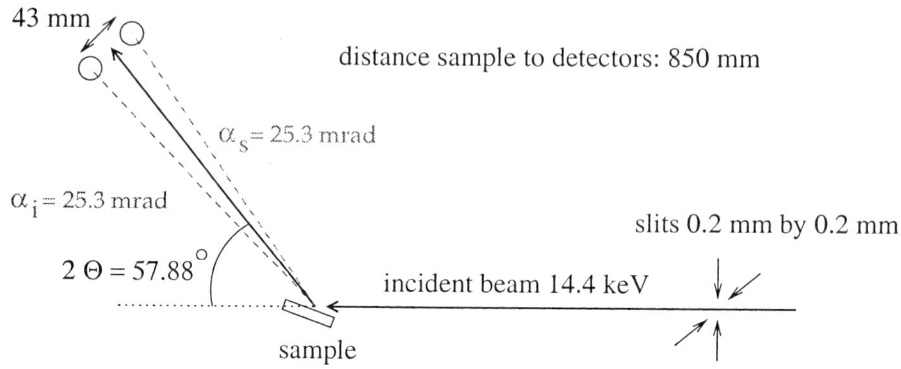

Figure 5.5: The experimental setup of ID18 at the ESRF during beam time MI-247, with the incident beam defining $\mathbf{k}_p$, the diffracted pump beam going upward and the detectors placed to register events with $\mathbf{k}_i$ and $\mathbf{k}_s$ lying within the vertical plane.

a schematic of the scattering geometry. Photos of a rather similar setup (see sec. 2.4) can be found on the CD in the directory chapters/pdc/photos/mi-311/. The pump photons of 14.4 keV energy hit the sample, oriented to $\Theta = 57.88°$ for the (400) refletion, at a flux of about $1.5 \cdot 10^{11}$ per second. Two Si drift diode detectors [132] were placed 850 mm from the sample and ± 21.5 mm from the Bragg reflected beam. These detectors have circular sensitive areas of 2.5 mm diameter. The angles of the down-converted photons relative to the Bragg-reflected beam were thus $\alpha_i = \alpha_s = 25.3 \pm 2.9$ mrad.[3] The detectors have an energy resolution of about 300 eV at moderate count rates below ca. 30000 photons of 10 keV per second, and the time resolution is about 100 ns. In a comparison to the experiment described in sec. 2.2, the superior energy resolution about balanced out against the inferior time resolution. Instead of measuring only coincidences, time correlation spectra of energy-discriminated detection events were obtained (see app. 6.5.1 for details). Figure 5.6 shows such a correlation spectrum on the left, obtained with a detuning angle $\delta\Theta = +0.0217°$,

[1]June of 1998 at the beamline ID 18 [130], beam time MI-247 (ESRF beam time allocation code).

[2]The classification of diamonds into types Ia, Ib, IIa and IIb is based on the impurity content, see for example [131], p. 669. Type I contains substantial amounts of nitrogen (up to about 0.3%) and is generally yellowish in color. They tend to have a better crystal quality (that is, for x-ray purposes) than the nicer-looking colorless type II stones.

[3]The following convention for the notation of angles shall be used here: Bragg angles Θ and their detuning $\delta\Theta$ shall be given in degrees because that is the usual unit used on diffractometers. The angles α_i and α_s are, however, given in milliradians because they have to be entered into eq. (5.1) in units of radians.

which is close to the value of 0.023° predicted by eq. (5.1). Coincidences oc-

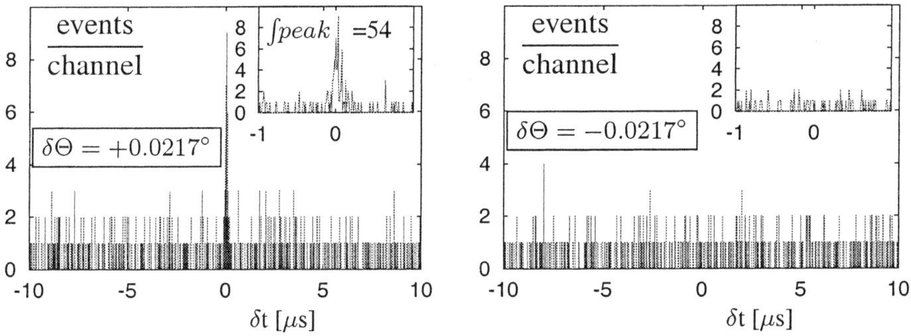

Figure 5.6: Time correlation spectra of beam time MI-247 at the ESRF, corresponding to figs. 4 and 5 in [113], but redrawn from the original data. The detuning angles are $+0.0217°$ on the left (maximimum event rate expected from eq. (5.1)) and $-0.0217°$ on the right (no events expected). The time scale is in microseconds and the channel resolution is 10 ns. The inserts show zooms into the central ± 1 μs. Both data sets were accumulated in 10500 s (2.92 h). At $\delta\Theta = +0.0217°$, a total of 54 coincidences occurred within ± 100 ns over a background of 7.3 in the same time interval and at $\delta\Theta = -0.0217°$, a total of 8 coincidences occurred within ± 100 ns over a background of 6 in the same time interval. The background-corrected peak event rate is thus about 1/225 s.

cur at zero time difference, and the rest of the spectrum is a measure of the background of accidental correlations, mainly due to stray photons at the pump energy being falsely recognized for being at the energy of the signal and idler. This is discussed at some more length in sec. 4.4. The recording of time correlation spectra made it unnecessary to introduce a delay to check the background level, as described in [112] (but see secs. 2.4 and 2.6 on that point). For comparison, another time correlation spectrum is shown in fig. 5.6 to the right. This one was taken at a detuning angle of $-0.0217°$, which is the right magnitude, but the wrong sign of detuning. In all other respects, the conditions were the same as with $\delta\Theta = +0.0217°$. A total of six such time correlation spectra were taken in that particular beam time at different detuning angles, and the coincidence rates show a clear peak at the expected detuning angle (fig. 6 in [113], not reproduced here because better data, obtained in subsequent beam times, is shown below).

2.4 High Event Rate

In another beam time[4], the experimental conditions were improved, and this beam time yielded the highest event rate to my knowledge ever observed so far [114]. The monochromator was set to 22 keV, and the ESRF was operating in the 2*1/3 fill mode (see ch. 1, sec. 4 and fig. 5.10). At an incident flux of about $2.5 \cdot 10^{12}$ photons per second, the rate of observed coincidences for ideal wave

[4]MI-311, April 1999 at ID18 of the ESRF

vector matching was about 1 in 13 seconds. The sample was a (100)-oriented, 0.4-mm-thick plate of natural diamond. The (111) reflection of that sample was used at $\Theta = 7.883°$, with the asymmetry chosen such that the sample volume illuminated by the incident beam was maximized. The detectors had apertures of ca. 2.5 mm and were spaced horizontally, ±19.5 mm from each other and 1420 mm from the sample, with 1285 mm of that length in He at ambient pressure and the remaining 135 mm in air, giving a transmission of 94.4% at 11 keV. This makes $\alpha_i = \alpha_s = (13.73 \pm 0.88)$ mrad. To find out if there is any marked polarization dependence (the radiation from the undulator is polarized in the horizontal plane), data were taken with the Bragg reflection lying within the vertical and in the horizontal scattering plane. Apart from the details given above, the experimental setup for the data shown in figs. 5.7, 5.8 and 5.9 was quite similar to that described in sec. 2.3. Photos of the setup can be found on the CD in the directory chapters/pdc/photos/mi-311/.

Figures 5.7, 5.8 and 5.9 summarize the data. Two representative time correlation spectra are shown in fig. 5.7 for detuning angles $\delta\Theta$ of 0.0215° and 0.024° (eq. (5.1) predicts 0.0208°). The close agreement of the former value of $\delta\Theta$ with that of fig. 5.6 is circumstantial. The coincidence peak is shifted away from zero because a 600 ns cable delay was inserted into one of the detector signal chains to catch artefacts in the data acquistion circuit. There are none.

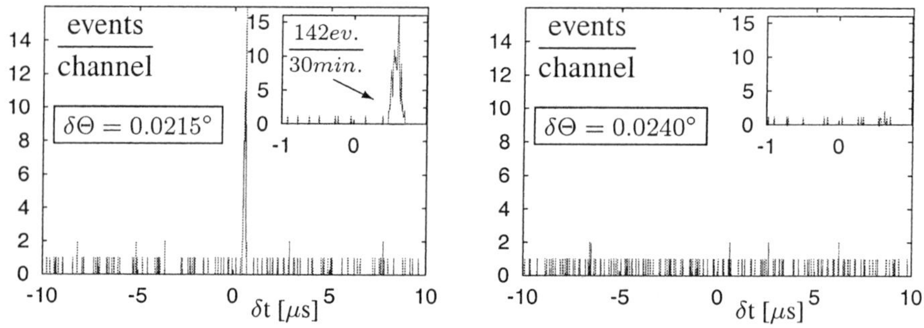

Figure 5.7: Two time correlation spectra from the vertical scattering geometry, each representing 1/2 hour of data acquisition at beam line ID18 of the ESRF during beam time MI-311. Left: $\delta\Theta = 0.0215°$, close to the value of 0.0208° predicted by eq. (5.1) and corresponding to the peak in fig. 5.8, right: $\delta\Theta = 0.024°$. The abscissae of the large graphs are time difference in μs, and the ordinates are events per time channel in 1/2 hour. The inserts show the central ± 1 μs. The peak appears at 600 ns due to a cable delay in one of the signal chains (see text). Its width is given by the detector time resolution.

Integrating over 50 channels (500 ns) around the peak position, subtracting the background and applying a monitor correction (i.e., a scaling to the incident intensity), yields the graphs in fig. 5.8. The background was determined by averaging over all of each spectrum except the peak region. For the monitor correction, the overall scaling factor was chosen to make the background-corrected and the background-and-monitor-corrected peak values of the "vertical" data agree closely: The peak of the event rate curve in fig. 5.8, left-hand side, corresponds to 146 raw events, 142.19 background-corrected and background-and-monitor-corrected events. The monitor correction factor lies within the bounds $0.9\ldots1.1$ for all points. Shown are the integrated event rates over the detuning angle for the vertical and the horizontal scattering geometry. Both curves are scaled with the same overall monitor correction factor. The peak "horizontal" event rate is lower by a factor of about 0.9, which agrees well with a polarization factor of $\cos^2 2\Theta$ with $\Theta = 7.88°$. The angular scale is relative to the Bragg diffraction peak, which is also indicated in the graphs. There are some glitches in the "horizontal" data at $\delta\Theta = 0.02°$, which are probably due to a mechanical inaccuracy in the horizontal circle of the diffractometer. The positions and widths of the down conversion peaks in fig. 5.8 are given by

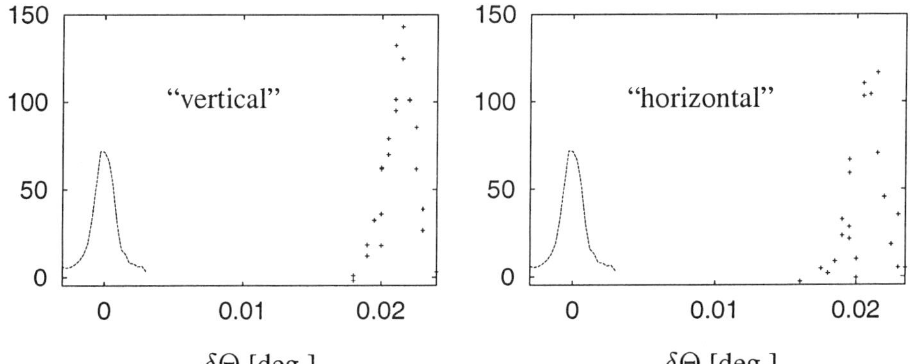

Figure 5.8: Number of down conversion events per half-hour, in vertical and horizontal diffraction geometry. The abscissa is the detuning angle $\delta\Theta$ and the ordinate is the number of events rate in 1/2 hour (see text for details). There are multiple data points for some values of $\delta\Theta$. Also shown is the Bragg reflection from the diamond crystal (the left curve in each of the graphs). No error bars are given because the points show actual numbers of events within about 10% (see text) and the background level can be estimated from fig. 5.7.

eq. (5.1) with $\alpha_{i,s} = (0.787 \pm 0.05)°$.

Figure 5.9 shows several time correlation spectra, restricted to $\pm 2.5\ \mu$s, and staggered according to the detuning angle. The event rates in this graph are raw, without background or monitor correction.

Finally figure 5.10 shows a sum over all time correlation spectra taken under comparable conditions with detuning angles $\delta\Theta = 0.016°\ldots0.024°$ and including vertical and horizontal scattering geometry data. No monitor or background correction was applied before taking the sum.

2.5 High Pump Photon Energy - 98.9 keV

One beam time[5] was given to measuring the conversion cross section at very high photon energies. Unfortunately, the level of background was much too high

[5] June 2000, beam time MI-393 at beam line ID-15A of the ESRF

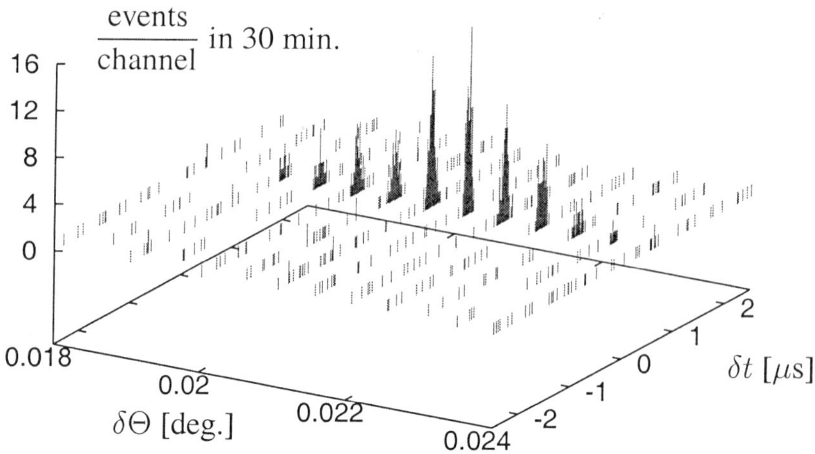

Figure 5.9: Time correlation spectra over detuning angle, from beam time MI-311 at the ESRF, corresponding to fig. 1 in [114], but redrawn from the original data. The background-corrected event rate peaks at $\delta\Theta = 0.021°$ with an integral of 142 events.

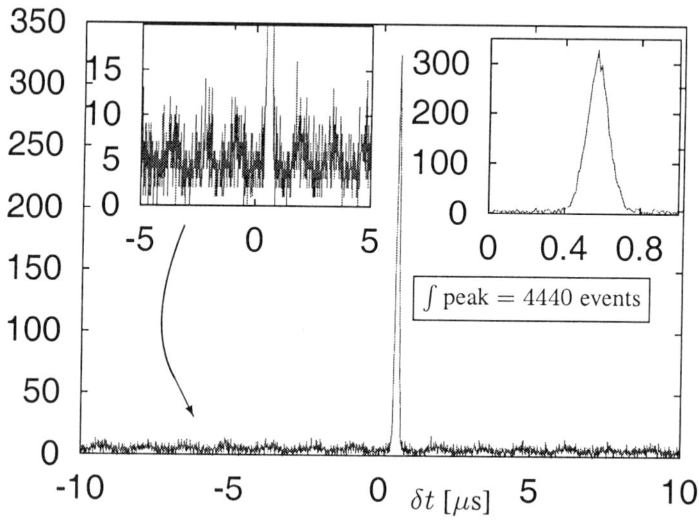

Figure 5.10: Sum over time correlation spectra with $\delta\Theta = 0.016° \ldots 0.024°$, without monitor or background correction. The background exhibits a slight modulation, which is the autocorrelation of the ESRF 2*1/3 fill pattern. As visible in the left-hand insert, its period is ca. 1.4 μs, corresponding to two blocks of filled bunches in the 2.84 μs ring period of the ESRF. The right-hand insert is a zoom into the central region. The peak position at 600 ns is due to the cable delay in one of the signal chains (see text), and its width is due to the detector time resolution.

with the particular setup used for that experiment, and the data show a marginal effect, if any at all. Nevertheless, the experiment shall be outlined briefly here. If nothing more, it can at least serve to illustrate background suppression issues, which are particularly challenging with high-energy photons. High-energy x-ray photons (say, above 50 keV) are highly penetrating and are readily Compton scattered multiple times. If the photon energy lies above the Pb K absorption edge at 88 keV, then Pb K fluorescence is excited, which penetrates rather well through lead because its energy is just below the Pb absorption edge.

Due to the very narrow width of x-ray reflections at high photon energies, it is hard to use a double-crystal monochromator and the beam from a single monochromator crystal takes off at the Bragg angle, instead of continuing straight as with a double-crystal monochromator. For geometrical reasons, the monochromator thus had to be placed inside the experimental hutch, instead of in a separate optics hutch. This setup is shown in fig. 5.11. The white beam from the wiggler entered a "pot" of lead through a narrow channel in a lead brick array. Inside that pot, the (3 3 11) reflection of a Ge crystal diffracted the x-rays of 98.9 keV (and harmonics) towards the sample, a Si crystal oriented to the (7 9 1) reflection. That beam path was also carefully shielded with lead bricks, leaving only tiny apertures for the x-rays. A liquid-nitrogen-cooled thirteen-

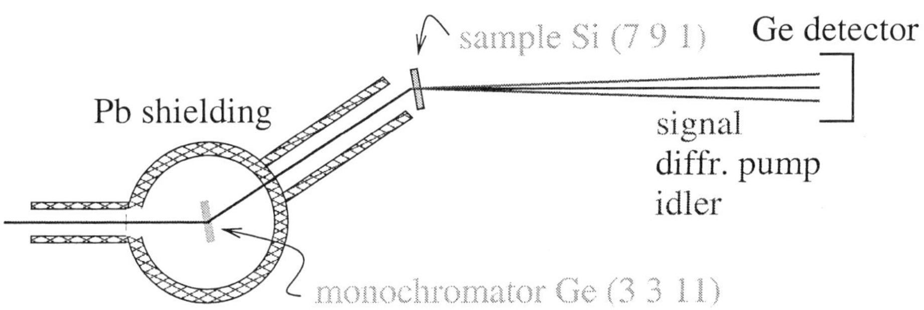

Figure 5.11: The setup of the high-energy experiment. The cross-hatched parts are lead shielding. The white beam, coming from the wiggler, enters the experimental hutch, and the selected energy of 100keV (plus harmonics) is diffracted by the monochromator crystal towards the sample, where the down conversion should occur. The detector is a cryogenically cooled thirteen-element germanium detector.

element Ge detector was used throughout the experiment, but the signals from only two elements were actually correlated.

Despite the impressive bulk of lead (see photos in chapters/pdc/photos/mi-393/ on the CD), the background radiation level in the hutch was still rather high. With the raw white beam entering the hutch, shielding is a major challenge, particularly considering the fact that the wiggler spectrum extends far beyond 100 keV. A multichannel analyzer (MCA) spectrum, taken in the process of background reduction, is shown in fig. 5.12. There is a strong peak at the incident photon energy and four peaks below it, which may be identified (in order of decreasing energy) as the Ge escape peak (due to the loss of a Ge K fluorescent photon leaving the detector during the detection process), Pb K_β, Pb $K_{\alpha 2}$, and Pb $K_{\alpha 1}$. A broad hump below the incident photon energy can be attributed to Compton scattering. The energy loss in Compton scattering depends on the scattering angle Θ as $E'/E = (1 + k(1 - \cos\Theta))^{-1}$, where k is the ratio of photon energy versus the energy in the electron rest mass. Here, $k \approx 0.2$. Given the collimated beam paths, the dominant source of the Compton hump is probably multiple Compton scattering within the sample. In contrast, single Compton scattering from the sample to the detector is geometrically restricted to roughly twice the Bragg angle at $2 \cdot 7.5°$. The resulting energy loss is ca. 70 eV, which might account for the shoulder on the low-energy side of the peak at 99 keV. The sharp peak at ca. 160 keV is unidentified and seems to be an electronic artefact.

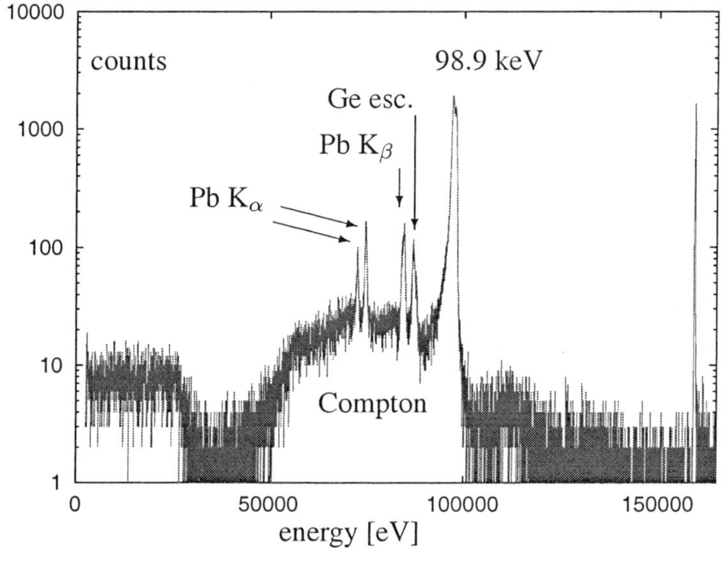

Figure 5.12: An MCA spectrum of 98.9 keV radiation, taken with a Ge detector. For details, see text.

After careful and massive shielding and collimation, the fluorescent peaks were eliminated (i.e., suppressed below the Compton hump). Of course, nothing could be done about the Compton scattering from the sample itself. Luckily, most of the Compton scattering was above the energy of 50 keV where the down-converted photons were expected. A time correlation spectrum of energy-selected events in two of the detector elements is shown in fig. 5.13. It was obtained by summing over the time correlation spectra of roughly one day of beam time, and then convoluting with the function $\exp(-(x/(30\text{ns}))^2)$. The time of $2 \cdot 30$ ns corresponds to the time resolution of the Ge detector, and, therefore, the convolution removes some of the statistical fluctuations in the original data without impairing the time resolution. There may be some coincidence signal, given by the peak at zero time difference. However, that peak is certainly not significant, being just slightly higher than the other ones.

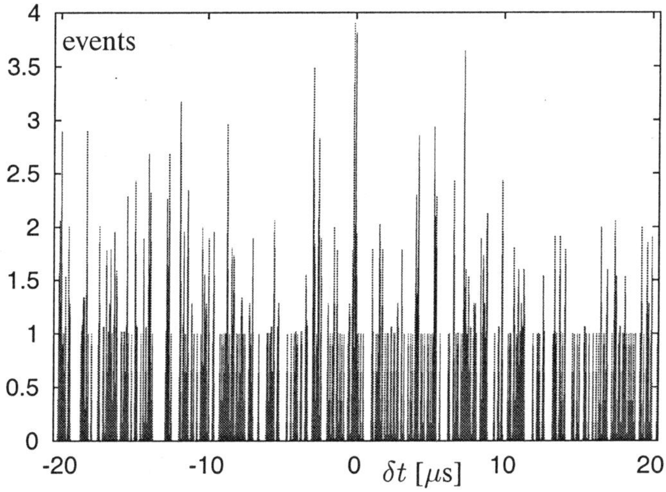

Figure 5.13: Histogram of a sum of time correlation spectra with a pump energy of 98.9 keV, convoluted with a low-pass filter function (see text).

2.6 Suppression of the Pump Photons with a Mirror

After careful shielding and collimation, most of the background signal is due to pump photons scattered within the sample towards the detectors (see sec. 4.3). Because the vast majority of them is close to the pump energy, one might use the wavelength dependence of total external reflection of x-rays to suppress them while leaving the down-converted photons unscathed.

The experimental setup of one such experiment[6] is shown in fig. 5.14. Putting a mirror into an upward-directed beam path (see fig. 5.5) does not seem very practical, and a horizontal scattering geometry was ruled out for a number of reasons, the most important one being the dispersion mismatch relative to the vertically diffracting monochromator and the relatively large horizontal divergence of the radiation. Instead, the incident beam of 25.514 keV photons was first deflected up by a Ge (220) reflection. That upward-going beam then hit the diamond sample. Due to the good match of lattice parameters between Ge (220) and diamond (111), the Bragg-reflected beam left the diamond almost in the horizontal direction (actually in a upward slope of $2 \cdot \Delta\Theta = 6.6$ mrad). The beams of down-converted photons with angles $\alpha_s = \alpha_i$ of 4.36 mrad within the vertical plane were then reflected off the surface of a vertically standing, 3-mm-thick glass plate of 356 mm length and 50 mm height. That plate was cut from a larger piece with defined surface flatness of 2-3 μm per inch.[7] The

[6] Jan. 2000 at station ID18 of the ESRF, beam time MI-362. Before that, the idea was tested in Oct. 1999 at Petra (DESY).

[7] Edmund Scientific beam splitter, stock no. A72-500 in the 2002 catalog

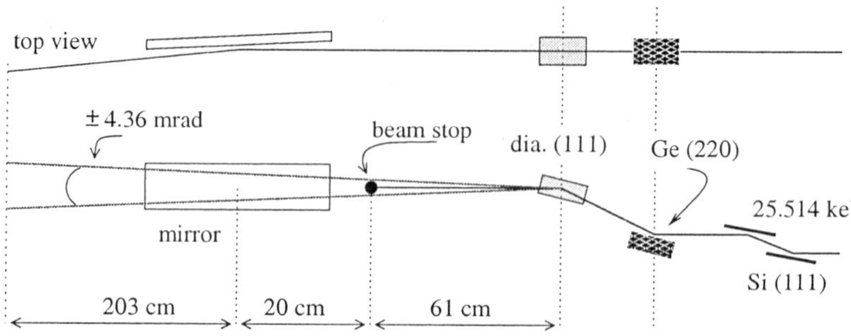

Figure 5.14: Mirror setup at beam line ID18 of the ESRF. Shown are a top view (above) and a side view (below) of the monochromator, Ge crystal, diamond sample, mirror, and detectors. The Bragg-reflected beam from the diamond crystal was stopped to prevent it from generating a strong background of stray radiation on the mirror. Most of the beam paths, including the mirror, went through He-filled tubes to reduce absorption and the background from air scattering.

glancing angle was chosen as 1.25 mrad, just above the critical angle of total reflection of 1.2 mrad for the pump photon energy. The Bragg-reflected beam from the sample was stopped before reaching the mirror to keep it from being scattered towards the detectors and generating a background much worse than that without any mirror.

The sample was a type Ib synthetic diamond from Drukker BV. [133], cleaved to a 1-mm-thick plate in the (111) orientation and cut to a rectangular outline of 3 mm by 4 mm. The raw synthetic stone from which this plate was made was selected out of a batch of three 10-to-15 mm sized stones after characterization with x-rays of 100 keV [134]. Its mosaic width was found to be better than 10^{-3} degrees throughout the material. The diamond is slightly yellowish in color (see photo P0001526.JPG in chapters/pdc/photos/mhatt on the CD) from nitrogen impurities which is typical for type Ib. For the purposes of x-ray diffraction, type I stones are considerably better than type II (compare fig. 3 in [113] with the 10^{-3} degree-rocking widths of the 1b-stone). Photos of that sample can be found on the CD in the subdirectories pertaining to this and the following down conversion experiments. This stone is also shown in fig. 3.5 and on the CD in the demonstration of the Poynting vector in dynamical diffraction.

The degree of angular accuracy necessary to match the Ge crystal to the monochromator is beyond the precision of the rotation stage used. Therefore, the Ge crystal, and also the diamond, were mounted on piezoelectric transducers. These were actually just bare tubes of piezoelectric material without any feedback (like a strain gauge). They proved, however, very reliable and stable if a change in the applied voltage was always done by running through the full hysteresis loop. The same kind of piezoelectric transducers was also used in the experiments described in secs. 2.7.2 and 2.7.3, and ch. 6 for the channel-cut crystal. Silicon drift detectors were used, as before, and the single channel analyzers in the respective signal chains were adjusted to 12.8 keV, where the down-converted photons were expected. The ESRF was operating in the uniform fill mode (i.e., all 992 buckets were filled). Due to equipment-related problems, only about one day's worth of beam time could actually be used. The incident intensity before the Ge crystal was about $2 \cdot 10^{11}$ photons per second at 25.5 keV. There was not enough space in the setup to place an ionization chamber into the upward-going beam from the Ge crystal, and, therefore, the intensity hitting the diamond crystal can only be estimated to about $1.5 \cdot 10^{11}$, taking into account the dispersion mismatch between the Si (111) monochromator and the Ge (220) crystal. A summary of the data is shown in fig. 5.15.

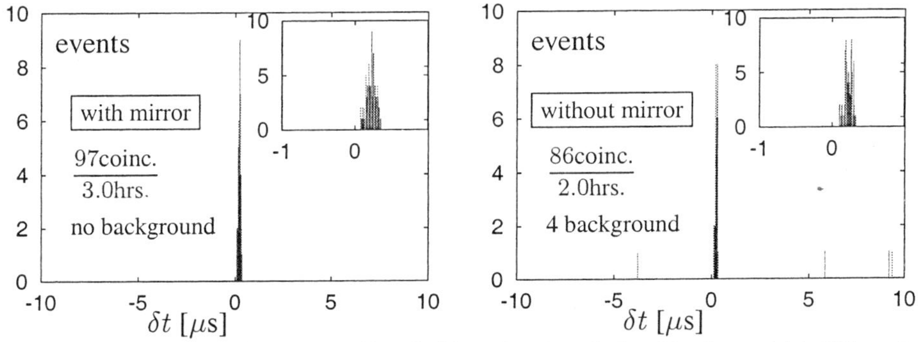

Figure 5.15: Comparison of event rates with and without the mirror. Left: with mirror, a total of 97 events in 3.0 hours and no background at all. Right: without mirror, a total of 86 coincidence events in 2.0 hours and 4 background events. The coincidence peak is shifted because of an intentional cable delay in one of the detector signal chains, see sec. 2.4.

Due to the relatively low incident intensity and the larger distance between the sample and the detectors, and perhaps also due to improved shielding, the background event rate was extremely low. Even without the mirror, there are only 4 background events in 2 hours, and there are none in the data "with mirror." The data are not monitor-corrected but taking the incident intensity into account would change the ratio of observed events by only about 10%. Within the level of uncertainty intrinsic to the extremely low event rates, the observed rate of roughly one in 80 seconds agrees well with the roughly 1 in 13 seconds at about six-fold higher incident intensity shown in sec. 2.4.

Barring games like "$4/0 = \infty$," the observed background suppression is not very striking. More importantly, however, the experiment did demonstrate that the down-converted photons can be reflected off a mirror with little loss. This may seem a bit trivial but is a necessary step towards taking advantage of the high quantum efficiency in x-ray detectors for tests of the quantum theory (see sec. 3.1). Nothing would be gained by having a high detection efficiency but losing down-converted photons on the way.

2.7 Small Angles

For a very small detuning angle $\delta\Theta$, dynamical diffraction of the pump wave comes into the play. One might also expect the rate of down conversion to go up, as α_i and α_s go to zero with decreasing $\delta\Theta$ (see eq. 5.1) because the relaxed geometrical constraints for the wave-vector matching result in a higher number of available final states (the so-called edge-enhancement [135]), illustrated in fig. 5.16. So far, three beam times have been devoted to the

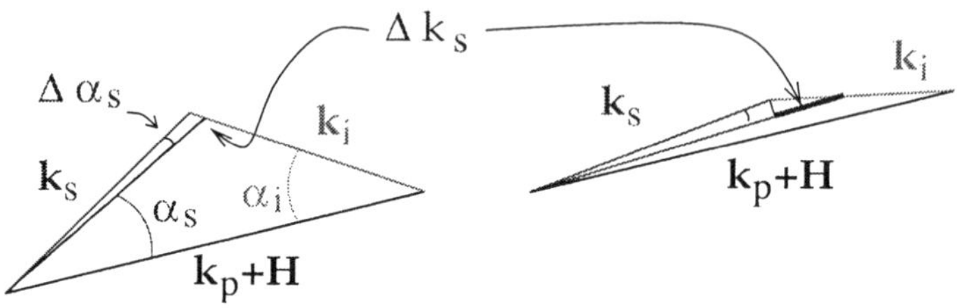

Figure 5.16: Illustration of the edge-enhancement effect. Left: at large angles $\alpha_{i,s}$, a given angular range $\delta\alpha_s$ results in a small wave number spread δk_s. Right: the same angular range $\delta\alpha_s$ gives a much larger wave number spread δk_s for small $\alpha_{i,s}$.

investigation of the small angle regime. The results were surprising because, instead of an enhancement, a suppression was observed. Although a definitive statement cannot yet be made, based on low event rates, suppressed even more, the evidence points to the suppression effect. A possible explanation, based upon dynamical diffraction effects, is given in sec. 2.7.4.

2.7.1 The First Small-Angle Experiment. A beam time was scheduled at the ESRF[8] to investigate the small-angle regime. The storage ring was operating in the uniform fill mode (see ch. 1, sec. 4) and the incident photon energy was 22 keV. In order to achieve very small angles $\alpha_{i,s}$, the scattering geometry with a primary upward-going beam, first tried in the mirror suppression experiment (sec. 2.6), was adopted. The setup (fig. 5.17) was similar in principle to the one shown in fig. 5.14, except that the pump photon energy and the distance between sample and detectors were different (given below), and the first crystal was InP, instead of Ge, as shown in fig. 5.14. The flux of pump photons at the sample diamond was measured to be ca. $6 \cdot 10^{11}$ photons per second.

The horizontal-going Bragg-reflected beam and the down-converted photons close to it were sent through an evacuated beam pipe from the experimental hutch, through the next hutch to the hutch at the very end of the beam line, 14

[8] Jan. 2001, MI-463

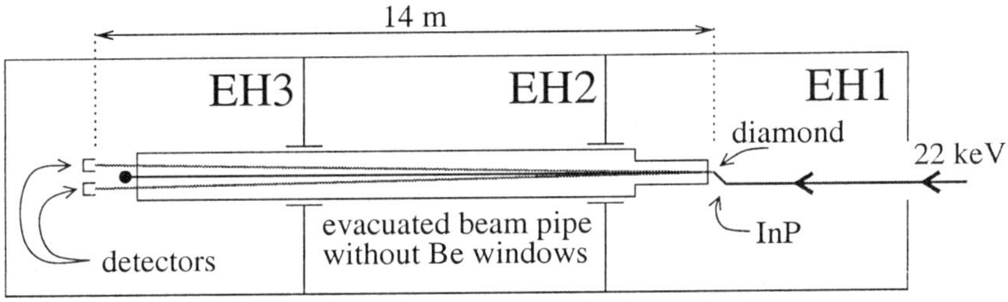

Figure 5.17: Small-angle experiment at the ESRF, side view of the three experimental huthces. The first crystal (InP) and the sample diamond were placed in the first experimental hutch "EH1" of beam line ID18. The Bragg-diffracted beam from the sample and the down-converted photons passed through the second hutch "EH2" in an evacuated beam pipe, into the third hutch 'EH3', 14 m downstream from the sample.

m downstream from the sample (see fig. 5.17). The beam pipe was sealed from the outside only by a 5-mm-diameter, 5-μm-thin Mylar foil at its entrance side and a Kapton foil at its exit side. No vacuum windows of any kind were present inside the pipe. This is important because the Bragg-diffracted beam from the sample is rather strong at small $\delta\Theta$, and small-angle scattering in windows would produce a high level of background in the detectors. The beam pipe has to begin as close to the sample as possible to minimize absorption of the down-converted photons and scattering of the Bragg-diffracted pump photons in air. Because of the small separation of the down-converted beams, the Bragg-diffracted beam cannot be stopped before hitting the entrance window, which must therefore be chosen to minimize the small-angle scattering from it. The exit window is not as critical: a cup-like beam stop of a few mm diameter right behind the window can efficiently catch the small-angle scattering and prevent it from reaching the detectors. The major source of background signal that remains is Compton and static and thermal diffuse scattering of the pump photons within the sample. Diffuse scattering, which is particularly strong at small $\delta\Theta$, might be reduced by use of a sample crystal of exceptionally high quality and by cooling it, but Compton scattering is unavoidable. However, the simplest way of reducing the background is probably to use detectors with a lower level of a low-energy tail in its energy spectrum (see fig. 5.34), or to use a mirror, as demonstrated in sec. 2.6.

Two Si drift detectors were placed 20.6 mm apart at the end of the beam pipe, 14 m downstream from the sample. The long beam path through apertures in the beam line necessitated a good match between the lattice plane spacings of the first, upward-reflecting, and the sample crystal. With germanium, the slope of the Bragg-reflected beam from the diamond would have been about 7.5 mrad at 22 keV pump photon energy, and, worse, the slope would be upward with an upward displacement by the double reflection (Ge-diamond). A much better choice is InP. Its (220) lattice spacing is better matched to diamond (111), giving a smaller and downward slope of the Bragg reflection from the diamond (1.34

mrad at 22 keV). Another advantage of InP is that the indium K absorption edge is at 27.9 keV, and a pump photon energy below that will not lead to a significant fluorescent background. Indium L fluorescence is below 4 keV and phosphorus K fluorescence is around 2 keV, making both irrelevant for background considerations for their small absorption lengths in air. In contrast, the germanium K fluorescence is at 10 to 11 keV, which is uncomfortably close to the expected energy of the down-converted photons (luckily below it, which is better than above, see fig. 5.34). The germanium K fluorescence is not a showstopper, as the low background in the mirror suppression experiment (sec. 2.6) demonstrates, but it is always better not have any at all.

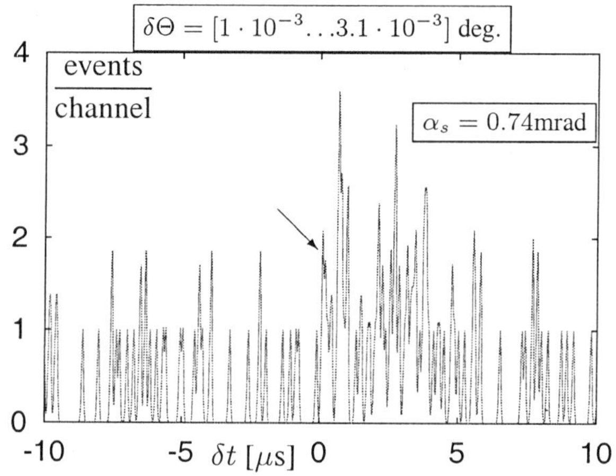

Figure 5.18: A time correlation spectrum with very small $\delta\Theta$, representing 6000 s of repeated scans over the interval $\delta\Theta = [1 \cdot 10^{-3} \ldots 3.1 \cdot 10^{-3}]$ deg. The original data are convoluted with a low-pass filter function (see text). The coincidence peak (arrow) is smaller than other, noncoincident peaks, and is most probably due to statistical fluctuations.

The data sets of this beam time show very small event rates, whose statistical significance is doubtful, at best. A sample correlation spectrum is shown in fig. 5.18, obtained by summing over repeated scans of $\delta\Theta = [1 \cdot 10^{-3} \ldots 3.1 \cdot 10^{-3}]$ deg., i.e., about the value of $1.4 \cdot 10^{-3}$ deg., expected by eq. (5.31). The original data were convoluted with the function $\exp(-(x/(50\text{ns}))^2)$, corresponding to the detector time resolution (see sec. 2.5). The background level is much higher than in the previous experiments because of the diffuse scattering from the sample, which is particulalry strong at small $\delta\Theta$. There are considerably more background correlations at positive δt, than there are at negative δt. This is due to to the high background count rates in the detectors, in combination with dead times and readout sequencing in the data acquisition electronics.

2.7.2 APS, 1-ID. The next beam time to study the extreme small-angle regime took place at the Advanced Photon Source (APS).[9] The setup is shown in fig. 5.19 and photos can be found on the CD in chapters/pdc/photos/sri/. The incident beam of 25 keV photons first hit a diamond plate in (111) orientation that diffracted a flux of about $3 \cdot 10^{10}$ photons per second upward towards the sample diamond crystal, placed ca. 12 mm after the first diamond. The sample, described in sec. 2.6, brought the (111)-diffracted beam back into the horizontal direction with a vertical displacement of about 3 mm. Use of a diamond as a first crystal and the close spacing were necessary because of the rather restrictive aperture sizes in the beam line (see below). In the absence of these restrictions, another material with more scattering power would have been used for the first crystal, such as InP in the experiments described in secs. 2.7.1 and 2.7.3.

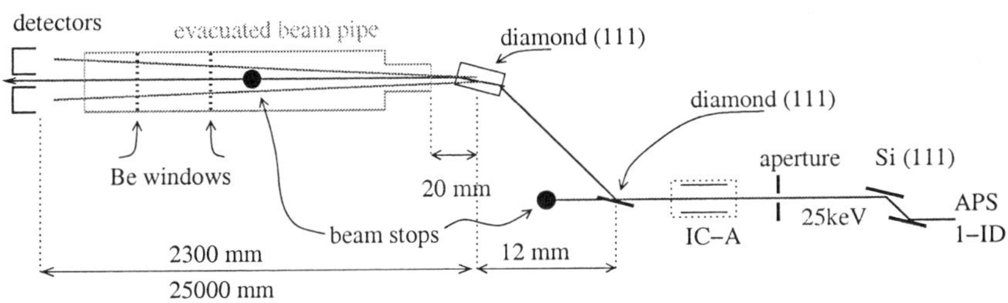

Figure 5.19: Experimental setup of the 1-ID beam line of SRI-CAT (APS) during the beam time of May 01.

Both diamond crystals were in the B hutch of beam line 1-ID. For the first part of the experiment, the detectors were also there at distances of 225 cm and 230 cm, respectively, from the sample. Figure 5.20 shows two time correlation spectra, taken at values of $\alpha_s = 5.3$ mrad and $\alpha_s = 3.6$ mrad, respectively.

In order to probe down conversion at very small α_s, the detectors were then placed into the C hutch, 24 m downstream from the sample. There is an evacuated beam pipe leading from B to C whose usual purpose is to transport the monochromatic beam to a sample in the C hutch. This beam pipe was used here to send the down-converted photons to the detectors. Due to small apertures in the beam pipe, the Bragg-diffracted beam from the sample had to be perfectly parallel and closely spaced to the line that the monochromatic beam usually takes through the beam pipe. This was possible only with a very close match of the lattice parameters in the two crystals—hence the choice of diamond for the first crystal and its proximity to the sample. The Bragg-diffracted beam from the sample had to be stopped before hitting the beryllium window at the entrance side of the beam pipe between the hutches, because it would have

[9]May of 2001, beam line 1-ID.

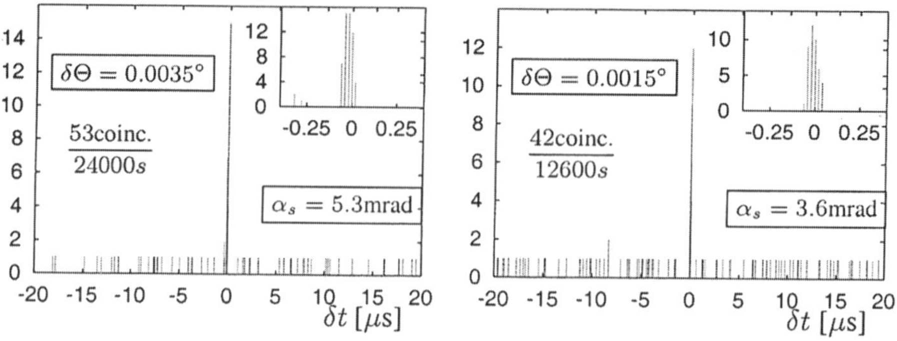

Figure 5.20: Time correlation spectrum for larger' $\delta\Theta$. Left: for $\alpha_s = 5.3$ mrad and integrated over $\delta\Theta = 0.0033° \ldots 0.0035°$ (denoted in the graph as "0.0035°") with a coincidence rate of $2.2 \cdot 10^{-3}\text{s}^{-1}$. Right: for $\alpha_s = 3.6$ mrad and integrated over $\delta\Theta = 0.00145° \ldots 0.00160°$ (denoted in the graph as "0.0015°") with a coincidence rate of $3.3 \cdot 10^{-3}\text{s}^{-1}$.

produced a disastrous background of small-angle scattering in the detectors. At that point, the down-converted beams were only about 1 mm apart.

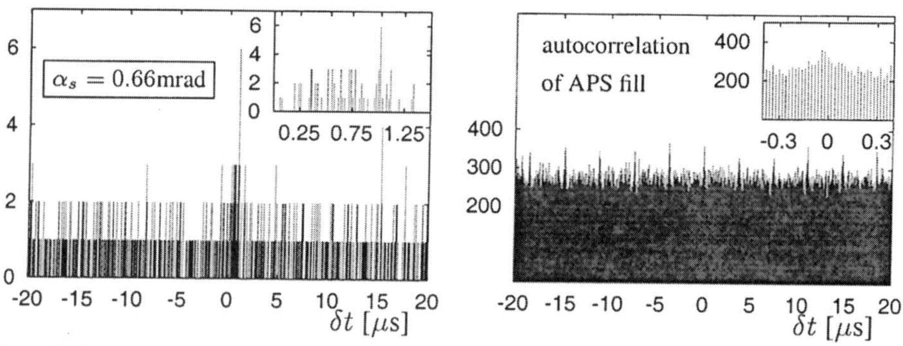

Figure 5.21: Left: time correlation spectrum for very small $\delta\Theta = 0.00045°$, representing 6000 s of data acquisition. Right: Autocorrelation of the APS fill pattern, showing that $\delta t = 0$ is no more than 100 ns away from zero in the graph. Therefore, the correlation peak at $\delta t \approx 1\mu\text{s}$ on the left-hand side is not a coincidence peak.

Figure 5.21 shows a time correlation spectrum for $\alpha_s = 0.66$ mrad and another one with a high background, which can be used for the purposes of calibrating zero time difference in the correlation in the following way: The APS fill pattern at the time of the experiment had six closely spaced bunches with 10% of the total current, showing up as narrow autocorrelation peaks in the time correlation spectrum. The one close to zero must be the real zero by the properties of the autocorrelation. It follows that the peak $\delta t = 1\mu\text{s}$ in the correlation spectrum on the left-hand side of fig. 5.21 is actually not due to down conversion, but is rather a statistical fluctuation. Therefore, no coincidence events above the background level are discernible in the small-angle data (and the same holds true for another data set, not shown here).

2.7.3 APS, 7-ID. The last down conversion beam time to date took place in July of 2002 at the 7-ID beam line (MHATT-CAT[10]) of the APS. Except for the event rate, which was much higher in the beam time described in sec. 2.4, this one gave (in my opinion) the best results so far. The setup was similar to the one used in the beam times of secs. 2.7.1 and 2.7.2 and is shown in fig. 5.22. The Si (111) double-crystal monochromator [136] of 7-ID was

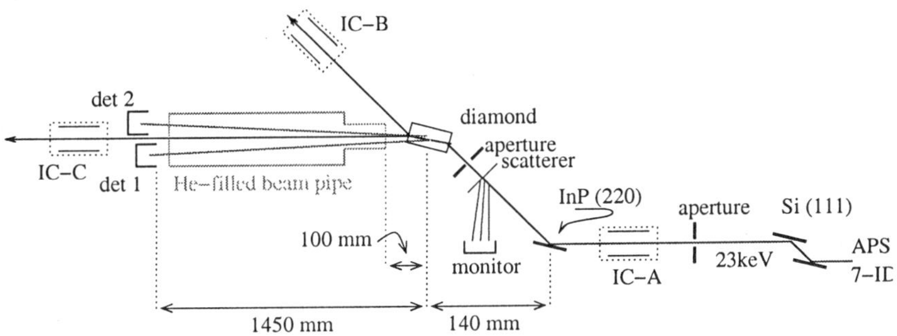

Figure 5.22: Experimental setup of beam line 7-ID (MHATT-CAT) of the APS for the beam time of Jul. 02.

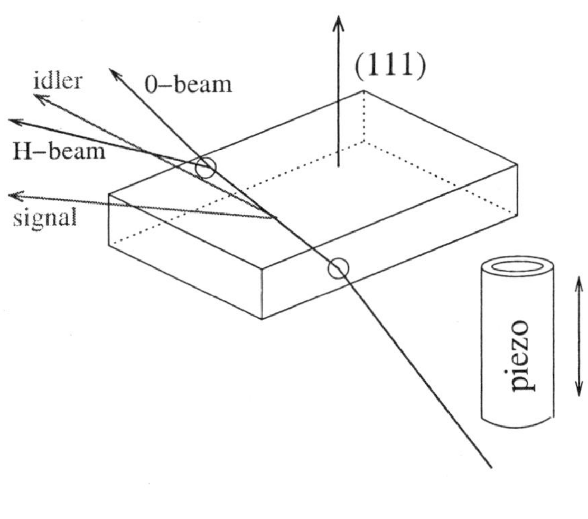

Figure 5.23: Closeup of the diamond sample. Compare with photo P0001526.JPG in the directory chapters/pdc/photos/mhatt of the CD.

adjusted to 23 keV. After passing through a 0.5*0.5 mm^2 slit and an ionization chamber (IC-A in fig. 5.22), the beam hit the cleaved (220)-oriented edge of an InP wafer at an intensity of about $1 \cdot 10^{11}$ photons per second. Thence,

[10]One of the Collaborative Access Teams of the APS, named after the contributors, Univ. of Michigan, Howard University and ATT (Lucent)

the upward-reflected x-rays encountered the (111) reflection (see fig. 5.23) of the sample diamond, 140 mm downstream. The advantagtes of InP over Ge for the first crystal were discussed in sec. 2.7.1. If the angular slope of the remaining mismatch between InP (220) and diamond (111) can be admitted by the apertures in the beam path, InP is also to be preferred over diamond (as used in the experiment of sec. 2.7.2), because diamond has a rather narrow rocking curve and reflects only a fraction of the flux offered by the monochromator (large diamonds are also harder to obtain). With the broader rocking curve of InP, long-term stability is also less of a concern. The InP crystal was mounted on a rotation stage (for the Bragg angle Θ_{InP}), which in turn was sitting on crossed translation stages, one moving horizontally in the transverse direction to position the InP crystal in the beam center and one horizontally along the beam to make the upward-reflected beam hit the diamond. The diamond was mounted only on rotation stages, one for the Bragg angle Θ_{dia} and one for the angle χ (about the direction of the beam line axis). The Bragg angles of both, the InP and the diamond crystal, could be fine controlled, each with a pair of piezoelectric actuators. Photos of the setup and further documentation can be found on the CD in the directory chapters/pdc/photos/mhatt. The piezos on the diamond sample were characterized by reflecting a laser beam off a small mirror, mounted next to the diamond (see photo P0001526.JPG in the directory chapters/pdc/photos/mhatt on the CD). Their response to an applied voltage is somewhat nonlinear, but a rough conversion factor can be given as 1 μrad per unit of the control signal (with 255 units corresponding to 500 V).

After going 100 mm through air, the refracted beam passed through a 5-μm-thick Mylar foil into a beam pipe filled with He at ambient pressure. At the other end of that beam pipe were two Si drift detectors [132] at distances of 145 cm and 147 cm from the diamond. placed such that the Bragg-reflected beam from the sample could pass between them and then on through an ionization chamber (IC-C in fig. 5.22). Their spacings from the Bragg-reflected beam were varied to select different angles α_s and α_i in symmetric ($\alpha_i = \alpha_s$) and asymmetric ($\alpha_i \neq \alpha_s$) configurations. The intensity in the 0-beam from the diamond (i.e., the beam diffracting in the diamond and continuing straight upwards) was measured with another ionization chamber (IC-B in fig. 5.22). Given the current-vs.-photon flux calibration curve of the ionization chambers and taking absorption within the diamond into account, this provided a quantitative measure of the intensity incident on the diamond. A piece of cardboard in the upward-directed beam from the InP crystal scattered a small fraction of the radiation, some of it picked up by a scintillation detector, which was used to determine the angle of maximum reflectivity of the InP crystal. The ionization chamber IC-B could not be used for that purpose because diffraction in the diamond would distort the measured reflectivity curve.

The data were recorded with an advanced data acquisition circuitry [137] (see also documentation in the directory chapters/pdc/equipment/evlog/ on the CD). It consists of five analog-to-digital converters (ADCs) and a digital event logger capable of recording each event separately with a time stamp, an eight-bit word from the respective ADC, representing the photon energy, and also some diagnostic information. This raw data can then be evaluated off-line to obtain energy-selective time correlation spectra. Only two of the five channels were actually used for the recording of data and two more for diagnostic purposes.

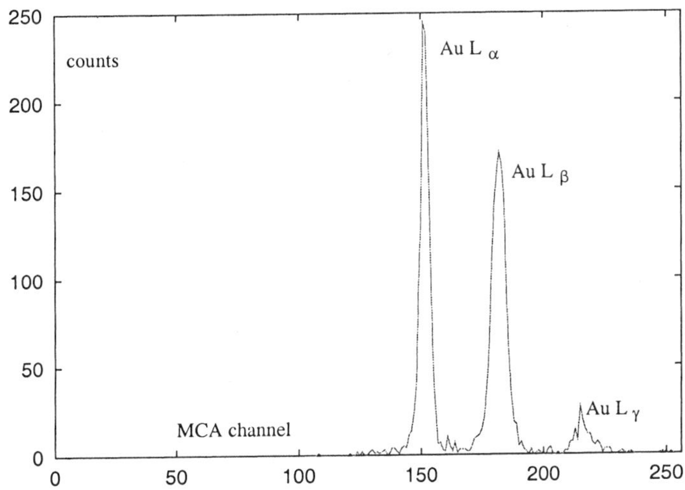

Figure 5.24: Energy calibration with Au fluorescence.

When the time information is discarded, one obtains an energy spectrum (commonly called an MCA spectrum) of the events in the respective detector. To calibrate the energy scale, one may take an MCA spectrum of some known x-ray fluorescent lines, such as Au in fig. 5.24.

Using the reference spectrum, one may now define limits on the photon energies and their sums to be admitted as a correlation event in the data evaluation. If only individual channel energy limits are imposed, the resulting time correlation spectra are equivalent to those obtained with single-channel analyzers (SCA) in the previous experiments. However, given the lack of live signal on which to calibrate SCAs, the possibility of deferring the decision on the limits to the data evaluation phase takes a lot of tension out of the experiment. Long-term drift of components, such as amplifiers, is also much less of a problem with the complete data sets from the event logger. With the possibility of imposing more complex filtering criteria for events to be admitted into the correlation spectra, one may obtain information that would be inaccessible with the simple SCA-based correlator. This proved important in some of the data shown below, where the individual energy bandwidths of the down-converted photons increase rapidly with decreasing α_s and α_i. This calls for a sum limit for the photon energies.

Data were taken for several settings of α_i and α_s (see fig. 5.25). In each of the graphs, a normalized event rate in down conversions per 10^{15} pump photons is given. The pump photon flux was obtained from the ion chamber readings for each data set separately. This individual normalization was particularly important during the first half of the beam time when the APS was running in the non-top-up mode,[11] and was therefore providing a varying flux of pump photons. The data exhibit two telltale signs of the edge-enhancement effect, namely an increase in the event rate and an increase in the energy spread of the individual photon energies, as α_s and $\delta\Theta$ become smaller. Actually, the increase in the event rate in fig. 5.25 is only partially due to the edge-enhancement and is partially due to the fact that, for decreasing $\alpha_{i,s}$, the detector apertures cover larger fractions of the cones (see fig. 5.1) into which the down-converted photons go. However, the edge-enhancement shows up in the increase of the normalized event rates (shown in framed boxes inside the graphs of fig. 5.25) beyond proportionality to $1/\alpha_{i,s}$ and in the increase of the energy spread (left inserts in fig. 5.1).

As expected, a dependence of the energy spread on $\delta\Theta$ with fixed $\alpha_{i,s}$ was not apparent (see fig. 5.26, right-hand side). The asymmetric geometry with $\alpha_i \neq \alpha_s$ was also tried once, with the expected result of an asymmetric energy splitting ration between the signal and idler photons. To test for possible artefacts in the time correlation spectra, data were taken at an intentionally wrong setting of $\alpha_s = 9.0$ mrad with $\delta\Theta = 0.005$. As expected, no coincidences were observed.

2.7.4 Suppression of Down Conversion at Small Angles. The reduction of the down conversion rate can be explained by two effects. The first of these can take away up to 3/4 of the down conversion rate and is based on dynamical diffraction. The other effect is purely geometrical in nature and is related to the edge-enhancement ("running against the edge").

As $\alpha_{i,s}$ become smaller, the pump wave enters the regime of dynamical diffraction. In the case shown in fig. 5.27 on the right-hand side, the incident pump wave has excited modes on both branches of the dispersion surface at the symmetry points. Only the waves with wave vectors $\mathbf{k}_0^{(\alpha)}$ and $\mathbf{k}_0^{\beta}$, each of which has one-half of the amplitude of the incident pump wave $(\mathbf{k}_p, \omega_p, \mathbf{A}_p)$, are relevant as pump waves because the coupling of the diffracted

[11]Most of the time, the APS runs in a mode in which the stored current is kept constant by injecting a small amount of charge every 2 minutes. Some of the beam time is, however, scheduled in non-top-up mode, with the current following the "sawtooth" pattern, as in other storage rings (ESRF, etc).

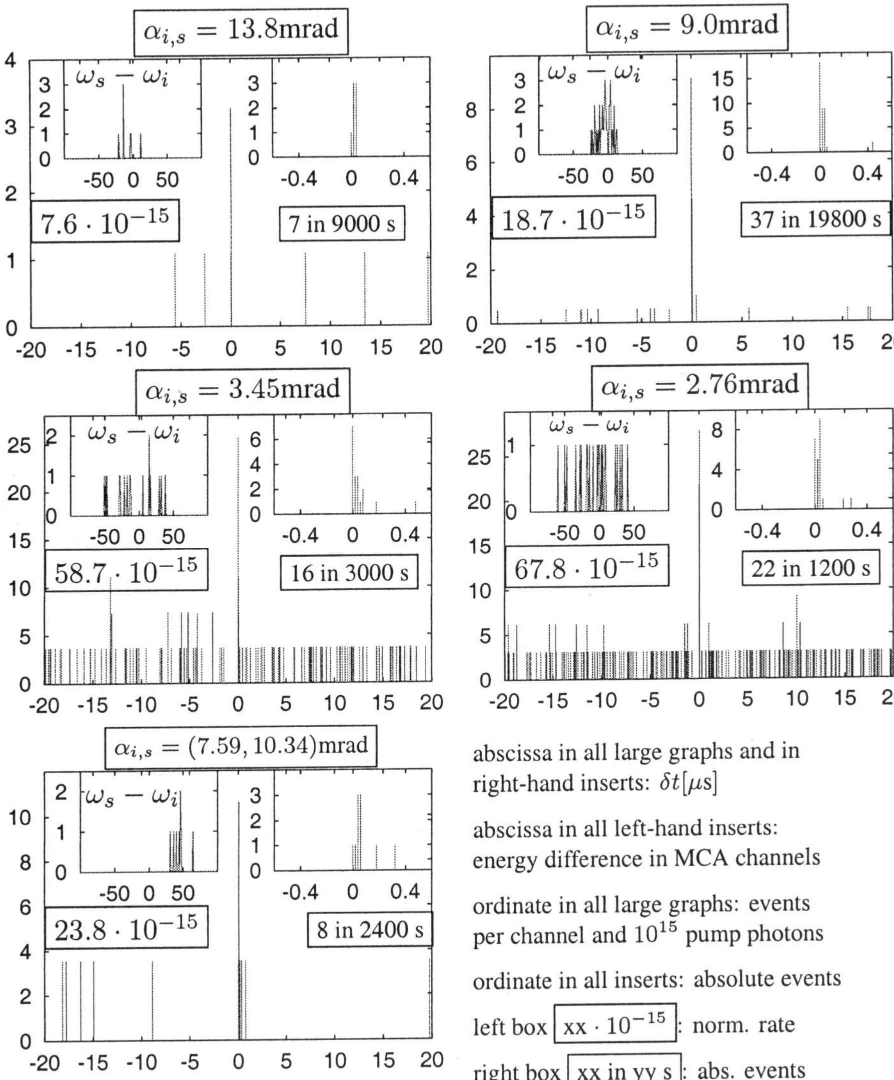

Figure 5.25: Time correlation and energy spread for several α_s. The main graphs give the correlation rate per 10^{15} pump photons over the time difference in μs, the right inserts are zooms into the coincidence peaks, given in absolute observed events, and the left inserts give the energy differences of the events in the respective coincidence peak in units of MCA channels (see fig. 5.24). The box to the left of the peak in each graph gives the coincidence rate per pump photon, integrated over the peak, and the box to the right of each peak gives integrated absolute coincidences within the respective data acquisition time. The upper four graphs show a sequence of decreasing $\alpha_s = \alpha_i$, with increasing normalized event rate and the energy spread of the individual down-converted photons (edge-enhancement, fig. 5.16). The lowest graph shows an asymmetric case $\alpha_s \neq \alpha_i$. Corresponding to the angles (α_i, α_s) of 13.8 mrad, 9.0 mrad, 3.45 mrad, 2.76 mrad and (7.59,10.34) mrad were detuning angles $\delta\Theta$ of 0.022°, 0.0097°, 0.00186°, 0.0008° and 0.01°, respectively.

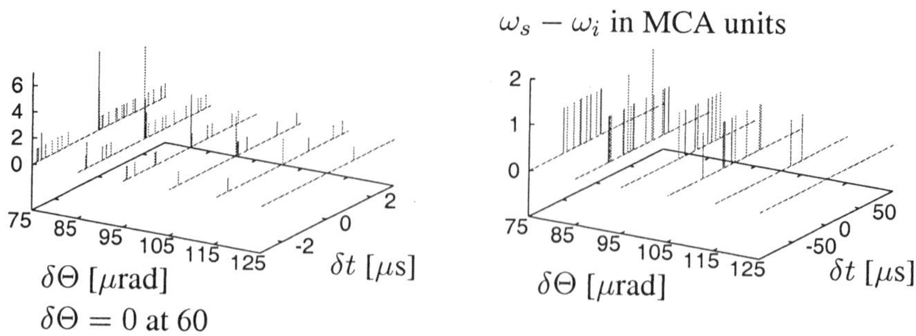

Figure 5.26: Time correlation spectra and energy spread for small $\delta\Theta$. Left: time correlation spectra for several values of the piezovoltage (in units of the digital-to-analog-converter driving the piezo), corresponding roughly to μrad, the Bragg peak is at 60. Right: difference of the down-converted photon energies in units of the MCA (see fig. 5.24). The time correlation spectrum and energy spread at a piezovoltage of 85 units are identical to those shown in the third graph (1 col., 2nd row) in fig. 5.25.

waves $(\mathbf{k}_\mathbf{H}^{(\alpha,\beta)}, \omega_p, \mathbf{A}_\mathbf{H}^{(\alpha,\beta)})$ to the down-converted modes is negligibly small (see ch. 2, sec. 1 and ch. 3, sec. 4.3). However, only the latter of the two pump

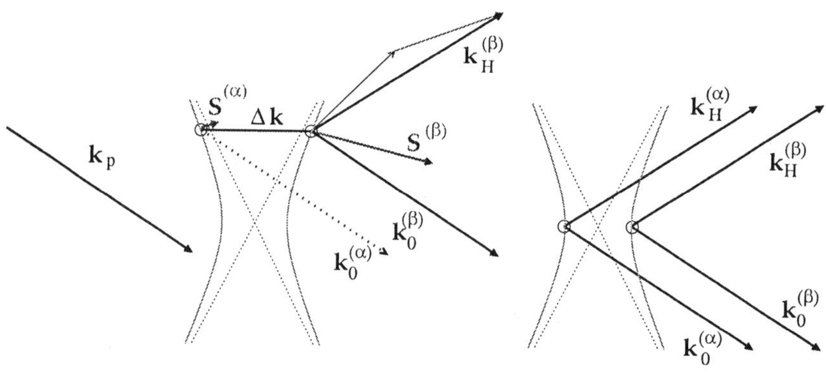

Figure 5.27: Dynamical diffraction of the pump wave in down conversion. Left: far off the Bragg condition, the wave $(\mathbf{k}_\mathbf{0}^{(\beta)}, \omega_p, \mathbf{A}_\mathbf{0}^{(\beta)})$ is almost equal in amplitude and intensity to the incident pump wave $(\mathbf{k}_p, \omega_p, \mathbf{A}_p)$ and dynamical diffraction effects need not be considered. Right: the incident pump wave meets the Bragg condition exactly. Of the four waves, only $(\mathbf{k}_\mathbf{0}^{(\beta)}, \omega_p, \mathbf{A}_\mathbf{0}^{(\beta)})$ is active as a pump. Its amplitude is one-half and its intensity 1/4 of that in the incident pump wave $(\mathbf{k}_p, \omega_p, \mathbf{A}_p)$.

waves $(\mathbf{k}_\mathbf{0}^{(\alpha)}, \omega_p, \mathbf{A}_\mathbf{0}^{(\alpha)})$ and $(\mathbf{k}_\mathbf{0}^{(\beta)}, \omega_p, \mathbf{A}_\mathbf{0}^{(\beta)})$ fulfills the wave vector matching condition. It has half the amplitude, i.e., one-quarter the intensity of the incident pump wave in it, and therefore leads to one-quarter the conversion rate one would expect from the incident pump in the absence of dynamical diffraction effects. The down-converted amplitudes generated by $(\mathbf{k}_\mathbf{0}^{(\alpha)}, \omega_p, \mathbf{A}_\mathbf{0}^{(\alpha)})$ are eliminated by destructive intereference within one pendellösung period of the pump because they "inherit" from it the difference wave vector $\Delta\mathbf{k}$, which determines the pendellösung period and at the same time represents the wave vector mismatch to the mode on the β branch that does fulfill the wave vector matching condition. Other than with linear diffraction, where photon number is conserved and destructive interference in one direction implies that the photons go somewhere else, destructive interference in down conversion means that the signal and idler photons are not generated at all.

Equivalently, one may argue that the pendellösung effect makes the energy flow of the pump wave within the crystal flip back and forth between the directions of the $\mathbf{k}_0$ and $\mathbf{k}_H$. Therefore, the pump wave is present in only half of the crystal volume, where the pendellösung goes in the direction of $\mathbf{k}_0$ (see above). This argument seem to predict a suppression by only one-half, in contradiction to the above prediction of up to 3/4 suppression. One must, however, also take into account the fact that the parts of the crystal volume that are active in down conversion by the above argument are spaced from each other by the pendellösung period and must contribute in phase with each other. The resulting interference condition accounts for another factor of one-half.

In the experiments shown in sec. 2.7, the suppression of down conversion due to dynamical diffraction is not as pronounced as described above because even with $\alpha_{i,s} = 0$ and diamond (111), the dynamical diffraction parameter η has a value of $\eta \approx 4/3$ (see app. 6.4.2), instead of $\eta = 0$ at the symmetry points of the dispersion surface.

It is therefore mostly due to the following geometric effect that the down conversion rate is so low in the extreme small-angle regime: As $\alpha_s \to 0$, a smaller and smaller range $\Delta\Theta$ corresponds to a given angular acceptance $\Delta\alpha_s$ of the detectors, as illustrated in fig. 5.28. As long as the divergence of the pump

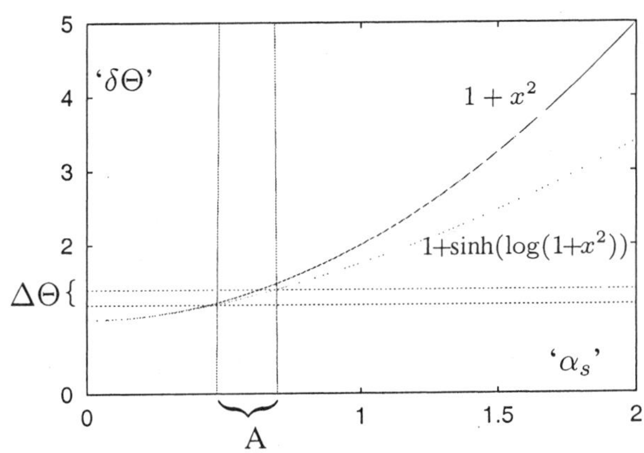

Figure 5.28: Limit of the edge-enhancement effect: plots of the functions $1+x^2$ and $1+\sinh(\log(1+x^2))$, representing $\delta\Theta$ versus $\alpha_s = \alpha_i$, according to eqs. (5.1) and (5.31).

wave, as offered by the source, is less than $\Delta\Theta$, all pump photons can contribute to the effect seen by the detectors. However, as $\alpha_{i,s} \to 0$, the source begins to overfill $\Delta\Theta$, and the observed event rate goes down. The pump divergence is comparable to the diamond rocking width in the experiment of sec. 2.7.2 (because the first crystal is also a diamond) and is much larger than the diamond rocking width in the experiment of sec. 2.7.1 (Si (111) monochromator, InP for the first crystal). In contrast, for both experiments, $\alpha_s^2/2$ is much less than $(3/2)\Gamma F_0$ in eq. (5.31). Therefore, the pump divergence overfills the angular range of α_s, and even more so the detector apertures (which are a fraction of α_s).

2.8 Wave Vector Matching by Dynamical Diffraction

The wave vector matching scheme shown in fig. 3.16 is not the only one possible. One might, for example, make use of the effective refractive indices of dynamical diffraction for the pump and the down-converted waves, as described below. Because this scheme requires Bragg's law to be fulfilled for all three of the waves involved, it is very restrictive and the energies and momenta of signal and idler are defined to within typically one part in 10^6, making them suitable for handling with crystal-based x-ray optical elements. The downside is that the number of final states for a down conversion is considerably lower than with the other wave vector matching scheme (fig. 3.16), and therefore the expected event rate is extremely low. I have tried this wave vector matching scheme several times, so far without success. The event rate to expect is much lower than with the wave vector matching scheme shown in fig. 5.1, because the energy splitting ratio is restricted to 1 part in 10^5 to 10^6. Because the down-converted photons are so well defined, it is however very attractive for applications that require coherent manipulation of the down-converted photons, such as tests of the quantum theory (see sec. 3.1).

Before going into the details of the wave vector matching, let us look at the cast of characters. There are now six waves directly involved in the nonlinear diffraction problem, namely pump, signal and idler, and also their diffracted counterparts. Again, the diffraction of the pump wave is essentially governed by the first line of eq. 3.63, the amplitudes of the down-converted waves being too weak to substantially influence it. For the signal and idler waves, the first line of eq. (3.63) describes the linear diffraction, and the other two give the nonlinear coupling to the pump. What are the relative weights of the linear and the nonlinear diffraction terms for the signal and idler waves? The answer is not immediately obvious because the nonlinear coupling is very small, but the pump wave is much stronger than the signal and idler. For an order-of-magnitude estimate, the ratio of nonlinear-to-linear terms is given by the expression $(r_e/q)|\mathbf{A}_p|$, where $|\mathbf{A}_p|$ is the amplitude of the pump wave, calculated in app. 6.1. Insertion of eq. (5.9) into that expression gives

$$\frac{r_e^2}{q^2}|\mathbf{A}_p|^2 = \frac{137}{3\pi}k_p dk_p r_e^2, \tag{5.3}$$

where $\hbar c/q^2 = 137$ was used and $k_p = \omega_p/c$ and $dk_p = \delta/c$ are the wave number of the pump and its spread. Even if an unrealistically large $dk_p = k_p$ is admitted (it is rather like $dk_p = 10^{-6}k_p$, see above), eq. (5.3) yields a value of $1.26 \cdot 10^{-7}$ at a pump photon energy of 24.8 keV ($\lambda = 0.5$ Å). Therefore, the dynamical diffraction of the signal and idler waves is governed by the linear diffraction terms in eq. (3.63), and the nonlinear coupling to the pump wave can be treated as a small perturbation.

Figure 5.29 shows a possible configuration. In the example shown here,

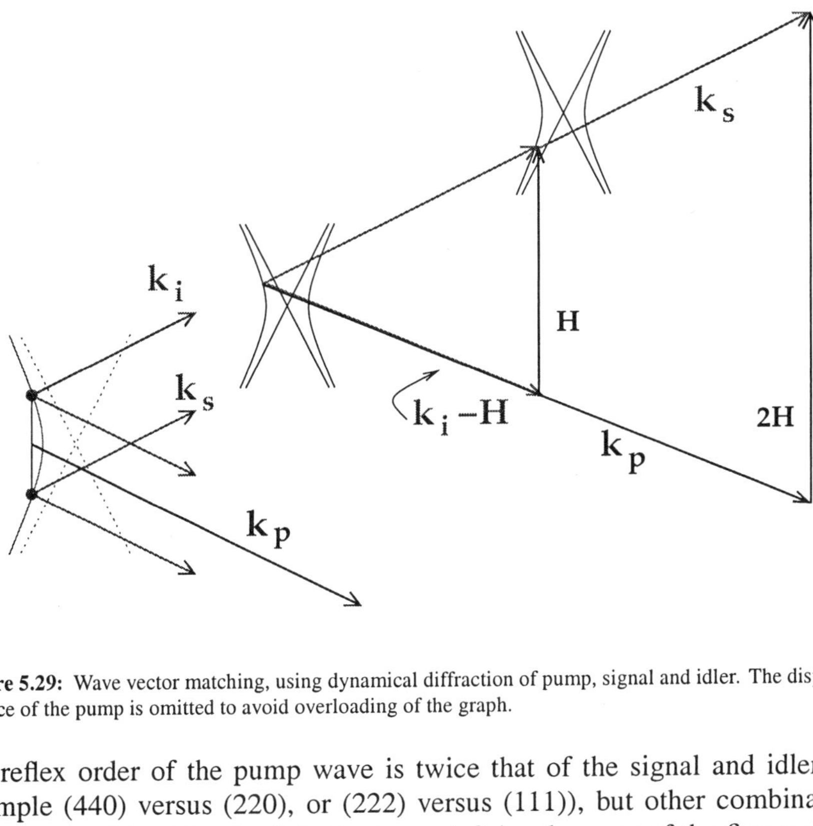

Figure 5.29: Wave vector matching, using dynamical diffraction of pump, signal and idler. The dispersion surface of the pump is omitted to avoid overloading of the graph.

the reflex order of the pump wave is twice that of the signal and idler (for example (440) versus (220), or (222) versus (111)), but other combinations are also possible. The detail in the lower left-hand corner of the figure shows the α branch of dynamical diffraction for the signal and idler waves and the pairs signal/diffracted signal, idler/diffrated idler. For the pump, no dispersion surfaces and only one wave vector are shown. The pump wave vector should be scaled down by a factor of $1/2$ because the dispersion surfaces of signal and idler must both be matched simultaneously. The scaled-down pump wave vector generally does not precisely hit the α branch of the dispersion surface for signal and idler. In most cases, it lies to the left of it, and the wave vectors are matched when the signal and the idler occupy different tie points on the dispersion surface, such that they split symmetrically away from the scaled-down pump wave, as shown in the figure. In particular cases, the dispersion splitting is strong enough to make the β branch of the scaled-down pump wave vector lie partially to the right of the α branch for signal and idler. In that case, signal and idler will occupy the same tie points at the intersection points of the scaled-down β-branch for the pump with the α-branch for signal and idler. For this intersection of pump-β with signal/idler-α to occur (not shown in the figure), the atom form factor may not depend too strongly on the momentum transfer. A more detailed discussion would go beyond the scope and intent of this section, so suffice it to say that this is possible with Si, Ge, GaAs, etc., but not with diamond.

3. Potential Applications

With event rates of the order of 5/s, it may be a bit bold to think of applications of x-ray PDC but nevertheless, a few possibilities shall be discussed here. Because the photons that do not undergo down conversion are left unchanged, a down conversion experiment could be integrated into a beam line and run in "parasitic mode."[12] This is discussed further in sec. 3.3.

3.1 Tests of the Quantum Theory

Parametric down conversion in the visible light regime is a prime source of entangled photons in experiments to test the quantum theory and in applications, such as quantum communication. In all these experiments, the quantum efficiency of the detectors plays a crucial role. Any lack thereof is equivalent to the coupling of a carefully prepared pure quantum state to external degrees of freedom ("heat bath"). The best commercially available visible light photon detectors reach roughly 70% efficiency [138]. In contrast, x-ray detectors can easily achieve better than 99%, the difference to 100% being mainly due to incomplete charge collection, as evident in fig. 5.34. Before x-rays can become a serious competitor in tests of the quantum theory, other aspects of the experiment must be addressed, too. It does not help to have a high detection efficiency but lose the photons in optical components. This was the intention of the mirror-suppression experiment described in sec. 2.6, but there is still a lot of work ahead. Instead of the wave vector matching scheme employed in all of the successful XPDC experiments (see fig. 5.1), it would be better to make use of a scheme based fully on dynamical diffraction, such as the one of fig. 5.29 because there, a well-defined relationship between the energies and momenta of the down-converted photons exists. This would greatly facilitate the use of low-loss diffractive x-ray optics.

---

[12] This unfriendly term was used for the first synchrotron radiation experiments on particle physics accelerators.

3.2 Sub-Poisson Absorption Spectroscopy

In the proposed application for tests of the quantum theory, the photons themselves are the object of investigation. In the application described here, namely x-ray absorption spectroscopy (XAS) of samples with little radiation tolerance, the sample is a conventional piece of matter, and the correlations between the down-converted photons are used to obtain information on it. The cross section in absorption spectroscopy is determined from the intensity transmitted through the sample, normalized with the incident intensity. Whereas the former can theoretically be measured with 100% quantum efficiency, the latter can be known only within the limits of Poisson statistics: some of the incident intensity must be split off and directed to the monitor detector. Regardless of how this is done (ionization chamber, scattering foil, etc.), the process of distributing the incident photons among the ones that go to the monitor and those that go to the sample is subject to statistical fluctuations. With correlated photons, one may instead use one for intensity monitoring and the other one for the actual measurement. Because of the low cross section of XPDC, this idea can never be competitive in applications where the statistical noise level is determined only by the available intensity. However, in applications where the radiation tolerance of the sample is the limiting factor, the use of correlated photons in the way described above can help to improve the signal quality at a given radiation dose. Particularly with a dilute elemental species, a large radiation dose is necessary to obtain a signal above the Poisson shot noise and a sub-Poisson statistic in the incident intensity can reduce the radiation dose drastically.

3.3 Integration into a Beam Line

Because of the low conversion cross section, it would be rather wasteful to dedicate a beam line to an application of XPDC. Luckily, such an experiment could be integrated rather transparently into a beam line whose main purpose is something altogether different. Figure 5.30 shows a schematic of a possible setup. A diamond crystal is placed into the beam, right after the monochromator. Some of the photons passing through will undergo down conversion, some will be diffracted out of the main beam, some will be absorbed, but the huge majority will just go on to the main experiment. Losses due to diffraction and absorption depend very much on details. If $\delta\Theta$ is made much larger than the rocking width, diffraction losses are negligible. To minimize absorption, the down conversion experiment should be run with photon energies above ca. 20 keV. Due to dynamical diffraction, the forward-diffracted beam is displaced: inside the crystal, the direction of energy flow is somewhere between the directions of $\mathbf{k_0}$ and $\mathbf{k_H}$ (see fig. 3.4), depending on the detuning angle $\delta\Theta$. This displacement can be kept smaller than the beam diameter if $\delta\Theta$ is made larger than a few rocking widths, such as in secs. 2.3 and 2.4.

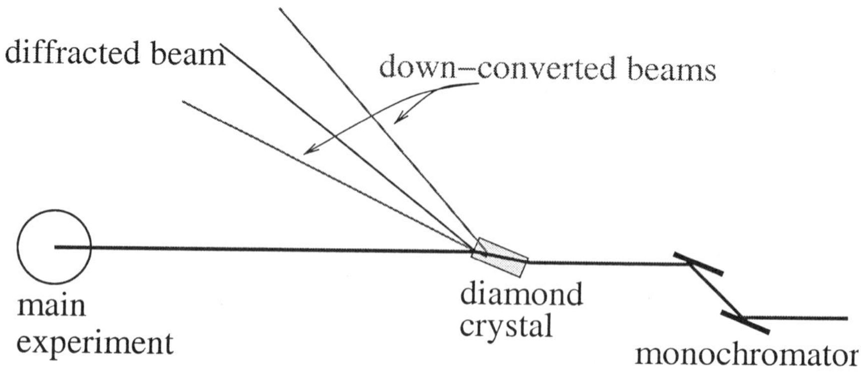

Figure 5.30: Integration of an XPDC experiment into a beam line. A diamond crystal is inserted into the beam going towards the main experiment. Some of the photons passing through it undergo down conversion. To keep the disturbance of the main experiment to a minimum, the detuning angle $\delta\Theta$ should be larger than a few rocking widths of the diamond (see text).

4. Experimental Issues

Due to the low cross section for the down conversion of x-rays, a number of experimental issues must be given careful consideration. The most important ones are stability of conditions, such as Bragg angles against long-term drift and the suppression of the statistical background. The highest event rate of x-ray parametric down conversion ever published is a bit less than one event in ten seconds. It takes therefore at least a good part of one hour to collect a significant number of events, and, obviously, Bragg angles of the monochromator and the sample must be stable throughout the data collection time. Unfortunately, x-ray monochromators on synchrotron radiation beam lines tend to drift by millidegrees over several hours, especially if the heat load on them changes due to the typical sawtooth pattern of the storage ring current. It is therefore important to have a monitor detector for every component that may drift, so that the alignment can periodically be verified and corrected (see fig. 5.22).

4.1 Background Suppression

Due to the small conversion cross section, the signal to be measured is very weak, and, therefore, suppression of background, i.e., stray radiation and electronic noise, is a major issue. A combination of several telltale signs can be used to distingiush down-conversion events from the background: The two down-converted photons must appear simultaneously, they must add up in energy to the pump, and the effect must show the right dependence on the wave vector matching. There is always a tradeoff between sensitivity towards various aspects of the signal: the more restrictive the wave vector matching condition is (which is good for a clear signal contrast), the fewer photons get to the detectors (which is bad due to shot noise) and the better the time resolution of a detector, the worse is generally its energy resolution.

An important distinction to make in time-correlation experiments is whether the noise in the detector signal chains is correlated or not. The former is mostly due to electric noise induced by strongly varying currents in power lines. The latter can be due to electronic noise, generated independently in the detector signal chains, and to stray radiation, statistically hitting the detectors. These types of background signal are discussed in detail below. A generic description of the time-dependent signal from a detector (A) whose output is to be correlated with another (B) is

$$S_i(t) = U_i(t) + D_i(t) + C_i(t) + T_i(t), \qquad i = A, B \tag{5.4}$$

where U means truly uncorrelated noise, for example generated within the signal processing electronics. D is the rate of statistical events that are derived from a common source. This might be photon shot noise, where the photons originate from a source that has a temporal modulation, such as the bunch structure of a synchrotron radiation source. C is the deterministic correlated noise, such as electric stray signals generated by power electronics, which couple into both signal chains (to be avoided at almost all cost). Finally, T is the true signal, such as the down conversion. C_A might differ from C_B due to different degrees of suppression, and T_A might differ from T_B due to absorption losses, etc. A time correlation spectrum will then show

$$\begin{aligned} R(\tau) = & \int U_A(t)\,[U_B(t+\tau) + D_B(t+\tau) + C_B(t+\tau) + T_B(t+\tau)]\,dt + \\ & \int D_A\,[C_B(t+\tau) + T_B(t+\tau)]\,dt + \int C_A(t) T_B(t+\tau) dt \\ & \int D_A(t) D_B(t+\tau) dt + \int C_A(t) C_B(t+\tau) dt + \int T_A(t) T_B(t+\tau) dt, \end{aligned} \tag{5.5}$$

where the integration is to cover the data acquisition interval. The first two lines contain only products of mutually uncorrelated rates and therefore produce a background which is statistically constant in τ ("uncorrelated noise") in fig. 5.31. The three integrals in the third line refer to the temporal correlations between signals of types D, C and T (defined above) in the two detectors which correspond to autocorrelation, corr. noise and down conv. in fig. 5.31.

4.2 Electric Noise

Electric noise is defined here as radiofrequency waves emitted from power lines that carry strongly varying currents. It is a particular problem in coincidence experiments because it generally couples more or less equally into the

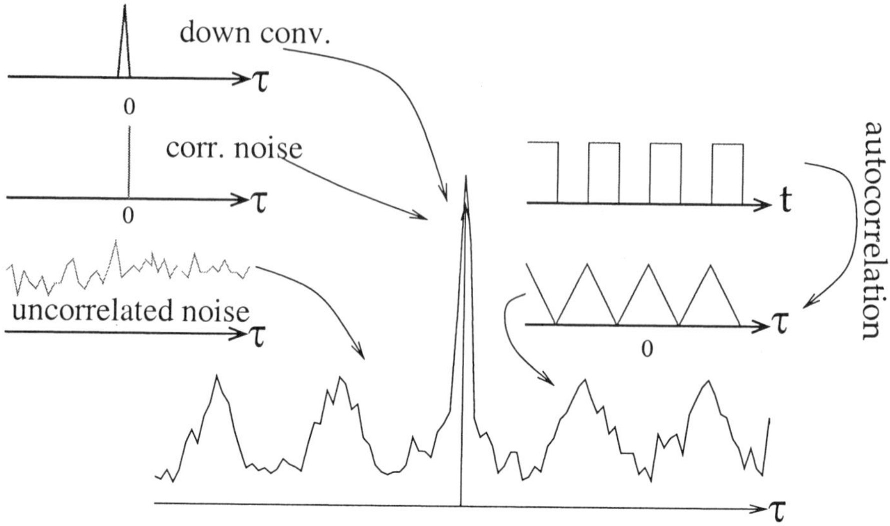

Figure 5.31: Contributions to a time correlation spectrum. Compare with fig. 5.10.

signal chains of both detectors and can thus generate false coincidences. Synchrotron radiation laboratories are electrically very noisy environments. There is a lot of machinery that operates on strongly varying currents, the worst being in most cases stepping motor drivers. A typical beam line at a third-generation facility contains tens to hundreds of stepper motors, necessary for remote control within the radiation shielding. In most cases, switched-current drivers are used, i.e., the motor phase currents are generated by switching a large current rapidly on and off. This saves power over truly continuous current drivers, such as the ones used at HASYLAB at DESY, but it is surpassed as a source of electric noise only by the servomotor drivers, which are found on a few beamlines. Due to x-ray shielding requirements, the signal cables from detectors to the data acquisition electronics are rather long (typically tens of meters) and can easily pick up the electrical noise. Using doubly shielded coaxial cables sometimes helps a bit, but more important is the elimination of ground loops, i.e., multiple electric ground references for the signals.[13] An example of an actually relatively harmless electric stray signal induction from a stepper motor is shown in fig. 5.32. The oscilloscope screen shot was obtained during the beam time described in sec. 2.7.3 and shows the output of one of the Si drift detectors [132], superimposed on two of the bursts that occur every 50 μs when the current in a stepping motor is switched while the motor keeps its position.

[13]On a personal note, I believe that issues such as noise suppression, shielding, grounding are receiving too little attention in synchrotron radiation laboratories.

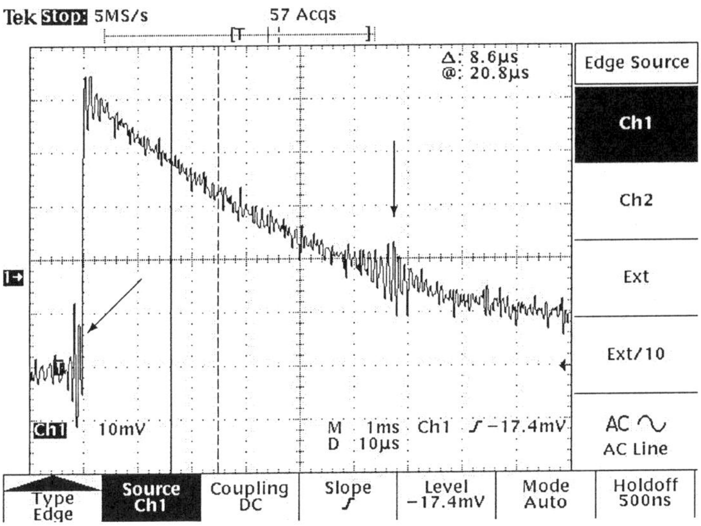

Figure 5.32: Detector output signal, stepping motor noise (arrows) superimposed on a pulse due to the detection of an x-ray photon. To achieve this rather harmless noise amplitude, some effort had to be expended at proper grounding.

The first burst comes fortuitously at the time when the detector output rises and the other one slightly to the right of the center of the screen. This is *not* the switching of currents when the motor is actually moving.

It is of utmost importance, especially for a coincidence measurement, to reduce the electric noise. This means (i) turning off stepper motors or filtering the phase currents, (ii) not using any servomotors at all, (iii) eliminating ground loops by making sure that the ground of the signal lines is distributed strictly from one point only (which is a major challenge in practice) and (iv) using well-shielded signal cables with solid ground. It may be advisable to place the main signal amplifiers as close as possible to the detectors and power them from the same AC power outlet (see (iii) above), so that the electric noise induced along the relatively long signal lines to the outside of the shielding hutch is not amplified with the signal. This is however not always a good choice (see discussion of this point in ch. 6, sec. 6.2, and in [139]) because very low and very high frequency noise is actually suppressed in a spectroscopic amplifier.

4.3 Stray Radiation

The highest event rate of parametric down conversion of x-rays that was ever published is a bit less than one event in ten seconds—at an incident x-ray flux of over 10^{12} photons per second [114]. This huge ratio illustrates the need to catch as much as possible of the stray radiation caused by the incident flux. Because the probability of a false coincidence is proportional to the square of the stray radiation background in the detectors, much in signal-to-background ratio can be gained by careful suppression of the stray radiation background. This shall be illustrated by looking in detail at one of the more complicated experimental setups, used in recent experiments on parametric down conversion, fig. 5.22. The x-rays, coming from the monochromator, are restricted to a beam of a cross section of 0.5*0.5 mm^2 to reduce the number of photons entering the experimental hutch but, for reasons of geometry, not eventually hitting the sample. The first solid object they hit is the InP crystal, which reflects the beam upward. Apart from the lattice spacing of the InP (220) reflection being matched very well to the diamond (111) reflection, InP produces no appreciable x-ray fluorescence at 22 keV incident energy: the threshold for the excitation of In K fluorescence is at 27.94 keV and the lower energy fluorescence (i.e., indium L and phosphorus K) is of no concern because it is almost totally absorbed by the air between the InP crystal and the detectors. Another material having a lattice spacing that is well matched to diamond is Ge. However, its K fluorescence at 9.9 keV and 11 keV is uncomfortably close to the expected energy of the down converted photons (here, 11 keV) and is not strongly absorbed in air. The scatterer in the upward-reflected beam is necessary to get enough intensity for the monitor detector, and there is also some amount of scattering from the air along the beam path. The purpose of the aperture just before the sample is to prevent radiation from the scatterer and upstream air paths from reaching the detectors.

The pump beam then enters the sample crystal. Some of the pump photons are scattered in it towards the detectors by Compton scattering, and static and thermal diffuse scattering. Not much can be done about Compton scattering, except perhaps the choice of a sample composed of heavier elements than carbon, which would however absorb more strongly. Diffuse scattering is particularly serious for small $\delta\Theta$. It might be reduced by a proper choice of sample crystal and by cooling it. None of these measures have been tried so far.

After leaving the sample, the Bragg-diffracted pump beam enters the beam pipe that must be installed to mimimize absorption of the down-converted photons. If evacuated, that pipe must have an entrance window, and, even in the case of a He-filled pipe, one would be hesitant to go without one.[14] Particularly for samll $\delta\Theta$, the Bragg-diffracted beam from the sample is strong and can produce a high level of background by small-angle scattering in any of the beam path windows it encounters. Therefore, windows are to be avoided or the Bragg-diffracted beam must be stopped before hitting them. The entrance window to the beam pipe was made of 5-μm-thin Mylar in all of the experiments described from sec. 2.4 onward, except for sec. 2.5. In those experiments

[14]Actually, there was no entrance window in the case of the first experiment, described in sec. 2.3; the upward tilt of the beam pipe helped to keep the helium inside.

in which the beam pipe was evacuated, the Mylar window had to be small in diameter (5 mm).

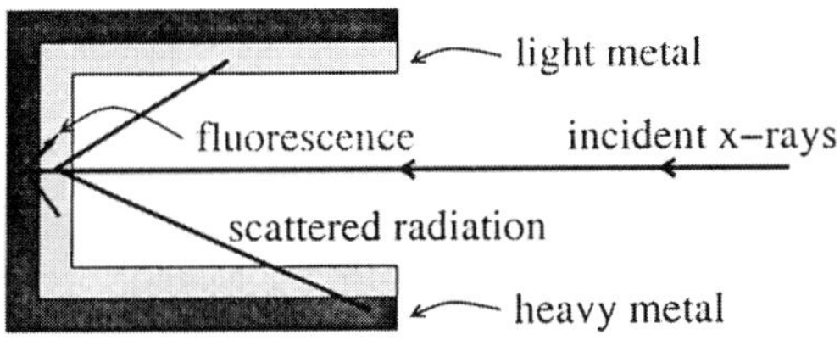

Figure 5.33: A beam stop for low background, showing a combination of two techniques. Sandwiching is employed to achieve good stopping power at low fluorescence and a cup-like shape restricts the solid angle into which any residual radiation from the beam stop can escape. One would usually apply only one of these techniques, i.e., cup or sandwich.

Attention must also be given to beam stops, lest they produce a background of their own. Generally, a beam stop must be thick enough to stop the pump photons and a possible higher harmonic contamination coming with them.[15] A cup-like shape helps to contain the photons emanating from it by scattering and fluorescence. Fluorescence can also be minimized by employing a sandwich structure: heavier elements, capable of stopping the x-rays, but producing energetic fluorescence are topped by layers of lighter elements, which absorb that fluorescence (see fig. 5.33). The fluorescence from the light elements is of lesser concern because it is absorbed in air (or a third sandwich layer).

4.4 Energy Resolution

Although, according to the wave vector matching diagram in fig. 5.1, photons at or near the pump energy should not reach the detectors, they do due to diffuse elastic and Compton scattering. In order to discriminate them from the signal and idler, energy-resolving detectors should be used. Because no detector has a perfect energy resolution, some of these photons will be erroneously registered as having the energy expected for the signal or idler, and these contribute to the background. Other than in spectroscopic applications, it is not so much the x-ray line resolution that counts—but the ability of a detector to discriminate between the pump and photons at roughly half the pump energy. Figure 5.34 shows an energy spectrum, taken with a Si drift detector [132], with the counts on a logarithmic scale over the *recorded* photon energy. The detector was exposed to elastically scattered stray radiation of 14.4 keV. The main peak at 14.4 keV has a shoulder towards lower energies at roughly 0.001 times lower

[15] Usually, the monochromator of a synchrotron beam line uses Si (111) crystals, which pass the 3rd harmonic of the nominal energy. The harmonic content can be reduced by slightly detuning the crystals relative to each other and by use of x-ray optics, such as lenses or mirrors.

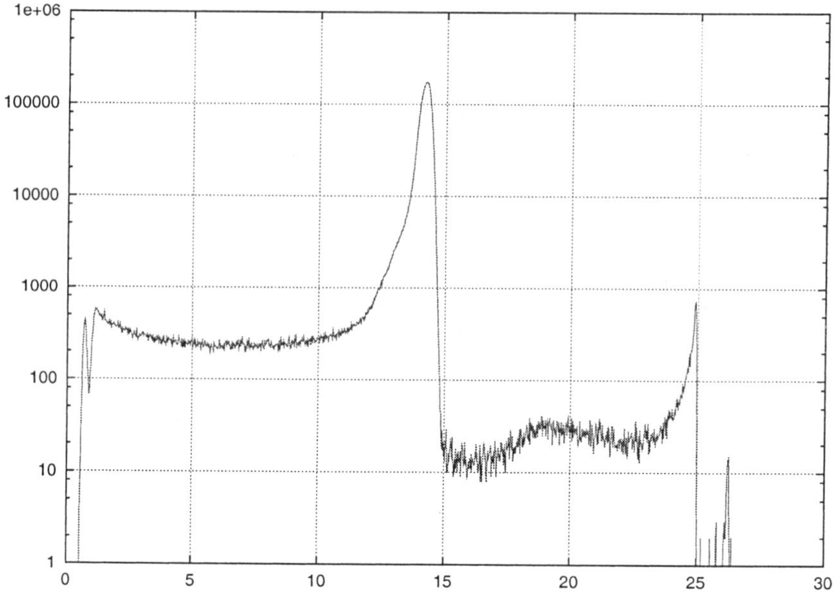

Figure 5.34: MCA spectrum of 14.4 keV photons, taken with a Si drift detector [132]. See text.

count rates. This is the so-called escape peak. At twice the incident energy, partially cut off for being at the scaling limit of the electronics, a peak is visible that is due to two photons of 14.4 keV hitting the detector simultaneously, i.e., within the 200 ns shaping time of the spectroscopic amplifier. The most striking feature is however the large plateau below 14.4 keV. It is due to events in which a photon generated a cloud of charges in the Si, but not all of them were collected by the pickup transistor. All solid-state x-ray detectors have this fault to some degree. When comparing fig. 5.34 to the MCA spectrum shown in fig. 5.12, one might think that the plateau shown here could also be due to Compton scattering (the explanation given there). This is not so because the Compton energy shift at 14.4 keV is very small and photoelectric absorption strongly suppresses multiple Compton scattering at lower x-ray photon energies. For details about the Si drift detector, see [140, 141, 142]. The energy-resolving power of the detectors, which is relevant for parametric down conversion experiments, is therefore the ratio of the peak to the counts in the plateau, integrated over the energy interval where signal or idler photons are expected.

4.5 Time Resolution

A coincidence is detected if both detectors register a photon within the expected energy range and at the same time within the time resolution of the signal chain. The probablity of a coincidence being registered due to two uncorrelated photons is thus proportional to the probability of one detector registering a photon within the set energy window *times* the probability of the other detector also doing so within a time interval given by the time resolution of the detectors and their signal processing chains. The measured rate of statistical coincidences due to actually uncorrelated photons depends on the detector energy resolution (discussed above), the time resolution and the time average of the square of the momentary intensity of stray photons. If a pulsed source is used (such as a synchrotron), then the detector time resolution needs be only as good as the reciprocal pulse repetition rate—or it must be better than the individual pulse length.

4.6 Time Structure of the Source

If the statistical background signal has a time structure, then its autocorrelation will appear in time correlation spectra (see fig. 5.10. This statistical background in the time correlation will *always* have a peak at zero time difference (coincidence). This is of course an obstacle to the detection of a faint true coincidence signal. Coincidence experiments should therefore be done with a source that has as little intensity modulation as possible on time scales that are resolvable by the detectors. For example, most of the down conversion experiments at the ESRF were done with the 2/3 or the 2*1/3 fill mode (see ch. 1, sec. 4), and some (secs. 2.6 and 2.7.1) in the uniform fill mode (i.e., all of the 992 buckets were filled). With the energy-resolving solid-state (Si drift and Ge) detectors used in all these experiments, the former two are more or less equivalent with regard to intensity autocorrelation, and the latter is obviously best.

4.7 Choice of Sample Material

The nonlinear material for x-ray down conversion must be crystalline because the wave vector matching works with a Bragg reflection and also because there must be a relatively large angle between the pump wave on the one hand and the signal and idler waves on the other hand. To keep the absorption of the down-converted photons reasonably low, a sample of light elements must be used. All experiments shown here were done with pump photon energies between 14.4 keV and 24 keV. With the down-converted photons having roughly half of that energy, only light elements can be present in the sample crystal. For that reason, beryllium was used by Eisenberger and McCall and in all other experiments (except the one described in sec. 2.5), diamond was used. Beryllium absorbs less than carbon, but diamond is probably still a better choice because with diamond, the statistical background is less for the following reasons: Due to the larger Z, the Compton scattering cross section is less, the static diffuse scattering is weaker because of the higher crystal quality and thermal diffuse scattering is less because of the strong covalent bonds.

There are of course a number of other candidate materials that fulfill the criteria of good crystalline quality and low absorption, such as lithium fluoride, lithium borate (LBO), organic crystals (sugar, etc.), but to my knowledge none of them have been tried yet. With higher energy radiation (like 100 keV in one experiment), absorption is less critical and materials, such as silicon are also usable.

5. Summary

The experimental results shown in the previous sections demonstrate the existence of the effect of parametric down conversion of x-rays and explore some of its features, such as the dependence on the wave vector matching, and the increase in event rate and energy spread due to the edge-enhancement effect. Possible applications include tests of the quantum theory, making use of the high quantum efficiency of x-ray detectors, and a drastic reduction of the radiation dose in the investigation of samples with little tolerance for radiation exposure. An important task on the way to potential applications will be the suppression of the background radiation. One possibility, namely the use of an x-ray mirror, was demonstrated.

A quarter-century passed before the pioneering experiment by Eisenberger and McCall was corroborated by Yoda et al. Since then, progress has been more rapid. This can be attributed largely to advances in x-ray sources, but even so, an x-ray down-conversion experiment remains challenging, and one has to pay attention to details, such as shielding against background, time and energy resolution of the detectors, and elimination of electronic noise as a possible source of artefacts in the time-correlation spectra.

6. Appendix

6.1 The Virtual Power Density of Vacuum Fluctuations

The zero-point motion of the electromagnetic field can be expressed in terms of a virtual power density, which is not directly observable or available as free energy. It appears only indirectly as a trigger to spontaneous emission, parametric down conversion, etc. Quantization is usually done by imposing periodic boundary conditions in a virtual box of the size $V = L^3$. The field amplitude can then be expressed through normal modes $\sqrt{V}\epsilon(\mathbf{k})\exp(\imath\mathbf{k}\cdot\mathbf{r})$. Due to the periodicity, a component k_j of $\mathbf{k}$ along one of the box edges fulfills $L = N2\pi/k_j$ with integer N. It follows that the modes have a spacing of $2\pi/L$, and therefore, the phase-space volume of one mode is $(2\pi)^3$. The gauge must be fixed to eliminate unphysical degrees of freedom, leaving two mutually orthogonal polarization states with $\mathbf{A}\cdot\mathbf{k} = 0$ in the transverse gauge. It follows that a phase-space volume of $(2\pi)^3$, situated at a phase-space point

$(\mathbf{r}, \mathbf{k})$ contains a zero-point energy of $2\hbar\omega/2$. The total volume in phase-space, corresponding to the infinitesimal frequency interval $[\omega - \delta/2 \ldots \omega + \delta/2]$ is given by a spherical shell in momentum space of radius ω/c and thickness δ/c, times a sphere in real space whose diameter[16] is the coherence length $2\pi c/\delta$. This gives a phase space volume of

$$\Pi = \left(\frac{4\pi^4}{3}\frac{c^3}{\delta^3}\right)\left(4\pi\frac{\omega^2}{c^2}\frac{\delta}{c}\right) = \frac{16\pi^5}{3}\frac{\omega^2}{\delta^2}. \tag{5.6}$$

With the above argument, Π contains a zero-point energy $\mathcal{E} = (2\pi^2/3)\hbar\omega^3/\delta^2$, which leaves that same volume through a real space surface of $\sigma = 4\pi^3 c^2/\delta^2$ (again, a sphere whose diameter is the coherence length) in a time $\tau = 2\pi/\delta$. The virtual power density of that flow into a solid angle of 4π is thus

$$\mathcal{P} = \frac{\mathcal{E}}{\tau\sigma} = \frac{\hbar\omega^3\delta}{12\pi^2 c^2}. \tag{5.7}$$

For a given situation with an admissible solid angle Ω, this value must be multiplied with a factor $\Omega/4\pi$.

The time averaged Poynting vector $\mathbf{S} = (c/4\pi)\mathbf{E} \times \mathbf{H}$ (see [1].(6.109)) in terms of the vector potential is:

$$|\mathbf{S}| = \frac{c}{4\pi}\left|\frac{\dot{\mathbf{A}}}{c} \times (\nabla \times \mathbf{A})\right| = \frac{|\mathbf{A}|^2\omega^2}{4\pi c}. \tag{5.8}$$

After equating eqs. (5.7) and (5.8), we obtain the virtual field strength of the vacuum fluctuations at a frequency ω within a frequency interval δ and into a solid angle of 4π:

$$|\mathbf{A}|^2 = \frac{\hbar\omega\delta}{3\pi c}. \tag{5.9}$$

6.2 Cross Section

The vector potential due to a current distribution $\mathbf{J}(\mathbf{r}, t)$ is [1].(9.13)

$$\mathbf{A}(\mathbf{r}) = \frac{e^{ikr}}{cr}\int \mathbf{J}(\mathbf{x})d\mathbf{x}. \tag{5.10}$$

Insertion of eq. (3.59) or (3.62), for the second-order current, replacing all vectors with scalars and taking $|\mathbf{h} - \mathbf{h}' - \mathbf{h}''| \approx k_n$ (within a factor of 2 or so) for order-of-magnitude estimates, gives for the contribution of a single electron

[16]The radiation passing through a point $\mathbf{r}$ along a direction $\mathbf{n}$ originates from points along $-\mathbf{n}$. After passage of the coherence time $2\pi/\delta$, the origination point is $\mathbf{r} - (2\pi c/\delta)\mathbf{n}$, which determines the antipode of $\mathbf{r}$.

(i.e., $\int \eta = 1$):

$$\mathbf{A}_s(r) = \frac{1}{r}\frac{e^3}{m^2c^4}|\mathbf{A}_i||\mathbf{A}_p| = \frac{r_e^2}{er}|\mathbf{A}_i||\mathbf{A}_p|. \tag{5.11}$$

Taking the square of eq. (5.11), inserting eq. (5.9) for $|\mathbf{A}_i|^2$ gives for the cross section for the scattering from a plane incident wave into a unit solid angle:

$$\frac{d\sigma}{d\Omega} = \frac{r_e^4\hbar\omega\delta}{3\pi ce^2} = \frac{137r_e^4k^2}{3\pi}\frac{\delta}{\omega}, \tag{5.12}$$

where $\hbar c/e^2 \approx 137$ was used (1/137 is the fine structure constant α). Because scalars were inserted for the vectors in $\mathbf{J}^{(2)}$, this is only an order-of-magnitude expression. For a typical x-ray photon energy of 12.4 keV ($\lambda = 1\text{Å}$) and a bandwidth δ/ω of, say, 1%, eq. (5.12) turns out to be $4.5 \cdot 10^{-9}r_e^2$, and thus much less than the Thomson scattering cross section r_e^2.

Equation (5.12) was obtained under the tacit assumption that $\mathbf{A}_s$ exhibits no angular dependence. It is justified because the second-order current $\mathbf{J}^{(2)}$ from a single electron is contained within a volume that is much smaller than the x-ray wavelength.

6.3 Amplitude Growth

With the cross section, eq. (5.12), one may describe incoherent scattering processes. For coherent processes, such as wave-vector-matched down conversion, it is more appropriate to determine the amplitude growth by use of a Takagi-style expression on the basis of eq. (3.64). For an order of magnitude estimate, it is sufficient to insert the scalar values of amplitudes and wave vectors. Furthermore, $\mathbf{h} - \mathbf{h}' - \mathbf{h}''$ is taken to have the same magnitude as the wave vectors. This gives then

$$2|\mathbf{k}_s||\nabla\mathbf{A}_s| = 4\pi\frac{r_e^2}{e}\tilde{\eta}_{\mathrm{H}}|\mathbf{A}_p||\mathbf{A}_i|. \tag{5.13}$$

The idler amplitude can be determined by the virtual power density of vacuum fluctuations. Insertion of (eq. 5.9) into eq. (5.13) and taking into account the limited solid angle Ω permitted by the wave vector matching, gives

$$|\nabla\mathbf{A}_s| = \frac{2\pi}{k_s}\frac{r_e^2}{q}\tilde{\eta}_{\mathrm{H}}\sqrt{\frac{\hbar\omega_i\delta}{3c\pi}\frac{\Omega}{4\pi}}|\mathbf{A}_p| = \frac{2\pi k_i}{k_s}r_e^2\tilde{\eta}_{\mathrm{H}}\sqrt{\frac{137}{12\pi^2}\frac{\delta}{\omega_i}\Omega}|\mathbf{A}_p|, \tag{5.14}$$

where $\hbar c/q^2 = 137$ was used. There is, of course, no distinction between signal and idler, and the above argument could be made with their roles reversed.

It is interesting to compare eq. (5.14) with Thomson scattering. Insertion of $\mathbf{k_O}$ fulfilling $|\mathbf{k_O}| = |\mathbf{k_h}| = k_r$ into eq. (3.30) leaves the term with $\tilde{\eta}_{\mathrm{H}}$ directly representing the Thomson scattering power. The scattering of $\tilde{\mathbf{A}}_0$ into $\tilde{\mathbf{A}}_\mathbf{H}$ is then given by

$$|\nabla \tilde{\mathbf{A}}_\mathbf{H}| = \frac{2\pi}{k} r_e \tilde{\eta}_\mathbf{H} |\tilde{\mathbf{A}}_0|. \tag{5.15}$$

For like amplitudes $|\mathbf{A}_p|$ and $\tilde{\mathbf{A}}_0$, one may take the ratio of eqs.(5.14) and (5.15) with $k = k_s$

$$r_e k_i \sqrt{\Omega} \sqrt{\frac{\delta}{\omega_i}}, \tag{5.16}$$

where, this being order-of-magnitude, $\sqrt{137/12\pi^2} \approx 1.0$ was approximated.

6.4 Wave Vector Matching

Wave vector matching must take into account the geometry, refraction, and, close to the Bragg condition in a crystal, dynamical diffraction effects.

6.4.1 Without Dynamical Diffraction. If dynamical diffraction effects are ignored, the wave vector matching becomes a relatively simple geometrical problem. The free-electron refractive index of a piece of matter is given by eq. (3.12) as $n \approx 1 - \delta(k) = 1 - (2\pi r_e/k^2)\tilde{\eta}_0$. If x is the energy-splitting ratio, i.e., a pump photon $\hbar\omega_p$ decays into signal $\hbar\omega_s = \hbar x \omega_p$ and idler $\hbar\omega_i = \hbar(1-x)\omega_p$, then

$$|\mathbf{k}_s| = \left(1 - \frac{\delta(k_p)}{x^2}\right) x k_p \quad |\mathbf{k}_i| = \left(1 - \frac{\delta(k_p)}{(1-x)^2}\right)(1-x)k_p, \tag{5.17}$$

where $k = \omega_p/c$ is the *vacuum* wave number of the pump, and $\delta = \delta(k)$ is the refractive decrement at the pump frequency.

We now abbreviate $k = k_p$ and $\delta = \delta(k_p)$. Taking the sum $|\mathbf{k}_s| + |\mathbf{k}_i|$ yields

$$|\mathbf{k}_s| + |\mathbf{k}_i| = \left(1 - \frac{\delta}{x(1-x)}\right) k < (1-\delta)\, k = |\mathbf{k}_p|. \tag{5.18}$$

The sum of the wave numbers of signal and idler is a bit short of the pump wave number, and collinear, or close-to-collinear wave vector matching will therefore not work (and, anyway, the down conversion cross section vanishes if the wave vectors are all parallel to each other).

Instead, one may use the wave vector matching scheme shown in figs. 3.16 and 5.1. In this scheme, a wave vector $\mathbf{k}_p + \mathbf{H}$ is obtained, which is associated with the pump but is shorter than $|\mathbf{k}_p|$ in eq. (5.18). One should keep in mind that the pump wave is still the one with $\mathbf{k}_p$, *and not* $\mathbf{k}_p + \mathbf{H}$. Leaving dynamical diffraction aside, the determination of $|\mathbf{k}_p + \mathbf{H}|$ is straightforward, as shown in fig. 5.35. The dotted lines represent circles about the reciprocal lattice points $\mathbf{0}$ and $\mathbf{H}$ with radii of $(1 - 2\pi r_e \tilde{\eta}_0)\, k_p$. For small detuning angles $\delta\Theta$, one

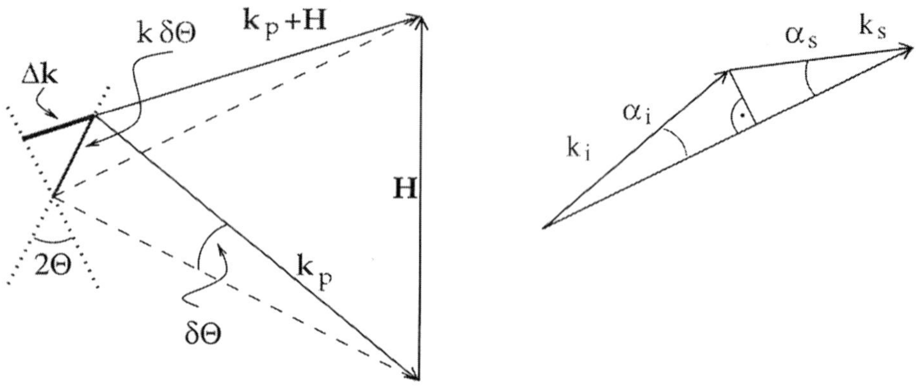

Figure 5.35: Wave vector matching.

may approximate them by straight lines, and the reciprocal length along the one belonging to $\mathbf{k}_p$ is $n_p k_p$, where $n_p = (1 - (2\pi r_e/k_p^2)\tilde{\eta}_0)$. The length of $\Delta k = |\mathbf{k}_p + \mathbf{H}| - (1-\delta)k$ can then be determined by simple geometry to be

$$\Delta k = -(1-\delta)k\delta\Theta \sin 2\Theta. \tag{5.19}$$

For $\mathbf{k}_i + \mathbf{k}_s = \mathbf{k}_p + \mathbf{H}$ to be true, two geometrical conditions must be fulfilled (see fig. 5.35, right-hand side):

$$\begin{aligned} |\mathbf{k}_i| \sin\alpha_i &= |\mathbf{k}_s| \sin\alpha_s \\ |\mathbf{k}_i| \cos\alpha_i + |\mathbf{k}_s| \cos\alpha_s &= |\mathbf{k}_p + \mathbf{H}| = (1-\delta)k + \Delta k. \end{aligned} \tag{5.20}$$

We have thus

$$\cos\alpha_i = \sqrt{1 - \sin^2\alpha_i} = \sqrt{1 - \frac{|\mathbf{k}_s|^2}{|\mathbf{k}_i|^2}\sin^2\alpha_s} \tag{5.21}$$

from the first line, and

$$\Delta k = k(1-x)\left(1 - \frac{\delta}{(1-x)^2}\right)\cos\alpha_i + kx\left(1 - \frac{\delta}{x^2}\right)\cos\alpha_s - k(1-\delta) \tag{5.22}$$

from the second line of eq. (5.20). For small angles α_i and α_s, several simplifying approximations can be made: First, δ is small, and we may approximate $\delta\cos\alpha_{s,i} \approx \delta$:

$$\Delta k \approx k\left[(1-x)\cos\alpha_i + x\cos\alpha_s - 1\right] - k\delta\left(\frac{1}{1-x} + \frac{1}{x} - 1\right). \tag{5.23}$$

Second, $|\mathbf{k}_s|^2/|\mathbf{k}_i|^2$ in eq. (5.21) may be replaced with $x^2/(1-x)^2$ and, $\sin^2\alpha_s$ being small, the root can be written as $\cos\alpha_i \approx 1-(x^2\alpha_s^2)/(2(1-x)^2)$. Third, $\cos\alpha_{s,i} = (1-\sin\alpha_{s,i})^{1/2} \approx 1-\alpha_{s,i}^2/2$. With these approximations, eq. (5.23) becomes

$$\Delta k \approx k\left[(1-x)\left(1-\frac{x^2}{(1-x)^2}\frac{\alpha_s^2}{2}\right)+x\left(1-\frac{\alpha_s^2}{2}\right)-1+\delta-\frac{\delta}{x(1-x)}\right]. \tag{5.24}$$

After comparing to eq. (5.19), and approximating $n_p = 1$, we obtain

$$\delta\Theta \approx \frac{1}{\sin 2\Theta}\left[\frac{x}{(1-x)}\frac{\alpha_s^2}{2}-\delta\left(1-\frac{1}{x(1-x)}\right)\right]. \tag{5.25}$$

In the symmetrical case $\alpha_i = \alpha_s$, the wave numbers $|\mathbf{k}_i|$ and $|\mathbf{k}_s|$ are equal to each other and therefore, $x = 1/2$. This simplifies eq. (5.25):

$$\delta\Theta \approx \frac{1}{\sin 2\Theta}\left[\frac{\alpha_s^2}{2}+3\delta\right] = \frac{\alpha_s^2+12\pi r_e\tilde{\eta}_0/k^2}{2\sin 2\Theta} = \frac{\alpha_s^2+12\pi\chi_0}{2\sin 2\Theta}, \tag{5.26}$$

in terms of the refractive index decrement δ (eq. (3.12)), the electron density η and the susceptibility χ (eq. (3.69)), respectively.

6.4.2 With Dynamical Diffraction of the Pump Only. Under conditions of dynamical diffraction, the length of a wave vector is $k(1-\delta)+\xi_h$, where ξ_h is defined in eq. (3.14). Only the pump wave is taken to fulfill the Bragg condition, and the considerations of sec. 6.4.1 are still valid for $\mathbf{k}_i$ and $\mathbf{k}_s$. Therefore, $-\xi_h$ can take the place of Δk in eq. (5.24), and (compare eq. (5.25))

$$\xi_h \approx -k\left[\frac{x}{(1-x)}\frac{\alpha_s^2}{2}-\delta\left(1-\frac{1}{x(1-x)}\right)\right]. \tag{5.27}$$

A parameter η is commonly used for ξ_h [73].(31b), shown here in the standard notation and the notation of this text (see ch. 3, sec. 5.1):

$$\xi_h = -\frac{k\Pi\Gamma F_{\mathbf{H}}}{2\sqrt{|b|}}\left(\eta\mp\sqrt{\eta^2+\frac{b}{|b|}}\right) = -\frac{2\pi\Pi r_e\tilde{\eta}_{\mathbf{H}}}{k\sqrt{|b|}}\left(\eta\mp\sqrt{\eta^2+\frac{b}{|b|}}\right), \tag{5.28}$$

where b is the asymmetry parameter [73] and $-$ in "$\mp$" is for the α and $+$ for the β branch.[17] The parameter η is given by [73].(32) (again in both types of notation)

$$\eta = \frac{b\Delta\Theta\sin 2\Theta+\frac{1}{2}\Gamma F_0(1-b)}{\Gamma\Pi\sqrt{|b|}|F_H|} = \frac{k^2 b\Delta\Theta\sin 2\Theta+2\pi r_e\tilde{\eta}_0(1-b)}{4\pi r_e\tilde{\eta}_{\mathbf{H}}\Pi\sqrt{|b|}}. \tag{5.29}$$

[17] When comparing with [73], please note that $-(\eta\mp\sqrt{\eta^2+1}) = (\eta\pm\sqrt{\eta^2+1})^{-1}$

By equating eqs. (5.27) and (5.28), and inserting eq. (5.29), one obtains the relation between α_s and $\Delta\Theta$.

In the symmetrical Laue case, $b = 1$. For large η, eq. (5.28) is then reduced to $\xi_h = -k\Pi\Gamma F_H \eta$. Upon equating this with eq. (5.27) and using $\delta = \chi_0/2$ and eq. (5.29) with $x = 1/2$:

$$\frac{\alpha_s^2 + 3\chi_0}{2} = \Delta\Theta \sin 2\Theta, \tag{5.30}$$

the result (5.26) of sec. 6.4.1 is recovered. For smaller η, one may substitute $\eta = \sinh \nu$ (see [73]), whereupon eq. (5.28) gives $\xi_h = -[k\Pi\Gamma F_H \exp(\nu)]/2$. Using also $\delta = \Gamma F_0/2$, it follows that

$$\frac{\alpha_s^2}{2} + \frac{3}{2}\Gamma F_0 = \frac{\Pi\Gamma F_H}{2} e^{\nu}. \tag{5.31}$$

To take an example, let us look at the case of vanishing α_s with the diamond (111) reflection in σ-polarization ($\Pi = 1$). Then, $F_0 = 48$ and $F_{(111)} \approx 12$ [115] and thus $\exp(\nu) = 12$ and $\eta = 5.96$, corresponding to a positive detuning $\delta\Theta$ by about 3 rocking widths.

6.5 Electronics

6.5.1 The Correlation Circuit. Energy-discriminated time-correlation spectra were obtained by use of analog/digital circuitry which I designed and built specifically for that purpose. It shall be discussed here because it has not yet been described in the openly available literature. Figure 5.36 shows a block diagram. The Röntec drift detectors have an output wave form as shown

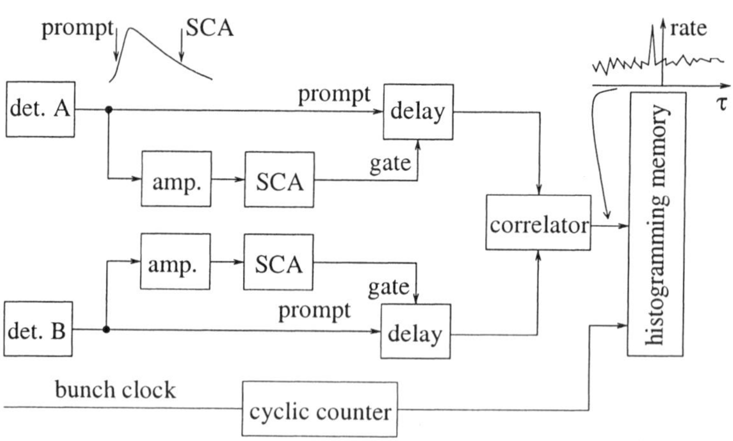

Figure 5.36: Block diagram of the correlator.

in fig. 5.32 with pulses rising to an amplitude of 5 mV per keV photon energy within 100 ns and falling off exponentially with a time constant of 50 μs. Pulses can stack up on each other. There is no need to reset the preamplifier, as with many other solid-state x-ray detectors. For each of the two detector channels, a comparator detects the rising edge and gives off a digital "prompt" signal. The detector output is also fed into a spectroscopic amplifier, which generates pulses whose height is proportional to the photon energy. This pulse is sent to a single-channel analyzer (SCA), which gives off a digital pulse if the photon energy falls between set limits. Only if the photon energy lies within set limits, and therefore the SCA gives off a pulse, should the event be used for the purposes of correlation. The SCA output comes several microseconds after the event (depending on the settings of the spectroscopic amplifier and the type of SCA). To be able to gate the prompt pulse with the SCA output, the prompt pulses are delayed in a digital counter, which counts down from a fixed value after receipt of a prompt pulse. Upon reaching zero and if a pulse from the SCA has arrived in the meantime, it generates a pulse that is passed to the digital correlator. The correlator is symmetric with respect to the two channels. Either one can start a digital counter, and the respective other one can stop it. If no pulse arrives in time to stop it before overrun, the circuit is reset. Otherwise, if a stopping pulse makes it in time through the gated delay, the count value and one bit to tell the start/stop polarity, are passed to a histogramming memory. The delay times (in units of the clock) and the maximum count of the correlator can be set by the host computer that controls the experiment. The digital part, that is the delay counters, the correlation counters and the histogramming memory, are implemented in Altera programmable logic devices. It is usually run at a clock speed of 100 MHz, giving a time resolution of 10 ns, which is quite sufficient for the Si drift detectors.

More information, including a sets of Altera design files for different versions of the device can be found on the CD. The time correlation circuit was later superseded by a more advanced device, described in app. 6.5.2 and [137].

6.5.2 The Event Logger. Having to set the SCA limits at the time of the experiment is a disadvantage, particularly so because, with the event rates of down conversion, there is no "live" signal to which to adjust them. Setting them too loosely will increase the background, and setting them too tightly cuts the observed event rate down. Furthermore, the extreme collinear geometry described in sec. 2.7 results in a rather broad distribution of individual energies of the down-converted photons (with a fixed sum of energies). It is much better to record all events with time and energy information, which fulfill rather relaxed selection criteria and to determine correlations off-line. This technique also permits finding correlation between any subset of detectors of a larger number than two. To meet that need, I designed an analog/digital circuit that, in its present implememtation, can record events from up to five detectors with time stamps of 20 ns resolution and eight-bit energy information for each of them. This circuit was used in the last experiment, described in sec. 2.7.3. It is described in the open literature [137], and a meaningful description would take too much space here. The Altera design files for that circuit are included on the CD.

Acknowledgments

The experiments described in the sections following 2.3 were done in teams of two to five investigators and also depended very much on the support by the respective beam line staff. My collaborators were: P. Fernandez (APS), G. Kuri (DESY), W.-K. Lee (APS), D.P. Lowney (Dublin City Univ.), G. Materlik (DESY), D.M. Mills (APS), Y. Nishino (DESY), D.V. Novikov (DESY), M. Oberparleiter (Univ. München), H. Schulte-Schrepping (DESY), S. Shastri (APS), and J. Würges (Univ. Hamburg). Affiliations are given for the respective times of the experiments.

I would also like to thank the beam line scientists who supported the experiments presented in this chapter and about as many more which gave inconclusive results but provided valuable experience. The beam line scientists not mentioned among the collaborators are: W. Drube at station BW2 and H. Franz at station Petra-I of DESY, T. Asthalter, A. Chumakov, B. Doyle, O. Leupold, and R. Rüffer at station ID18 of the ESRF, K.-D. Liss, and T. d'Almeida at station ID15A of the ESRF, A. Mashayeki at station 1-ID of the APS, D.A. Arms, E.M. Dufresne, and D. Walko at station 7-ID of the APS. Many thanks to all of you for good team work and enjoyable science!

The printed circuit boards of later versions of the correlator and the event logger were produced at the DESY electronic shops.

Use of the APS was supported by the U.S. Department of Energy under the contract W-31-109-ENG-38. Operation of the MHATT-CAT sector 7 beamline at the Advanced Photon Source was supported by DOE grant No. DE-FG02-99ER45743.

Chapter 6

LASER PUMP, X-RAY PROBE SPECTROSCOPY ON GAAS

B.W. Adams
Advanced Photon Source, Argonne National Laboratory, Argonne, IL 60439, USA
adams@mail.desy.de

M.F. DeCamp, E.M. Dufresne, D.A. Reis
mattydee@umich.edu, dufresne@umich.edu,dreis@umich.edu
Univ. of Michigan, Department of Physics, 500 E. University Drive,
Ann Arbor, MI 48109, USA
dreis@umich.edu

Abstract A laser pump, x-ray probe spectroscopic experiment and its results are presented. We measured the Ga K_α x-ray fluorescence following x-ray absorption at the Ga K absorption edge and determined its increase due to excitation with sub-picosecond pulses of laser light at 4.6eV photon energy. Within about 200 ps after the laser excitation, additional possible final states for the x-ray absorption process are present in the valence band, leading to an increased rate of x-ray absorption/fluorescence processes. The technique could eventually lead to femtosecond-resolving pump-probe spectroscopy with an absolute reference energy level and also to a femtosecond-resolving x-ray detector. The latter is of particular importance to future short-pulse x-ray sources, such as free-electron lasers.

1. Introduction

Elementary processes in solid-state physics and chemistry take place at electron-volt energies, corresponding to a few femtoseconds on the time scale and the elementary length is given by the interatomic spacing of 2-3 Ångströms (see fig. 6.1). Whereas the energy and time scales can be addressed with short-pulse lasers, the resolution of individual atoms does require x-rays. With recent technological advances, it has become possible to combine short-pulse laser

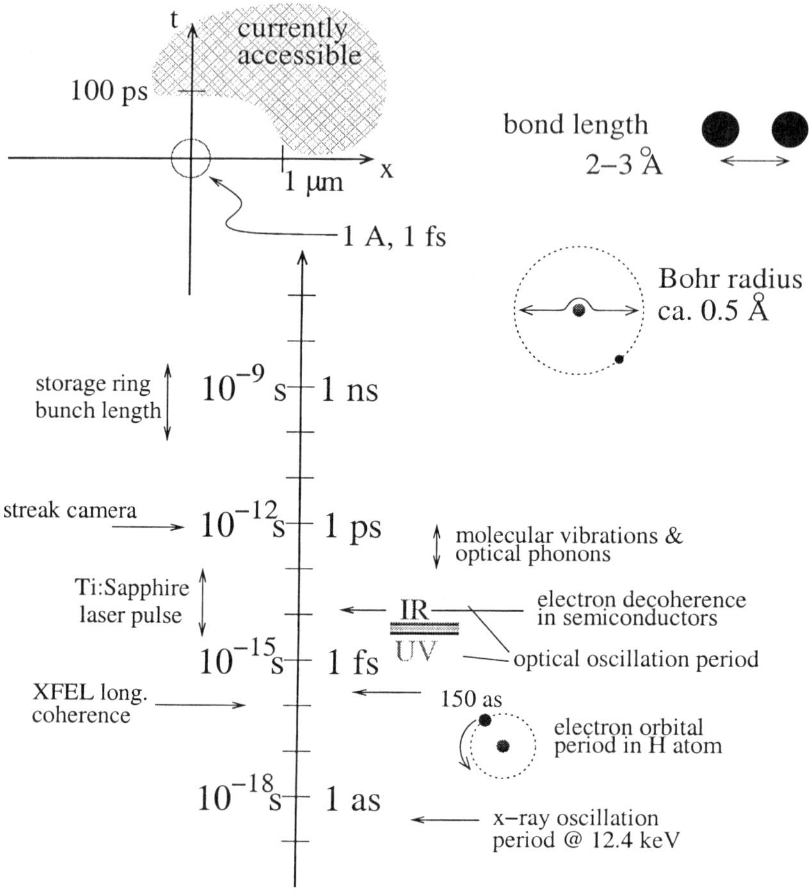

Figure 6.1: Time and length scales of elementary processes in chemistry and solid state physics and resolution of available instruments. The Bohr radius is about 0.5 Å, interatomic distances in condensed matter are a few Å and the elementary time scales of condensed matter dynamics are upwards of one femtosecond. In the upper right hand corner is a map of the region in space and time that is currently accessible with lasers and x-rays (and all at longer time/length scales, too). The goal is to resolve elementary processes at about 1 Å and 1 fs resolution.

radiation with x-rays and join their specific strengths. The first pioneering experiments (see [123, 143, 144] and chapters 9 and 8) were done in a pump-probe configuration, with x-ray diffraction showing laser-induced structural changes, such as ultrafast melting. Extended x-ray absorption fine structure (EXAFS) has also been used to measure changes in interatomic distances after ultrafast laser excitation.

This chapter is about another type of laser pump, x-ray probe experiment, in which laser-induced changes in the *electronic structure* of a semiconductor are probed with x-ray spectroscopy. The motivation for this experiment was initially to provide the operating principle for a femtosecond-resolving x-ray detector [145] and laser/x-ray correlator (discussed further in sec. 7.2). However, a probably more interesting application would be to develop it into a spectroscopic technique for charge-carrier dynamics in photoexcited semiconductors, which refers transitions within the band structure to an absolute energy reference level.

2. Physics Background

X-ray absorption near-edge structure (XANES) spectroscopy is a powerful tool to obtain chemical information, such as the oxidation state of a particular elemental species in a molecule [146]. In the context of semiconductor physics, the band structure and localized states can be probed. In combination with ultrafast laser excitation, the technique can give insights into charge-carrier dynamics. Solely laser-based spectroscopic techniques can provide information on the band structure and carrier dynamics only in the form of a convolution because there is no absolute energy reference within the band structure. That is, however, available with x-ray spectroscopy because the core levels are discrete and are almost totally unaffected by processes involving outer-shell electrons. X-ray absorption and fluorescence spectroscopy can thus be used to probe laser-induced changes and their relaxation dynamics, relating all energies in the band structure to a core level. There is one other technique that can also put spectra on an absolute energy scale, namely electron photoemission spectroscopy. However, due to the small mean free path of electrons in condensed matter, that technique is limited to surfaces. X-rays, on the other hand, can be used to study bulk *and* surface properties, with the information depth under control through incidence and exit angles of the radiation. Details on the spectroscopic application can be found in sec. 7.1.

In the particular case of the experiment described here, a sample of GaAs was illuminated with pulses of frequency-tripled light from a Ti:sapphire laser. The laser photon energy of 4.5 eV corresponds to a maximum in the GaAs absorptivity due to almost parallel branches in the valence and conduction bands (see fig. 6.2). Photons at electron-volt energies induce mainly direct transitions, and the absorption cross section is therefore given by the parallelism of branches in the band structure. In contrast to that, an x-ray photon at 10.367 keV does carry a momentum that exceeds the Brillouin zone diameter. This momentum is transferred to the atom that absorbs the x-ray photon and thence to the surrounding lattice. An x-ray absorption process will therefore excite a phonon wave packet, and the transfer of momentum to the photoelectron is smeared out.

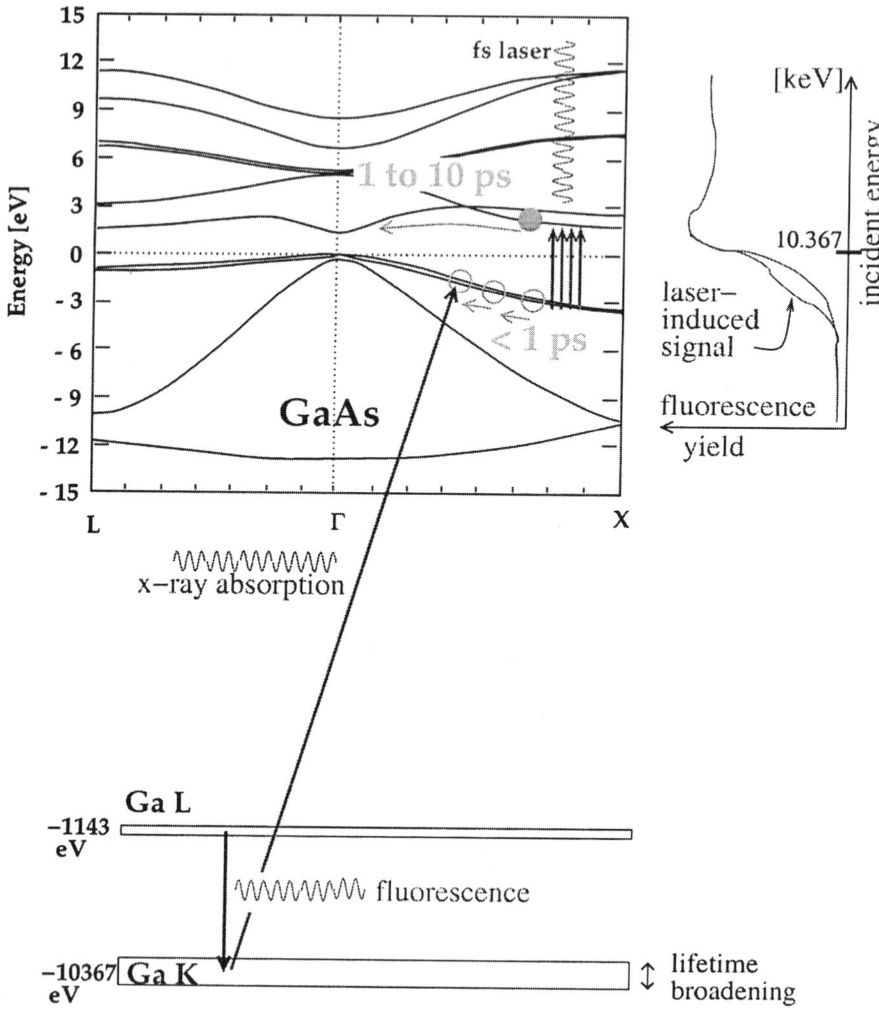

Figure 6.2: GaAs bandstructure with laser and x-ray transitions.

Once created ca. 3 eV below the band edge, the holes drift to the top of the valence band within typically 100 fs to 200 fs (see fig. 6.2). The electrons are initially deposited into a secondary minimum in the conduction band and therefore take a few picoseconds to relax into the conduction-band minimum. Recombination across the band gap takes typically one to several hundred picoseconds, depending on sample purity, proximity to a surface, etc. X-rays which are monochromatized to a photon energy corresponding to a transition from the Ga K shell to a few eV below the lower edge of the GaAs conduction

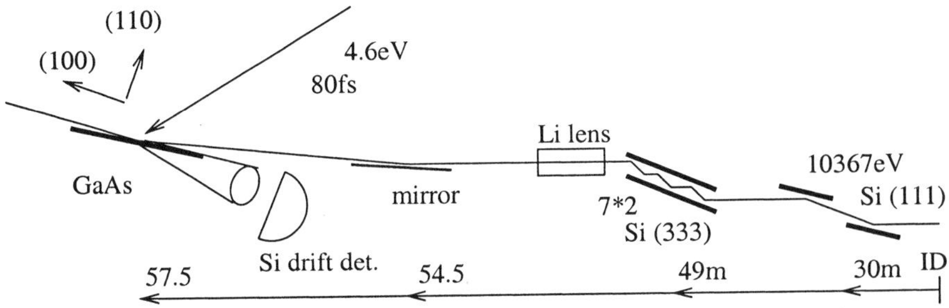

Figure 6.3: Schematic of the beam line.

band (i.e., just below the Ga K absorption edge) experience a strong change in the absorption cross section when states are emptied within the valence band due to laser excitation. The hole in the Ga K shell can be filled by a number of different relaxation processes, the dominant one being the emission of Ga K_α fluorescence. The intensity of Ga K_α fluorescence is therefore a measure of the concentration of holes in the valence band. Different depths within the valence band can be addressed by appropriate monochromatization of the x-rays and the time dependence of the fluorescence gives information on the laser-induced carrier dynamics within the valence band.

3. The Experiment

The experiment was carried out at the 7-ID beam line [136] of MHATT-CAT at the APS. Apart from the Ti:sapphire laser system that is installed there, the beam line is rather typical for a 3rd-generation synchrotron radiation facility. Figure 6.3 shows a schematic of the x-ray optics of the setup used in the experiment described here. The cryocooled Si (111) double-crystal monochromator was adjusted to the energy of 10.367 keV, corresponding to the Ga K absorption edge. At a storage ring current of 100 mA, it provided a flux of roughly $5 \cdot 10^{12}$ photons per second within a bandwidth of ca. 1 eV (see below). In order to narrow the energy width of the incident beam and particularly to suppress the wings of the bandpass, the x-rays were made to go through 14 reflections (up-horizontal-up-....) of the order (333) in a channel-cut Si crystal. They then passed through a horizontally focusing lithium lens [147] and were reflected onto the sample by a plane mirror whose angle was adjusted to suppress the harmonics of 10.367 keV in the beam and thus minimize the fluorescence background from the sample. The lens, besides increasing the flux on the sample, also helped a bit in the harmonic suppression (because the focal length depends on the photon energy). Both, the lens, and the mirror came after the monochromatizing elements and any divergence introduced by them had no adverse effect on the energy bandwidth (more on that in sec. 6.1).

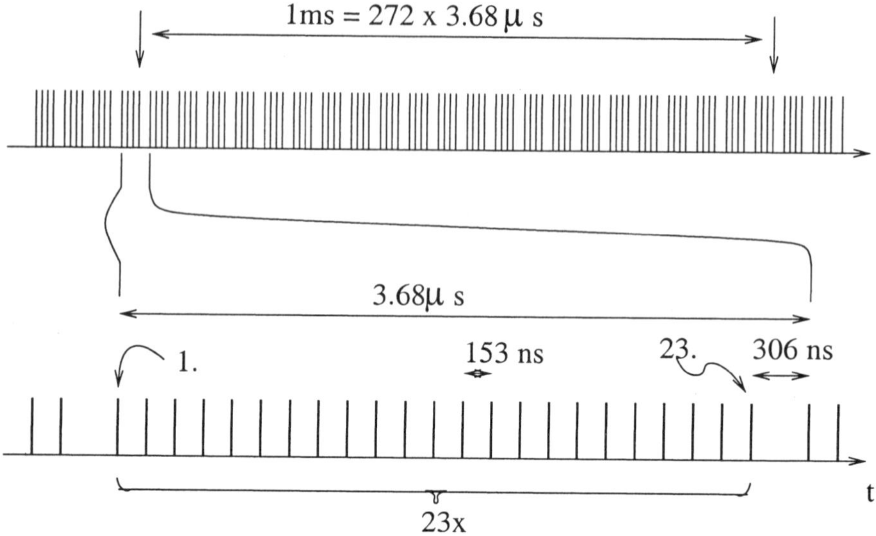

Figure 6.4: Bunch fill pattern of the APS. Upper trace: one millisecond, corresponding to 272 ring periods. Lower trace: a zoom into one ring period, showing bunches nos. 1 to 23, spaced 153 ns and the gap of 306 ns. Compare with the oscilloscope screen shot in fig. 6.5 and with fig. 9.2. Output of the laser amplifier coincides with the 23rd bunch, once in each millisecond.

A Ti:sapphire laser system and electronics for the synchronization of the laser to the storage ring are installed in the D hutch of the MHATT-CAT beam line. Figure 6.6 shows a schematic of the laser system and the synchronization electronics. A continuously pumped Ti:sapphire oscillator, emitting 50 fs, 4 nJ pulses at a repetition rate of 88 MHz, is phase-locked [124] to the 352 MHz radio frequency (rf) of the storage ring. The timing of the laser pulses relative to the rf, and thus the x-rays, can be changed in steps of 19 ps through a digital delay in the phase lock loop (PLL). A Pockels cell selects one pulse each millisecond out of the pulse train emanating from the oscillator. These pulses are then amplified in an eight-pass Ti:sapphire amplifier, which is pumped by an YLF laser just before each one of the pulses from the Pockels cell. The kilohertz repetion rate of the pump is also derived from the ring rf. Figure 6.4 shows the timing of the experiment: The APS operated in its standard mode in which out of 24 buckets (see p. 3), spaced evenly at 153 ns intervals, 23 were filled with electron bunches, and one was kept empty. The laser timing was adjusted to make the pulses from the Ti:sapphire amplifier coincide with the last bunch before the 306 ns gap (henceforth the 23rd bunch), once each millisecond.

In the experiment described here, the oscillator was tuned to a center wavelength of 820 nm. The output of the amplifier was then first doubled in frequency by a crystal of potassium dihydrogen phosphate (KDP), out of which a crys-

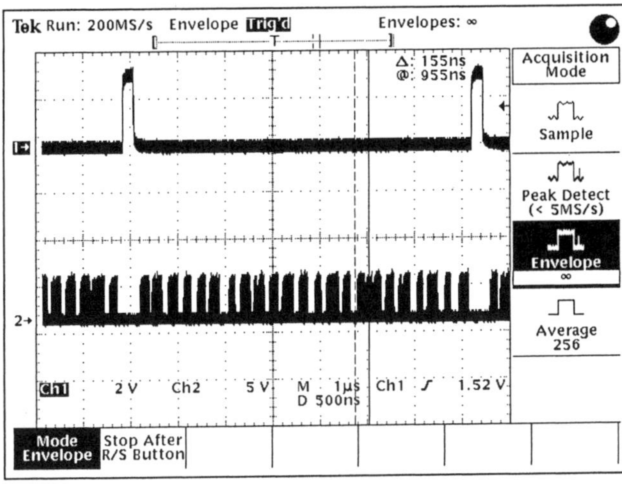

Figure 6.5: Oscilloscope screen shot of the APS bunch fill pattern, generated from the SCA output in the signal processing chain (see text) in the digital oscilloscope's envelope mode. The upper trace shows the P0 bunch clock signal, occuring once per storage ring period. The lower trace is an envelope of the SCA output. Actually, the probability of an SCA pulse is below one in a period.

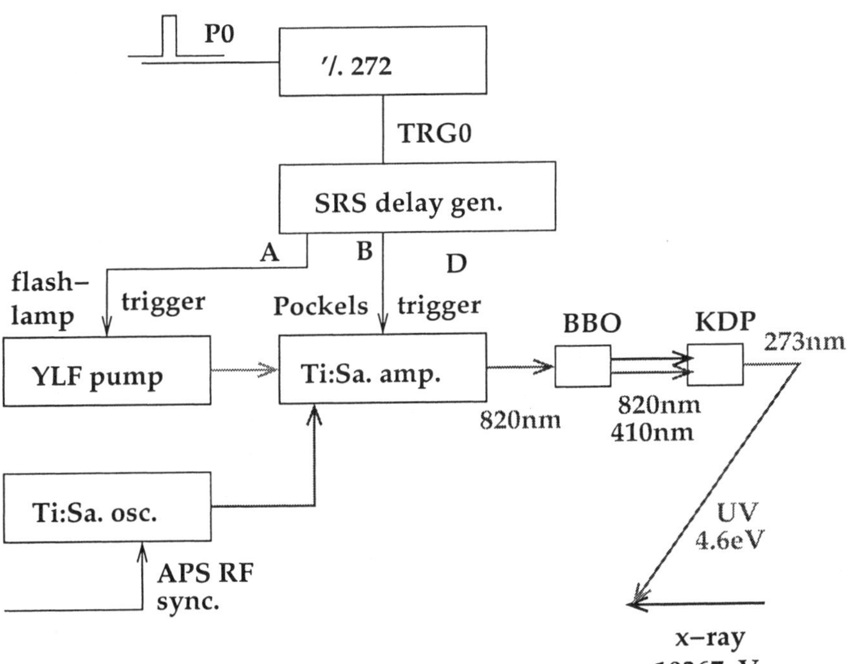

Figure 6.6: Laser and synchronization. See also fig. 9.4.

tal of beta barium borate (BBO) generated frequency-tripled light by adding fundamental frequency photons to the doubled ones.

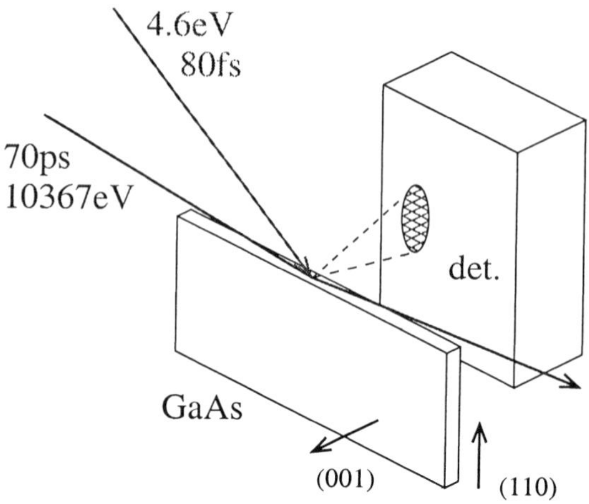

Figure 6.7: Schematic of sample and detector.

The sample was a (001)-oriented wafer of GaAs, broken along a $(1\bar{1}0)$ line to reveal a fresh (110) cleavage surface, ca. 20 mm by 0.5 mm in size. The laser light hit the surface at an angle of ca. 15°, giving an elongated footprint of ca. 0.5 mm by 5 mm (the laser focus itself was not circular). In order to match the x-ray penetration depth to that of the laser light, the x-rays were incident at an angle of 3.9 mrad, i.e., just above the critical angle $\alpha_c = 3.65$ mrad of total reflection,[1] along the long side of the surface. Having a narrow strip of sample surface greatly facilitated the adjustment of spatial overlap between laser light and x-rays. Temporal overlap had been adjusted before placement of the sample, using the laser-induced modulation of the Borrmann effect in a Ge sample (see ch. 9).

A Si drift diode detector [132] of the type used in most of the down conversion experiments described in ch. 5 was placed next to the sample, so that it would detect x-rays leaving the sample at a small angle of ca. 5 degrees to the surface and at close to a right angle to the incident x-rays. Figure 6.7 shows a schematic of this placement, and photos can be found on the CD in chapters/psg/photos. The detector is capable of resolving x-rays of 10.367 keV versus Ga K_α fluorescence at 9.25 keV, but not versus Ga K_β fluorescence at 10.26 keV (see figs. 5.24 and 5.34 for representative energy spectra taken with this kind of detector). The grazing exit angle helped in limiting the effective sample depth and the close-to right angle put the direction of the x-rays coming from the sample roughly parallel to the polarization of the incident x-rays, and thus reduced the background from elastic scattering of incident x-rays. A lead

[1]The refractive index with x-rays is slightly less than unity, see eq. (3.12). Therefore, total reflection occurs for a beam coming from outside of the sample.

mask (not shown in fig. 6.7) was attached to the sample to make sure that only x-rays from the region of overlap between the laser light and the x-rays would reach the detector (see photographs on the CD)).

Figure 6.8 shows the signal processing chain. The detector output (see

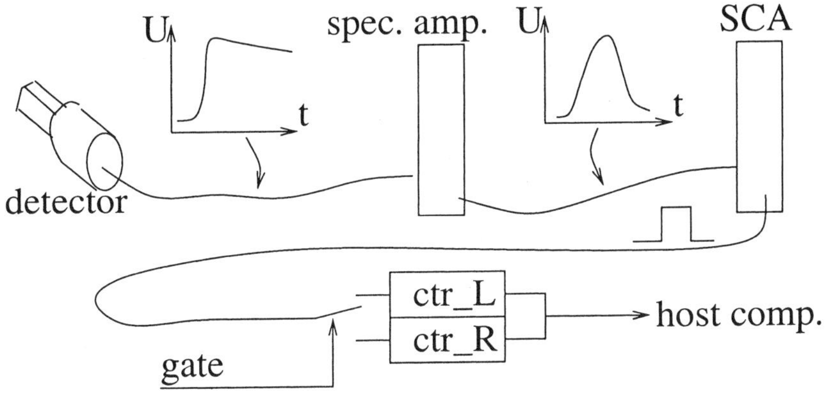

Figure 6.8: The signal processing chain. The output of the detector (compare fig. 5.32) is fed into the spectroscopy amplifier and thence into an SCA. The digital output of the SCA goes into one of two counters ("ctr-L" for "with laser", "ctr-R" for "reference"), depending on the state of the gate signal which is logic-1 for the bunch which is related to a laser pulse, and logic-0 otherwise.

fig. 5.32 for an example) was taken outside the hutch with a doubly shielded coaxial cable and fed into a Canberra 2025 [148] spectroscopic amplifier, set to a shaping time of 500 ns. Despite the shielding on the coaxial cable, the switching noise of the Pockels cell was clearly visible in the input to the amplifier (see fig. 6.12). Issues arising from this noise, particularly concerning the placement of the spectroscopy amplifier, are discussed in sec. 6.2. The output of the spectrocopy amplifier was connected to a single-channel analyzer (SCA), which was adjusted to give off a digital pulse for inputs whose height corresponded to a detected photon energy of the Ga K_α fluorescence. This digital signal was fed into a set of gated digital counters. The gate signal was logic-1 at times at which an SCA pulse from that particular bunch would occur, to which the laser was timed. At all other times, it was logic-0. Depending on the state of that gate signal, the SCA pulses were made to increment one of two counters, which thus represented "photons from the bunch with laser" and "photons from the bunch without laser."

With the Si drift detector, the spectroscopy amplifier, and the SCA used in this experiment, the count rate should not exceed about 50000 per second. It is not so much the loss of energy resolution with increasing count rate that imposes this limit, but the timing jitter of the SCA pulses due to overlapping pulses in the detector output. With too large a jitter, the SCA pulses, which come rather close to each other (see fig. 6.5), would not be clearly separable any more. Because the bunch repetition rate with the fill pattern shown in figs. 6.4

and 6.5 is about $6.5 \cdot 10^6$, the probability of a photon from any given bunch hitting the detector must be no greater than ca. 0.8%. This means that the count rate from the bunches that coincide with laser pulses must be limited to 8 per second, and it therefore takes a while to collect a sufficient number of photons.

4. Results

Figure 6.9 shows the Ga K_α fluorescence yield over the incident photon energy, recorded with and without the channel-cut crystal in place. The slope of the two curves at the absorption edge is the same and is therefore not due to instrumental effects but rather due to the lifetime broadening of the K shell. The channel-cut crystal does however help to suppress the fluorescence at the lower end, presumably because the multiple reflections suppress the wings of the energy bandpass (see fig. 6.11).

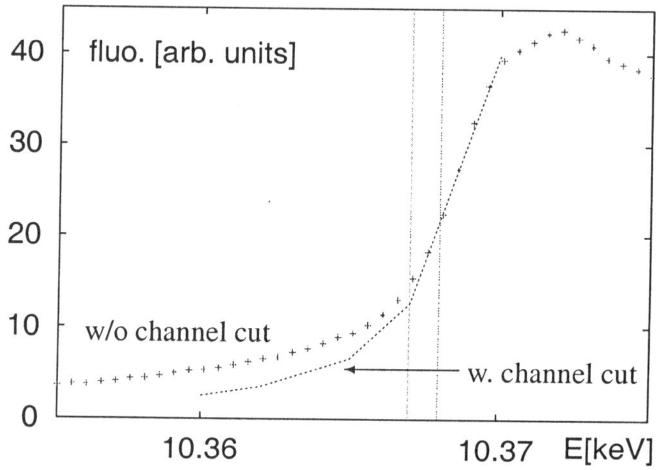

Figure 6.9: Ga K_α fluorescence yield at the Ga absorption edge. Crosses: Monochromator scan without the channel cut in place. Dashes: Monochromator scan with the channel cut in place and throughput maximized at each energy point. The vertical lines indicate the two energies chosen for the timing experiment (fig. 6.10).

Two incident energies, namely 10.367 keV and 10.368 keV, were chosen for the experiment by the following reasoning: the inflection point of the curve in fig. 6.9 is taken to correspond to the lower edge of the conduction band, i.e., where the lowest empty states are. Laser-induced changes in the x-ray absorption/fluorescence must occur at least 1.5 eV below that point, at the upper edge of the valence band. The energies of 10.367 keV and 10.368 keV bracket that point. If more beam time had been available, lower incident energies might have been probed, too. However, the strongest signal is to be expected right at the upper edge of the valence band because the photoexcited holes relax to that point within tens to hundreds of femtoseconds but stay there tens to hundreds of picoseconds, before recombination eliminates them.

For both of the above incident energies, the laser timing relative to the x-rays was scanned repeatedly and data taken at delays of $\{-300, -150, 0, 150, 300\}$ ps, where a positive delay means that the laser pulse comes before the x-rays. The fluorescence yield from the bunches "with laser" was normalized with the fluorescence yield from those "without laser" at each point in each scan separately, and the data were added up to produce the graphs in fig. 6.10. We estimate that the fluence of UV light was 10^{-4}J/cm^2, and the initial carrier density was about 10^{19}cm^{-3}.

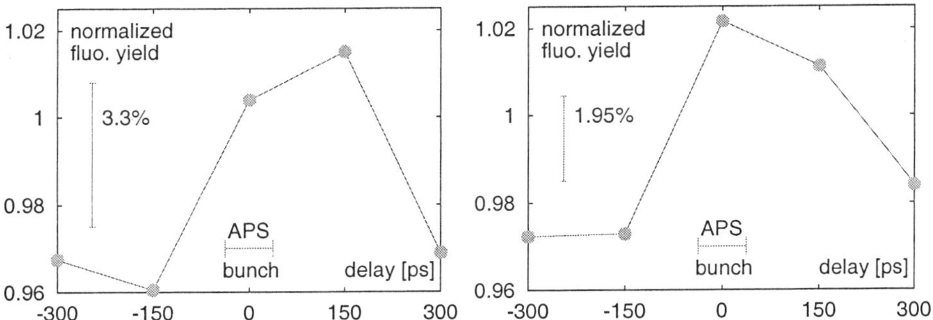

Figure 6.10: Fluorescence yield in laser timing scans. Left: at an incident photon energy of 10.367 keV, right at 10.368 keV (vertical lines in fig. 6.9). The error bars indicate the Poisson counting statistics. A positive delay means an earlier laser pulse. Time in the sample therefore runs from left to right.

5. Discussion

Details of the carrier dynamics in GaAs depend strongly on the strength of the excitation because of the multitude of contributions from phonon interactions and multicarrier effects. Nevertheless, the general picture emerges that relaxation of electrons within the X and L valleys of the conduction band occurs roughly within one picosecond, intervalley scattering on time scales from below one picosecond to ca. 10 ps, and full recombination within hundreds of picoseconds (see [149], fig. 7). The dominant dissipation mechanisms are carrier-carrier scattering for high, and carrier-phonon scattering for low excitation densities, respectively. The elementary time constant of the former is given by the plasma frequency $\omega_p^2 = 4\pi c^2 r_e \rho$ (ca. 105 fs at a carrier density of $\rho = 1 \cdot 10^{18}$ cm^{-3}), but with some modification due to screening effects. For the latter, the elementary time constant is typically 100 fs to 200 fs (the LO phonon in GaAs has a period of ca. 115 fs).

The signal in fig. 6.10 must therefore be due to photoexcited holes drifting to the top of the valence band, where they persist until recombination removes them. Ideally, the base line in the two graphs should be at unity because the fluorescence at the bunch "with laser" should be equal to that "without laser" if the laser excitation comes too late. Although it is not possible to determine the reason for the lower baseline with certainty, it is quite possible that the 23rd bunch had a lower than average charge. In the non-topup mode, it is quite normal for the the APS to have a bunch-to-bunch charge variation of 5% or more [150].

6. Experimental Issues

Some of the experience gained in the course of the experiment described here may be useful for similar projects. Two points turned out to be crucial and shall be discussed in more detail.[2]

6.1 Monochromatization

The quality of the monochromatization is critcal to the success of the kind of spectroscopic experiment shown here. It is not sufficient to specify only a width of the incident energy band because the intensity in the wings of the bandpass curve contributes to the amount of background. Figure 6.11 shows a comparison of a single versus multiple reflections of spectrally broad perfect plane-wave-type radiation from a perfect crystal.

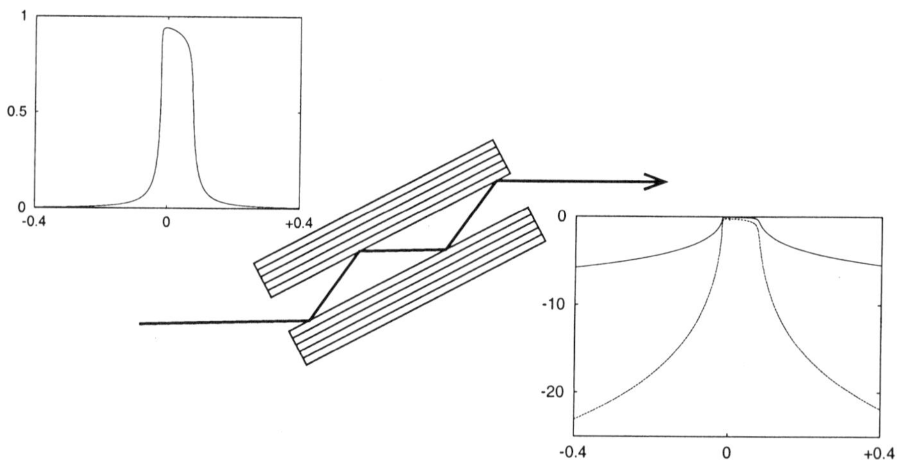

Figure 6.11: Improving the energy bandpass with multiple reflections. Upper left: the Si (333) reflectivity at 10.367 keV, plotted over relative detuning of ±0.4 eV, plot data generated with the program XOP [115]. Center: Schematic of a channel cut, here with two double bounces. Lower right: The same Si (333) reflection on a logarithmic scale. The comparison of the single- and the four-bounce reflectivity curves shows suppression of the wings.

Another effect that can broaden the energy bandpass is the Bragg-"rainbow" from a divergence due to a combination of the source divergence and a "bump" on the first monochromator crystal induced by the heat load of a few hundred Watts. This effect can both be measured and eliminated by the combination

[2]Please see footnote on p. 144.

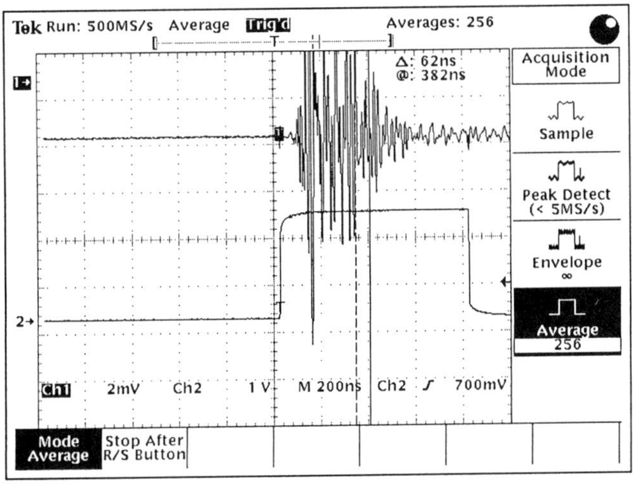

Figure 6.12: Pockels cell noise in the detector output (upper trace) and the counter gate signal (lower trace, see text).

of two reflections, such as here (111) in the monochromator and (333) in the channel cut. In the experiment described here, the combined source/thermal bump divergence was found to be less than ten percent of the natural width of the (111) reflection (details in [139]).

A focusing mirror before the monochromatizing elements can also introduce a divergence that contributes to an energy spread. Even if the mirror is meant to focus only horizontally and the monochromator diffracts in a vertically standing plane, optical aberrations do introduce a noticable effect [151]. This was however no concern at MHATT-CAT because the lens and the mirror came after the crystal optics.

6.2 Electronic Noise

X-ray photon detectors must be highly sensitive to be able to pick up events with energy contents of roughly 10^{-15} J, or even to resolve the energy within 10^{-17} J. Unfortunately, this makes them very susceptible to electronic noise pickup. The situation is of course even worse with photon detectors for visible light. Particularly in the context of the study of ultrafast phenomena, there are ample sources of high-frequency electronic noise that might be picked up by a sensitive amplifier in a photon detector. One such example, namely the switching noise from the Pockels cell in the Ti:sapphire amplifier and how to deal with it, are discussed here. Figure 6.12 shows the output of the Si drift detector used in the experiment described here, after passage along several ten meters of doubly shielded coaxial cable. This is the signal amplitude observed after careful analysis and improvement of current paths, grounding, etc. A question that comes up whenever a filter amplifier is used with a detector (see fig. 6.8) is whether to place it close to the detector inside the experimental hutch or rather outside. The latter choice is often preferred because of easier accessibility. When noise suppression is critical, the amplifier is in most cases kept close to the detector, so that coupling into the signal before amplification (where it is relatively stronger) is minimized. Here however, the opposite seems to be the case, and placement of the amplifier outside of the hutch is the better choice by the following reasoning: Being an eight-pole filter [148], the amplifier has an asymptotic high-frequency amplitude suppression of -24 dB ($= 4 \cdot 10^{-3}$) per octave. The lowest frequency in the Pockels cell noise was about 20 MHz which is 3 to 4 octaves above the characteristic frequency of the filter amplifier, given by its shaping time constant. Therefore, the suppression of the Pockels cell noise is theoretically 3*24 dB and thus much stronger than the amplifer gain of about 22 dB. One might ask if the signal in fig. 6.10 is perhaps an artefact induced by electronic noise. Corresponding to the persistence of the signal of ca. 250 ps would be a frequency of roughly 1.5 GHz, which is suppressed even more (theoretically by more than 20 dB), well below the 2% shown in the figure. As annoying as the Pockels cell noise might be, one good side to it was that it could be used for a rough adjustment of the gate signal to the counters.

7. Potential Applications

Two possible applications of the experiment described here are presented below. One is science-oriented, with the aim of providing information on carrier dynamcis in photoexcited semiconductors. The other one is instrumental, providing a way to measure the relative timing of laser and x-ray pulses on the femtosecond level.

7.1 Spectroscopy with an Absolute Energy Reference

The experiment presented here may be developed into a powerful spectroscopic technique to measure the dynamics of photoexcited carriers in semiconductors and other materials. Laser spectroscopic methods provide information only in the form of a convolution over the band structure because of the continuous nature of the density of states. In contrast, core levels are discrete and they are affected only very slightly by processes taking place within the band structure. Spectroscopy that involves core states can therefore provide direct, unconvoluted information. The core states should be at least several keV deep, so that transitions to/from them involve x-rays that can penetrate beyond the immediate surface (hard x-rays can be made surface specific by going to grazing incidence). Unfortunately, deep core levels are subject to considerable lifetime broadening (about 2.5 eV for the Ga K level) and can therefore not be used for high-resolution studies of the band structure. There is however a way out of this dilemma, namely the combination of x-ray absorption with x-ray fluorescence, using an analyzer crystal to narrow the detected fluorescence energy bandwidth [152], see schematic in fig. 6.13. In the ideal case of an infinitely narrow incident energy bandwidth and an infinitely narrow analyzer bandpass, one would be left only with the lifetime broadening of the upper level of the fluorescent transition, i.e., the Ga $L_{2/3}$ levels in the experiment described in this chapter. In real life, this must of course be convoluted with the instrumental resolution of the monochromator and the analyzer. This combination of x-ray absorption and fluorescence with an analyzer, and the choice of a narrow final state of the fluorescence, combines thus a narrow energy reference from a core state with little lifetime broadening with the penetrating power of hard x-rays.

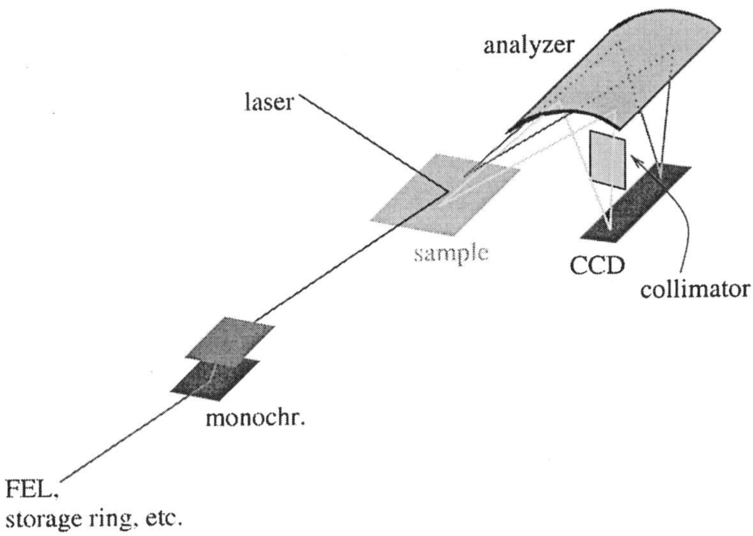

Figure 6.13: X-ray spectroscopy with fluorescence analyzer to eliminate the core hole lifetime broadening.

With the Si drift detector used in the above experiment, the choice of fluorescent lines was limited to Ga $K_{\alpha_{1,2}}$ because other lines would be too close to the incident energy to be resolved. However, $K_{\alpha_{1,2}}$ is not the best choice for high resolution spectroscopy because the $L_{2/3}$ levels themselves have a considerable lifetime width of ca. 0.8 eV [153]. With an analyzer, proximity of the fluorescent to the incident energy is of no concern, and one may choose a line with narrower final state, instead (here Ga K_{β_1}).

Another potentially interesting type of spectroscopy would be to set the analyzer to select a final state actually within the band structure. With a well-defined incident and analyzer bandpass energy, there would still be an energy reference, but now one would be probing carrier correlations within the band structure.

7.2 A Femtosecond Detector and X-Ray/Laser Correlator

As stated in the introduction, some of the elementary processes in chemistry and solid-state physics take place within a few femtoseconds. X-ray instrumentation for that time scale is not yet available. The effect studied in this chapter could be used for ultrafast x-ray detection and correlation with visible light. With an analyzer for the fluorescent radiation, it is possible to make detection of x-ray fluorescence dependent on the simultaneous presence of x-rays and *deep*-valence holes. With a few-cycle laser pulse, these can be created within femtoseconds, and their relaxation towards the top of the valence band, and thus out of range for the x-ray absorption/fluorescence process, takes place within roughly 100 fs. Figure 6.14 shows a schematic of a correlator. The x-rays

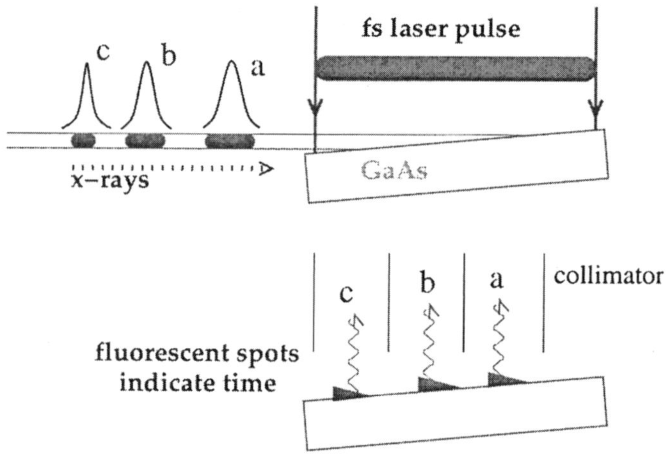

Figure 6.14: A laser/x-ray correlator. X-rays with a photon energy just below the Ga K edge and with a femtosecond time structure generate fluorescent spots on a GaAs surface when/where they coincide with a femtosecond laser pulse.

strike the surface of the converter material (such as GaAs) at a grazing angle, while a short laser pulse is incident at a large angle. Due to the crossing beam geometry, time is converted to location on the surface. A high x-ray intensity occurs where the x-rays coincide with the laser pulse. The fluorescent spots have tails to the right in the figure because of the asymmetry in the times for hole creation (few-fs) and hole relaxation (100 fs).

Acknowledgments

Operation of the MHATT-CAT sector 7 beamline at the Advanced Photon Source was supported by DOE grant No. DE-FG02-99ER45743. Use of the APS and one of us (B.W.A.) were supported by the U.S. Department of Energy under the contract W-31-109-ENG-38. We want to thank D. Arms for his support of the experiment.

Chapter 7

ULTRAFAST STRUCTURAL CHANGES INDUCED BY FEMTOSECOND LASER PULSES

H.O. Jeschke

Department of Physics and Astronomy, Rutgers University, 136 Frelinghuysen Road, Piscataway, NJ 08854-8019, USA

jeschke@physics.rutgers.edu

M.E. Garcia

Institut für Theoretische Physik der Freien Universität Berlin, Arnimallee 14, 14195 Berlin, Germany

garcia@physik.fu-berlin.de

Abstract A theoretical approach to treat laser induced femtosecond structural changes in covalent solids is described. Our approach consists in molecular dynamic simulations performed on the basis of a time-dependent, many-body potential energy surface derived from a tight-binding Hamiltonian. The shape and spectral composition of the laser pulse is explicitly taken into account in a non-perturbative way. We show a few examples of the application of this approach to describe the laser excitation of bulk diamond and ultrathin graphite and silicon films.

1. Introduction

The interaction of femtosecond laser pulses and solids gives rise to a large variety of novel and interesting effects. An important part of the laser induced ultrafast phenomena observed recently is concerned with structural changes. Laser pulses of durations ranging from a few femtoseconds up to half a picosecond and intensities above 10^{12} W/cm^2 may induce lattice instabilities which lead to thermal and nonthermal structural changes.

In the last years different types of lase induced ultrafast structure effects were observed, like excitation of coherent phonons [154, 155, 156, 157, 158] or nonequilibrium phase transitions, the latter including order-disorder [159, 160], solid-solid [161, 162] and solid-liquid [163, 164, 165, 166, 167, 157, 168, 169,

158, 170, 171] transformations. While the first experimental studies were based on optical methods (measurement of the reflectivity or dielectric function), several interesting x-ray experiments have been recently performed to analyze laser induced melting and crystalline-crystalline structural changes. In most experiments the rocking curves (changes in time of the Bragg angle) were measured [157, 168, 169, 158, 170, 171]. All these results show clear evidences of nonthermal structural changes.

Regarding the systems studied experimentally, a group of materials that merit particular attention are semiconductors because of their technological significance. Of special interest are materials like silicon (Si) and gallium arsenide (GaAs) which are widely used in semiconductor devices; here short time behavior becomes ever more important as the frequencies at which these devices are operated increase [172, 167]. The study of electronic relaxation time scales is very important for the development of nanodevices. A semimetal like germanium-doped antimony (GeSb) or carbon nanotubes are likely candidates for future technological applications like ultrafast optical storage media or cluster assembled materials [173, 174]. Finally, covalent solids like diamond, graphite and other carbon materials like DLC are of enormous technological interest because of their physical properties including extreme hardness and very high thermal conductivity. For these materials there is a large field of research concerning ultrashort laser-induced modification, annealing, ablation and machining [175].

Covalent systems are also interesting from the viewpoint of basic research, since their structural response differs dramatically from that of metals. For low or moderate laser pulse intensities, the different processes which take place during and after excitation of metals can be described by a simple physical picture, sketched in Fig. 7.1.

The femtosecond pulse creates hot electrons, which transfer their energy to the lattice via electron-phonon coupling and to the environment via diffusion. No laser induced bond breaking occurs. These processes are well described by the well known two-temperature model [176], which consists of a system of coupled differential equations for the electronic temperature $T_{\mathrm{el}}(t)$ and the lattice temperature $T_{\mathrm{lattice}}(t)$. Equilibrium parameters, like the lattice and electron specific heats, are included in the model. In contrast to the rather simple case of metals the femtosecond laser excitation of covalent solids exhibits new and interesting features. First, absorption of photons in covalent solids leads rapidly to bond-breaking processes. As soon as bonds are broken, the picture of electron-phonon coupling is no longer valid. Moreover, the concept of phonons becomes meaningless, since the potential landscape cannot be represented any more by a superposition of harmonic potentials. As a consequence, the two-temperature model breaks down and a different theoretical approach is necessary. Such an approach must take into account explicitly the atomic coor-

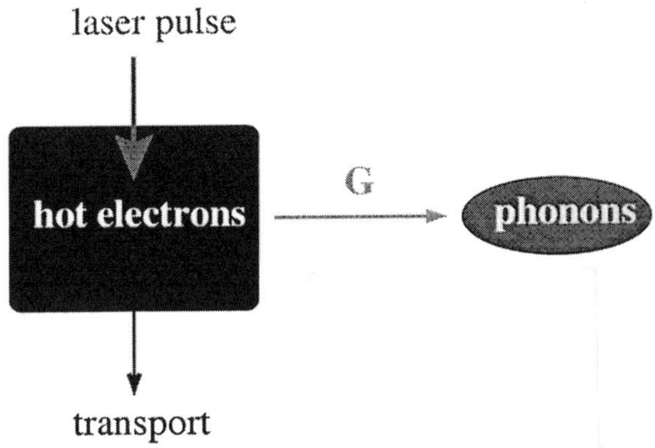

Figure 7.1: Schematic illustration of the processes taking place upon laser excitation of metals for low laser intensities. These processes can be well described by the two-temperature model.

dinates and must yield a correct description of the bonding. In this chapter we will describe in detail a microscopic theory for laser induced structural changes in covalent solids.

Good candidates for exhibiting interesting properties upon laser excitation are the different solid phases of Carbon, described in detail by Bundy *et al.* [175]. The open questions related to the phase diagram are still of fundamental interest. In particular, the question of whether several distinct liquid phases of carbon exist [177, 178], and if these phases can be induced by laser pulses attracts considerable attention.

Despite the intense experimental activities, only a few theoretical methods were derived in the last years to describe laser induced structural changes. Theoretical descriptions of laser induced structural changes in the different systems under experimental study like metals, covalent solids and ionic crystals are needed which are able to predict the effect of the influence of pulse duration and intensity on the lattice dynamics. As mentioned above, this chapter will deal with a complete theoretical description for the case of covalent solids.

In the literature, different theoretical approaches to the response of covalent solids to strong laser excitation can be found, which we mention in the following. P. Stampfli et al. [179, 180, 181] studied for the first time laser-induced lattice instabilities in Si, Ge and C. They analyzed the instability caused by the electron hole plasma in terms of one or two phonon degrees of freedom, and assuming that the density of the excited material remains constant during and after laser heating. This method can only give insight into the first femtoseconds after the laser pulse. By assuming the instant creation of an electron-hole plasma of a given density ξ at a time $t = 0$, the duration of the laser pulse is set to zero ($\tau = 0$). Herrmann, Gerlach and Campbell [182] have performed MD simulations of the laser ablation of silicon. They use modified Stillinger-Weber potentials to model the interaction between the atoms. As we show in section 2, the use of model interatomic potentials which do not include the electrons as degrees of freedom do not allow for a theoretical treatment of laser induced structural changes.

Parrinello and coworkers employed first principles MD simulations [183, 184, 185]. These are based upon a treatment of the electrons and ions in the lattice with density functional theory in the local density approximation and with plane waves as basis functions. The volume of the MD cell was kept constant. The numerical cost of such calculations is very high [185], so that these methods would not permit a study of many different laser intensities and durations on various materials as was the intention of this work. Concerning the results obtained for the laser melting of silicon and graphite, respectively [184, 186], it seems possible that the choice of a constant volume MD method imposed too severe a restriction on the relaxation channels available to the material. Likewise, the results of a study of the liquid state of carbon [183] should be quite sensitive to the density, and thus a calculation at constant (graphitic) density might not answer all questions on the nature of molten carbon. Following Stampfli et al., these authors also assume for their melting studies the instant creation of a high electron temperature T_e at a time $t = 0$. Thus, the laser pulses are taken to be delta functions of duration $\tau = 0$ which makes it impossible to analyze the influence of the pulse duration. A similar approach as the one described above, but using a tight-binding basis instead of Kohn-Sham functions has been used by K. M. Ho and coworkers [187].

The method which will be presented in the next section was developed by us to study laser excitation of diamond, graphite, C_{60}-crystal and silicon in the bulk [161, 188, 189, 190, 191] and in ultrathin films [192, 193], and of C_{60} clusters [194].

Independently of our studies, R. Allen and coworkers developed a similar method to describe the excitation of coherent phonons and laser induced melting [195, 196, 197, 198]. This method is suitable for the study of the first few femtoseconds after excitation, since the assumption of constant volume is made and it is assumed that the electronic system does not undergo decoherence and thermalization processes after laser excitation.

The aim of our investigations is to understand on which time scale do laser-induced structural transitions occur, which relaxation channels are present and wheter these relaxation channels be controlled by laser parameters like intensity, duration, and light frequency.

An important question which we attempt to answer is if there are fundamental differences between laser-induced and thermally induced bond breaking and phase transitions.

Important aspects taken into account in our theory are:

(a) the admission of all relaxation channels in the response to an electron-hole plasma, including the calculation for all degrees of freedom and the admission of changes in sample volume and geometry,

(b) the handling of the electronic nonequilibrium caused by strong laser excitation and its subsequent equilibration and relaxation, and

(c) the treatment of laser pulses with finite durations $\tau > 0$.

The following section gives a brief account of the method applied in this work which will show how these extensions of previous theoretical work will be performed.

2. Theory

Our model for the femtosecond laser excitation of covalent solid is based on the following physical picture. Since light couples primarily with the electronic system, a laser pulse of moderate intensity produces the excitation of electrons from occupied to unoccupied levels. This creation of electron-hole pairs occurs with a time-dependent probability which is proportional to the (time-dependent) intensity of the laser field. As a consequence of the extremely fast excitation process, a nonequilibrium distribution of electrons is created. Through electron-electron collisions this electron distribution thermalizes to an equilibrium (Fermi-like) occupation of the electronic levels. Parallel to this electron thermalization process, hot electron transport from the excited region into the rest of the material sets in. During this complex electron dynamics the lattice does not remain frozen, but might undergo bond breaking and restructuring processes. These structural changes may lead to another crystal phase, to melting or to ablation. We show in this section that the reason why the lattice structure undergoes changes upon laser heating is related to the rapid and nonadiabatic change in time of the potential energy landscape $U(\{\mathbf{r}_{ij}\}, t)$ of the system when a large fraction of the electrons is excited. Note, that $U(\{\mathbf{r}_{ij}\}, t)$ is usually a very complex function of the interatomic distances $\mathbf{r}_{ij} = |\mathbf{r}_i - \mathbf{r}_j|$ and, for most materials, cannot be written as a sum of pair potentials.

We describe laser induced structural changes with the help of molecular dynamics simulations (MD). This means, that we treat the nuclear degrees of freedom of the problem classically. However, the electron dynamics must be treated quantum mechanically. The forces acting on the atoms are obtained as $\mathbf{F}_i(t) = -\vec{\nabla}_i \; U(\{\mathbf{r}_{ij}\}, t)$ and cause the lattice dynamics, described by Newton's equations of motion.

As mentioned in the previous section, special attention has to be given to the choice of the external parameters under which the MD simulations are performed. Simulations at constant volume would not be able to describe laser induced structural transformations involving density changes. The assumption

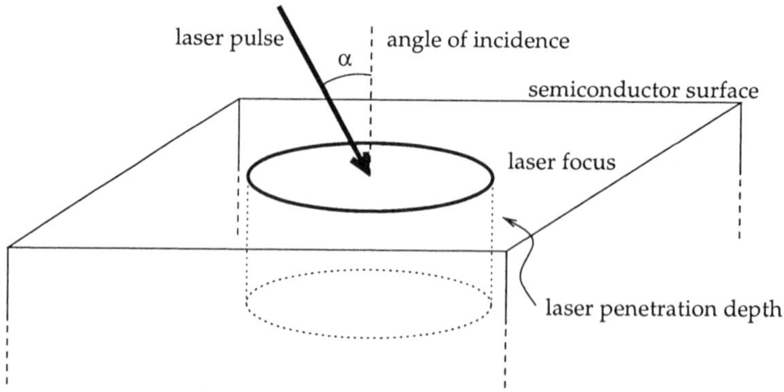

Figure 7.2: Sketch of the geometry of a laser pulse acting upon a solid surface. Note that this figure does not show the length scales in proportion; see text for dimensions.

of constant volume would artificially suppress such relaxation channels and would not describe correctly the laser heating process.

It is known [199] that the laser excited material reacts by a very fast expansion, which will be stronger in areas that have absorbed more laser energy; the laser intensity distribution can be expected to be Gaussian in space [200]. Thus, a surface profile as shown schematically in Fig. 7.2 is possible after a few picoseconds. For this reason, one possible approach to describe the situation is to make the volume and shape of the MD supercell variable, depending on the forces that prevail within. On the other hand, the considered MD supercell will because of its small extensions be surrounded by material that is subject to approximately the same degree of excitation. Therefore, locally a translation invariance is existent that is the prerequisite for the use of periodic boundary conditions.

Basically, we consider two different MD approaches, which will now be explained in some detail.

2.1 Lattice motion: molecular dynamics simulations

When we deal with the situation of an ultrafast laser pulse impinging on the surface of a semi-infinite solid, the geometry is as shown in Fig. 7.2. The typical lateral extension of the laser spot is 15 μm [165] and the penetration depth of the laser light is of the order of 100 nm, depending, of course, on the material and the frequency of the laser pulse. Consequently, the volume of the material excited by the laser is approximately $7 \times 10^{13} \mathring{A}^3$, while the volume of a typical MD supercell is $\Omega = 10^3 \mathring{A}^3$. Thus, the theory presented here can only deal with a very small sector of the entire system of laser excited material plus surrounding cold lattice. Taking this into account, we perform the following

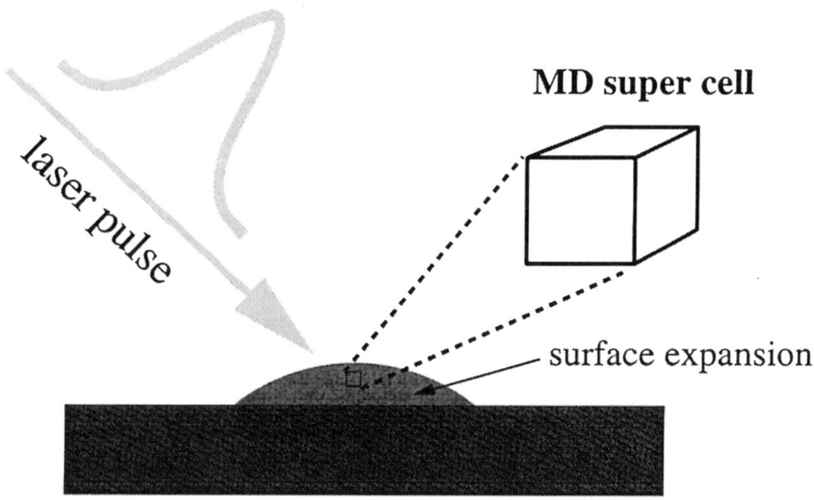

Figure 7.3: Illustration of the model used for the MD supercell to describe the time evolution of a piece of bulk solid upon laser excitation. The laser excited region of the covalent material expands into the vacuum due to femtosecond laser heating. We place the MD supercell inside the expanding region and use periodic boundary conditions in all three directions, simulating a small region of translational invariance. Shape and size changes of the MD supercell are permitted.

MD studies:
(i) We consider a subregion of volume $\Omega_{MD} \sim 10^3$ Å^3, represented by a MD supercell with periodic boundary conditions all directions, inside the much larger irradiated volume ($\Omega_{irr} \sim 10^{13}$ Å^3). Fig. 7.3 shows schematically where the MD cell is placed and which part of the whole system it represents. Clearly, the cell is expected to undergo homogeneous rapid volume and shape changes upon laser heating. The only constraint acting on the MD cell is the external pressure on the solid, which remains constant.

(ii) We consider ultrathin films made out of a few number of layers. In this case, as shown in Fig. 7.4, the system can be described by a MD supercell which is periodically continued in the horizontal directions. The MD cell is rigid in order to account for the inertial confinement. However, it contains vacuum above and below the material, so that expansion of the system in the vertical direction is allowed. No assumptions on the external pressure are made.

Both techniques (i) and (ii) can be mathematically formulated as follows.

(i) Variable MD cell inside the heated region (bulk). We perform simulations at constant pressure [201, 202], which are based upon a Lagrangian of

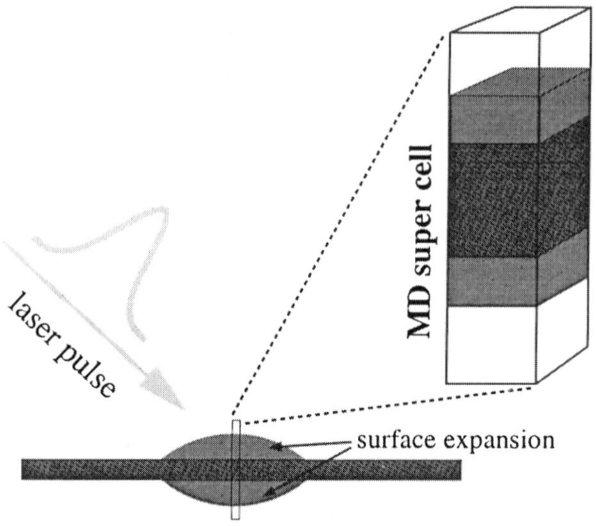

Figure 7.4: Modeling of the MD supercell for femtosecond laser heating of a thin film. The laser excited film expands into the vacuum on both sides. The MD supercell contains vacuum above and below the film in order to allow for an unhindered expansion perpendicular to the film. In the direction perpendicular to the surface, periodic boundary conditions are assumed.

the form

$$L(t) = \sum_{i=1}^{N} \frac{m_i}{2} \, \dot{\mathrm{s}}_i^T \, h^T \, h \, \dot{\mathrm{s}}_i + K_{cell} - U(\{r_{ij}\}, t) - P\Omega. \tag{7.1}$$

Here, the coordinates s_i of the N atoms are taken relative to the vectors $\mathbf{a}$, $\mathbf{b}$ and $\mathbf{c}$ that form the MD supercell and constitute the matrix $h = (\mathbf{a}\ \mathbf{b}\ \mathbf{c})$. Note that the primitive vectors $\mathbf{a}$, $\mathbf{b}$ and $\mathbf{c}$ can move as a consequence of the laser excitation, changing the shape and the volume of the MD cell. The absolute coordinates of the atoms are obtained from $\mathbf{r}_i = h\mathrm{s}_i$. The volume of the MD cell, which is allowed to change in time is given by $\Omega = \det(h)$, and P is the external pressure, which remains constant. The first term of the Lagrangian accounts for the kinetic energy arising from the motion of the N atoms relative to the MD cell. The superscript T denotes the transposition of cell matrix h and relative velocity vectors $\dot{\mathrm{s}}_i$. The second term (K_{cell}) represents the contribution of the motion of the primitive vectors, i.e., the time-dependent shape changes of the MD cell, to the kinetic energy of the system [201, 202, 203]. There are different approximations for K_{cell}, which will be discussed below. The third term represents the potential energy surface $U(\{r_{ij}\}, t)$, which is determined from a microscopic theory as described below. The quantities r_{ij} are the inter-atomic distances. Note that, since periodic boundary conditions are assumed in all directions, the interatomic distances are calculated as

$$r_{ij} = \sqrt{[h(\mathbf{s}_i - \mathbf{s}_j + \mathbf{z})]^2}\,, \qquad z_\alpha \in \{-1, 0, 1\}\,. \tag{7.2}$$

Using the Euler-Lagrange formalism, $N + 9$ equations of motion for the atomic coordinates $\mathbf{s}_i$ and for the MD cell coordinates h_{kl} can be derived from Eq. (7.1) [201, 202].

Before writing the equations of motion we point out that their form will depend on the explicit form of K_{cell}. For the form of K_{cell} there are a number of proposals. One of the simplest is that by Parrinello and Rahman [202], which we will denote with K_{PR} here:

$$K_{\mathrm{PR}} = \frac{w_{\mathrm{PR}}}{2}\,\mathrm{Tr}(\dot{h}^{\mathrm{T}}\dot{h})\,. \tag{7.3}$$

In this case, the Euler-Lagrange formalism yields the equations of motion

$$\ddot{\mathbf{s}}_i = -\frac{1}{m_i}\sum_{j\neq i}\frac{\partial U(r_{ij})}{\partial r_{ij}}\frac{\mathbf{s}_i - \mathbf{s}_j}{r_{ij}} - g^{-1}\dot{g}\dot{\mathbf{s}}_i\,, \tag{7.4}$$

and

$$\ddot{h} = (\Pi - p_{\mathrm{ext}})\frac{\sigma}{w_{\mathrm{PR}}}\,, \tag{7.5}$$

with

$$\Pi = \frac{1}{\Omega}\sum_{i=1}^{N} m_i\mathbf{v}_i\mathbf{v}_i^{\mathrm{T}} - \frac{1}{\Omega}\sum_{i=1}^{N}\sum_{j>i}\frac{\partial U(\{r_{ij}\},t)}{\partial r_{ij}}\frac{\mathbf{r}_{ij}\mathbf{r}_{ij}^{\mathrm{T}}}{r_{ij}}\,. \tag{7.6}$$

Here, the definitions $g = h^{\mathrm{T}}h$, $\mathbf{v}_i = h\dot{\mathbf{s}}_i$ and $\mathbf{r}_{ij} = \mathbf{r}_i - \mathbf{r}_j + h\mathbf{z}$ have been used and $\sigma = (\sigma_{\alpha\beta})$ with $\sigma_{\alpha\beta}$ calculated as $\sigma_{\alpha\beta} = \partial\Omega/\partial h_{\alpha\beta}$.

Improved choices of K_{cell} avoid the problem that K_{PR} does not fulfil the virial theorem [204, 203] and is not invariant with respect to the form of the MD supercell [203, 205]. The proposal by Cleveland reads [203]

$$K_{\mathrm{C}} = \frac{w_{\mathrm{C}}}{2}\,\mathrm{Tr}(\dot{h}\,\sigma^{\mathrm{T}}\,\sigma\,\dot{h}^{\mathrm{T}})\,, \tag{7.7}$$

where σ is defined as in Eq. (7.5). With this term for K_{cell} inserted into the Lagrangian (Eq. (7.1)), the equations of motion for the relative atomic coordinates $\mathbf{s}_i$ do not change with respect to Eq. (7.4), while those for the cell coordinates h become

$$\begin{aligned}\ddot{h} &= (\Pi - p_{\mathrm{ext}})\frac{h}{w_{\mathrm{C}}\,\Omega} \\ &\quad - \frac{2\dot{\Omega}\,\dot{h}}{\Omega} + \mathrm{Tr}(\dot{h}\,\sigma^{\mathrm{T}}\,\sigma\,\dot{h}^{\mathrm{T}})\frac{h}{\Omega^2} + \frac{\dot{h}\,\sigma^{\mathrm{T}}\,\dot{h}}{\Omega} + (\dot{h}\,\sigma^{\mathrm{T}}\,\sigma\,\dot{h}^{\mathrm{T}} - \sigma\,\dot{h}^{\mathrm{T}}\,\dot{h}\,\sigma^{\mathrm{T}})\frac{h}{\Omega^2}\,.\end{aligned} \tag{7.8}$$

Here, the microscopic stress tensor Π is again calculated according to Eq. (7.6). (Note, in the last term of this equation a misprint in Eq. (18) of Ref. [203] has been corrected.)

In $K_{\rm PR}$ and $K_{\rm C}$ the parameters $w_{\rm PR}$ and $w_{\rm C}$ appear, and they play a role in the equations of motion for h that is analogous to that of the masses m_i in the equations of motion for the s_i. Thus, while $w_{\rm PR}$ and $w_{\rm C}$ can be chosen as convenient for the calculation if only statistical averages over trajectories are determined, they must be chosen with care if dynamical properties are calculated. According to Andersen [201], these parameters should be adjusted so that the fluctuations of the MD supercell volume Ω take place on a time scale of $\sqrt[3]{\Omega}/c_M$, where c_M is the speed of sound in the material. For graphite and diamond, we have found $w_{\rm PR} = M/25.5$ to fulfil this condition, where M is the total mass of all atoms in the MD supercell. For silicon, we use $w_{\rm PR} = M/1.6$.

(ii) Ultrathin films. In this case the simulations are performed according to the geometry illustrated in Fig. 7.4. The MD cell is rigid, does not change its shape or volume in time, and defined by the constant primitive vectors $\mathbf{a} = (a, 0, 0)$, $\mathbf{b} = (0, b, 0)$ and $\mathbf{c} = (0, 0, c)$. Periodic boundary conditions in the lateral directions (defined by $\mathbf{a}$ and $\mathbf{b}$) are used. Since the primitive vectors are rigid, this means that we do not allow expansion of the system in the lateral directions. This accounts for the inertial confinement which is produced by the rest of the film. In contrast, expansion of the system along the vertical directions $\mathbf{c}$ and $-\mathbf{c}$ are, of course, allowed. This is achieved by setting a large value for the constant c, so that enough vacuum is present above and below the film. No assumptions are made regarding the external pressure. In view of the above discussion the second and the fourth terms of the Lagrangian (Eq. (7.1)) are dropped and the Lagrangian L_{film} for simulations on ultrathin films is given by

$$L = \sum_{i=1}^{N} \frac{m_i}{2} \dot{\mathrm{s}}_i^{\mathrm{T}} h^{\mathrm{T}} h \dot{\mathrm{s}}_i - U(\{r_{ij}\}, t)\,. \tag{7.9}$$

Using the Euler-Lagrange equations we can again calculate the equations of motion for the atoms

$$\ddot{\mathrm{s}}_i = -\frac{1}{m_i} \sum_{j \neq i} \frac{\partial U(\{r_{ij}\}, t)}{\partial r_{ij}} \frac{\mathrm{s}_i - \mathrm{s}_j}{r_{ij}}\,. \tag{7.10}$$

These equations describe the motion of atoms in a MD supercell with periodical boundary conditions for the case that the cell size and shape does not change during the calculation. For the cell there are no equations of motion, since $\ddot{h} = 0$.

2.2 Potential energy surface: laser induced electron dynamics

The molecular dynamics techniques described above constitute a standard method to study structural properties of solids in thermodynamical equilibrium. In those cases, the potential energy surface $U(\{\mathbf{r}_{ij}\})$ is usually approximated by a model potential, like, for example, Lennard-Jones [206], Tersoff [207] or Stillinger-Weber [208], which consist of functions of the interatomic distances $\mathbf{r}_{ij}$. For the description of laser induced structual changes, however, such model potentials are not useful, since they do not contain the electrons as degrees of freedom. Therefore, they do not allow any reasonable description of the excitation process. As mentioned at the beginning of this section, the laser pulse couples primarily with the electronic system. Hence, for the studies to be presented in this chapter, the many-body potential energy surface $U(\{\mathbf{r}_{ij}\}, t)$ must be derived from a microscopic electronic Hamiltonian.

We assume that the valence electrons of the covalent solid can be described by an effective single-particle Hamiltonian H_{eff} in a restricted Hilbert space spanned by M functions. H_{eff} depends, of course, on the electronic degrees of freedom, but also on the interatomic distances $\mathbf{r}_{ij}$, since the properties of the electrons change if $\mathbf{r}_{ij}$ change. The effect of the rest of the core electrons and the internuclear repulsion is approximated by a repulsive potential $E_{\text{rep}}(\{\mathbf{r}_{ij}\})$. As effective Hamiltonian we use

$$H = \sum_{i\alpha} \epsilon_{i\alpha} n_{i\alpha} + \sum_{\substack{ij\alpha\beta \\ j\neq i}} V_{ij}^{\alpha\beta}(r_{ij}) c_{i\alpha}^{+} c_{j\beta}. \tag{7.11}$$

Here, $\epsilon_{i\alpha}$ is the on-site energy of atom i and orbital α. $c_{i\alpha}^{+}$ and $c_{j\alpha}$ are the creation and annihilation operators, and $V_{ij}^{\alpha\beta}(r_{ij})$ the hopping integrals. For the description of carbon and silicon, the $2s$, $2p_x$, $2p_y$ and $2p_z$ orbitals are taken into account. For the radial part $R(r)$ of the hopping integrals and for E_{rep} we employ the forms proposed by Xu et al. [209]:

$$R(r) = \begin{cases} \left(\frac{r_0}{r}\right)^n \exp\left\{n\left[-\left(\frac{r}{r_c}\right)^{n_c} + \left(\frac{r_0}{r_c}\right)^{n_c}\right]\right\}, & r < r_1 \\ d_0 + d_1(r-r_1) + d_2(r-r_1)^2 + d_3(r-r_1)^3, & r \geq r_1 \end{cases} \tag{7.12}$$

$$E_{\text{rep}} = \textstyle\sum_i f\Big(\sum_j \phi(r_{ij})\Big) \quad \text{with} \tag{7.13}$$

$$\phi(r) = \begin{cases} \phi_0\left(\frac{d_0}{r}\right)^m \exp\left\{m\left[-\left(\frac{r}{d_c}\right)^{m_c} + \left(\frac{d_0}{d_c}\right)^{m_c}\right]\right\}, & r < d_1 \\ c_0 + c_1(r-d_1) + c_2(r-d_1)^2 + c_3(r-d_1)^3, & r \geq d_1 \end{cases} \tag{7.14}$$

Thus, for distances $r > r_1$ and $r > d_1$ respectively, the hopping integrals and $\phi(r)$ are replaced by polynomials that go to zero and are fitted smoothly to the functions valid for $r < r_1$ and $r < d_1$. The repulsive potential $E_{\rm rep}$ is taken as a polynomial of the function $\phi(r)$. For all parameters in Eqs. (7.12)-(7.14) see Ref. [209].

The effective single-particle Hamiltonian $H_{eff}(\{\mathbf{r}_{ij}\})$ [Eq. (7.11)] yields M eigenvalues ϵ_m, which clearly depend on $\{\mathbf{r}_{ij}\}$.

The time-dependent potential energy surface $U(\{\mathbf{r}_{ij}\}, t)$ which governs the motion of the atoms is equal to the free energy of the electrons $F_{el}(\{\mathbf{r}_{ij}\}, t)$, and given by

$$U(\{\mathbf{r}_{ij}\}, t) = F_{el}(\{\mathbf{r}_{ij}\}, t) = \sum_m n(\epsilon_m, t)\epsilon_m + E_{\rm rep}(\{\mathbf{r}_{ij}\}) - T_{el}(t)S_{el}(t), \tag{7.15}$$

where $n(\epsilon_m, t)$ are the occupations of the corresponding electronic levels ϵ_m. Due to the presence of the laser pulse, these occupations are time-dependent, as we show below. The first term of Eq. (7.15) represents the attractive contribution from the valence electrons. The third term contains the electronic temperature T_{el} and the electronic entropy S_{el}, which is given by

$$S_{el} = -k_B \sum_m \left[n(\epsilon_m, t) \log\left(n(\epsilon_m, t)\right) + (1 - n(\epsilon_m, t)) \log\left(1 - n(\epsilon_m, t)\right)\right]. \tag{7.16}$$

Note that the functional dependence of $U(\{\mathbf{r}_{ij}\}, t)$ on the interatomic distances $\{\mathbf{r}_{ij}\}$ is strongly dominated by the electronic occupations $n(\epsilon_m, t)$ present in the first and third terms of Eq. (7.15). If the electronic occupations undergo strong changes, then the potential energy landscape $U(\{\mathbf{r}_{ij}\}, t)$ will change significantly.

In view of the above discussion one can understand the mechanisms for laser induced structural changes by using a simple physical picture sketched in Fig. 7.5. Before laser heating the solid is in thermodynamical equilibrium at a low initial temperature ($T_i \leq 300K$). This means that electrons fill mainly the states below the Fermi level. The resulting ground state potential landscape $U_i(\{\mathbf{r}_{ij}\})$ shows minima at the crystal lattice sites. Atoms are sitting on these sites and therefore no forces act on them. However, the situation changes dramatically when the solid becomes heated by a laser pulse. For moderate and high fluences the laser creates a considerable amount of electron-hole pairs, as illustrated in Fig. 7.5. This means that the electronic occupations $n(\epsilon_m, t)$ change. Consequently, the functional form of $U(\{\mathbf{r}_{ij}\})$ undergoes quantitative and qualitative changes, which result in the dissappearance or shifting of the minima. Since these modifications occur on a time scale which is much shorter than the reaction time of the lattice, the atoms remain in their positions, most of them no longer minima of the new potential landscape. Therefore, forces act now on the atoms, which start to move. This is the way laser induced structural changes take place.

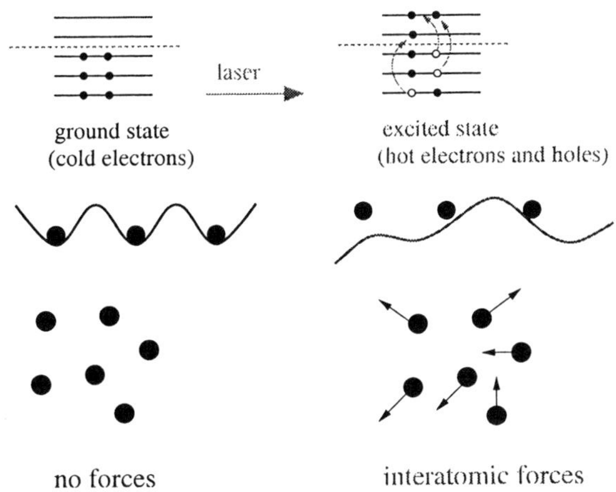

Figure 7.5: Illustration of the microscopic physical picture derived from Eq. (7.15). a) Situation before the action of the laser pulse: the solid is in thermodynamic equilibrium. For low temperatures electrons fill the states up to the Fermi level. The atoms are at the equilibrium positions of the ground state potential energy surface. b) The laser pulse changes the electronic occupations. This leads to rapid changes in the potential energy landscape. As a consequence, the lattice becomes unstable and forces on the atoms appear, driving a structural change.

Fig. 7.5 does not show how the electron occupations change in time. This is a fundamental point, which must be taken carefully into account by any theory aimed at the description of laser induced structural effects. In order to obtain an equation for the rate of change of the electron occupations $n(\epsilon_m, t)$ we consider the equation of motion for the density matrix $\hat{\rho}$, which reads

$$\dot{\hat{\rho}} = -\frac{i}{\hbar}\left[H_{TB} + V_{laser}, \hat{\rho}\right] + \left.\frac{\partial \hat{\rho}}{\partial t}\right|_{coll} . \qquad (7.17)$$

The first term refers to the coherent motion of the electrons, which is driven by the laser field and which involves optical transitions between the energy levels of the Hamiltonian H_{TB}. Since the tight-binding Hamiltonian is a single-particle one, it does not describe interactions between the excited electrons, which are essential for the thermalization processes. Thus, we treat them by including a second, dissipative term in Eq. (7.17).

Eq. (7.17) represents a system of coupled differential equations for the diagonal and nondiagonal elements of the density matrix. However, since the electron-hole plasma created by the laser undergoes rapid dephasing, the relevant equations of motion will be those describing the diagonal elements. Therefore, neglecting the contribution of the nondiagonal elements one obtains [210]

$$\frac{dn(\epsilon_m, t)}{dt} = \int_{-\infty}^{\infty} d\omega\, g(\omega, t-\tau)\Big\{[n(\epsilon_m - \hbar\omega, t-\tau) \quad (7.18)$$
$$+n(\epsilon_m + \hbar\omega, t-\tau) - 2n(\epsilon_m, t-\tau)]\Big\} - \frac{n(\epsilon_m, t) - n^0(\epsilon_m, T_{el})}{\tau_1}.$$

Here, the first (integral) term describes the laser excitation and creation of a nonequilibrium electron distribution. $g(\omega, t)$ is the intensity function of the laser pulse, which describes the intensity as a function of frequency and time as a product of the time envelope and its Fourier transform. Thus, the electronic distribution is at each time step folded with the current intensity function $g(\omega, t)$. This means that at each time step, the occupation of an energy level ϵ_m changes in proportion to the occupation difference with respect to levels at $\epsilon_m - \hbar\omega$ and at $\epsilon_m + \hbar\omega$. In Eq. (7.18), the optical matrix elements are assumed to be equal to unity, so that the band structure plays the main role in the absorption process. This simplification makes sense, as the calculation of optical matrix elements is at the same time tedious and known to be rather inaccurate in the TB approximation. Eq. (7.18) can be straightforwardly generalized to include the dipole matrix elements. The second term of Eq. (7.18) describes the electronic thermalization resulting from electron-electron collisions through a relaxation time τ_1. Thus, with a time constant τ_1, the distribution $n(\epsilon_m, t)$ approaches a Fermi-Dirac distribution $n^0(\epsilon_m, T_e)$, which is given by

$$n^0(\epsilon_m, T_e) = \frac{1}{\exp\{\beta(\epsilon_m - \mu) + 1\}}, \quad \text{with } \beta = \frac{1}{k_B T_e}. \quad (7.19)$$

Here, T_e is the electron temperature, and μ is the chemical potential. This simple approach works very well because for dense electron-hole plasmas in covalent solids extremely low relaxation times τ_1 have been found. Chemla and coworkers have reported a carrier thermalization faster than 10 fs in GaAs [211]. Thus, we use $\tau_1 = 10$ fs throughout this chapter. Note, for such a short thermalization time the exact electronic dynamics leading to electronic equilibrium do not play a significant role for the structural changes we are studying here.

The electronic temperature T_e and the chemical potential μ, which appear in the Fermi-Dirac distribution and which are not determined by Eq. (7.18), need to be fixed by an additional principle. We will demand that the nonequilibrium distribution $n(\epsilon_m, t)$ approaches the Fermi-Dirac distribution while conserving the total energy of the system. Energy loss mechanisms, which are of course already present as soon as some laser energy has been absorbed, will be treated in the following section in a way that does not interfere with this principle of

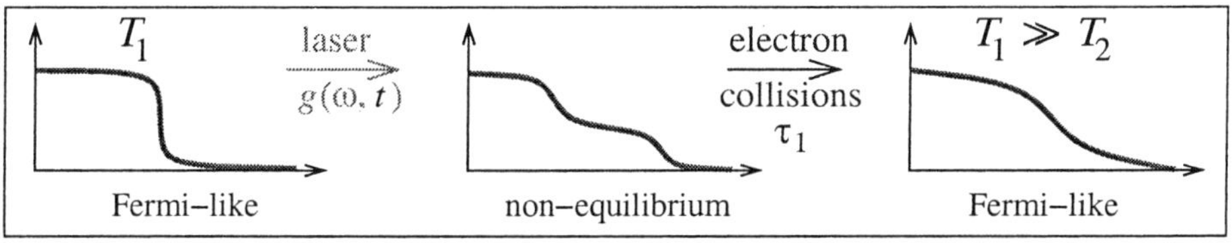

Figure 7.6: Illustration of the electron excitation and thermalization processes described by Eq. (7.18). The temperature T_2 reached after thermalization is much higher than the initial temperature T_1, since the electron system has absorbed energy from the laser pulse.

energy-conserving equilibration. The total energy at time t is required to be

$$E_{\text{tot}}(t) = U(\{r_{ij}(t)\}, t) + E_{\text{kin}}(t) \stackrel{!}{=} E_{\text{tot}}(t=0) + E_{\text{abs}}(t) - E_{\text{loss}}(t)\,, \tag{7.20}$$

where $E_{\text{kin}}(t)$ is the kinetic energy of the ions, and $E_{\text{abs}}(t)$ is the energy that has been absorbed from the laser pulse up to the time t. A further, obvious demand which is necessary to determine T_{e} and μ is that the number of electrons N_{e} stays constant over the entire calculation:

$$N_{\text{e}}(t) = \sum_m n(\epsilon_m, t) \stackrel{!}{=} N_{\text{e}}^0\,. \tag{7.21}$$

Thus, by enforcing conservation of energy during the electron thermalization we make sure that we get physically meaningful results. In Fig. 7.6 the processes of excitation and thermalization are illustrated.

With this approach the effect of a laser-induced electronic nonequilibrium on the structural response can be studied. This theory for the time dependence of the occupation numbers would be too simple if electron dynamics during the first few femtoseconds after laser excitation were in the centre of our study. However, the main focus of this work are laser-induced structural changes on a time scale of a few picoseconds. These are not expected to be very sensible to fine details of the nonequilibrium distribution created by the laser pulse. Thus, Eq. (7.18) is used for the study of the dependence of structural relaxation on duration, frequency and intensity of the exciting pulses.

2.2.1 Summary of the numerical approach. Our theoretical description of laser induced structural changes requires to solve in a parallel way the classical equations of motion for the nuclear degrees of freedom (and also of the lattice vectors, in the case of simulations at constant pressure) and the equations of motion for the electronic occupations [Eq. (7.18)]. The equations of motion are, of course, coupled. Changes in the occupations lead to changes in the potential energy surface. These lead to changes in the atomic coordinates (due to the presence of forces on the atoms) which cause changes in the

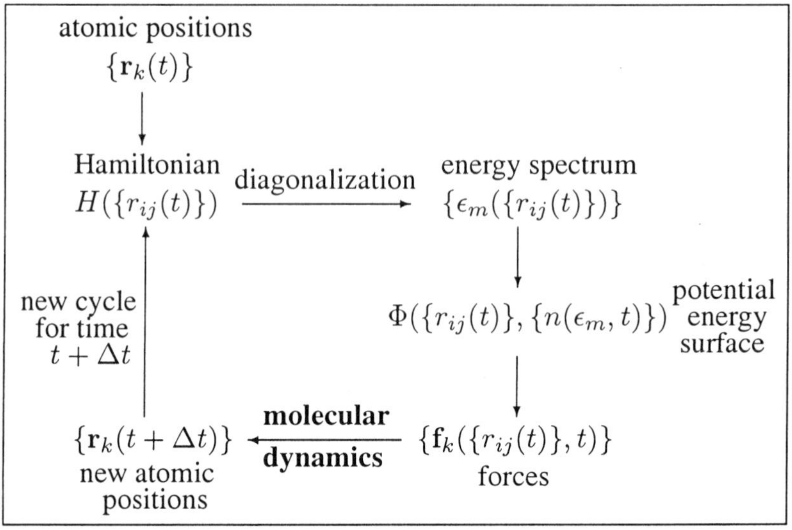

Figure 7.7: Molecular dynamics scheme for the determination of changes of the atomic structure in response to electronic excitation. An explicitly time dependent potential energy surface (PES) is obtained on the basis of an electronic Hamiltonian and with the help of a time dependent nonequilibrium electronic occupation. This PES yields forces for the MD. Then these are used to work out the relaxation of the atomic structure.

eigenvalue-spectrum of the Hamiltonian H_{TB} (due to changes of the hopping matrix elements).

In Fig. 7.7, our calculation procedure is summarized. It is important to point out, that for each time-step of our simulations we have to diagonalize H_{TB} and also solve Eqs. (7.20) and (7.21) to determine the instantaneous electron temperature and chemical potential.

A further technical point which is important to mention is that the calculation of the forces from the expression (7.15) for the potential energy surface $U(\{\mathbf{r}_{ij}\}, t)$ does not involve the gradients of the electron occupations. It has been shown [212, 213] that the correct lattice dynamics is obtained by taking only the gradients of the eigenvalues of H_{TB} and of the repulsive potential in Eq. (7.15) without further terms involving derivatives of the occupation numbers $n(\epsilon_m, t)$. This is because the terms arising from the gradients of $n(\epsilon_m, t)$ in the first and third terms of (7.15) approximately cancel out (except for negligible corrections).

2.2.2 Pair correlation function. For the characterization of the structural changes during a solid to liquid phase transition, a meaningful quantity is needed. The pair correlation function (or radial distribution function) $g(r)$ is the density of other particles around a given particle. This function has a sharply peaked structure in a perfect crystal, while in a liquid, it shows structure only for small r. It is normalized so that at large r, $g(r) = 1$. Thus, it is given by [214, 215]

$$g(r) = \frac{\langle \Delta N(r) \rangle}{n_a \, 4\pi \, r^2 \, \Delta r} \tag{7.22}$$

where $\Delta N(r)$ is the number of atoms situated at a distance between r and $r + \Delta r$ around a given particle, and $n_a = N/\Omega$ is the atomic density of the material. The denominator of Eq. (7.22) corresponds to the average number of particles in the spherical shell of radius r and thickness Δr. The average $\langle \ldots \rangle$ is performed over all atoms in the material sample.

The pair correlation function will show a strong peak at the nearest neighbour distance r_{nn}. For higher distances, peaks of decreasing amplitude will indicate more distant neighbours around a particle.

3. Ultrafast nonequilibrium graphitization of diamond

A most interesting process observed in diamond between damage and ablation thresholds is shown in Figs. 7.9 and 7.10. [1] It is an ultrafast nonequilibrium graphitization of diamond. In Figs. 7.9 and 7.10 we show detailed snapshots of a graphitization trajectory. The laser pulse of $\tau = 20$ fs duration had an energy of $E_0 = 1.0$ eV/atom which is very close to the damage threshold of diamond. In Fig. 7.8 the time evolution of the corresponding MD supercell volume is shown. The material was initially thermalized to a temperature of $T = 300$ K. Upon incidence of the ultrashort laser pulse at $t = 0$ fs, the material expands sharply due to the weakening of the bonds by the laser excitation. The cohesion of the lattice is still strong enough so that the expansion peaks at $t = 20$ fs, and the material contracts again. This ultrafast "breathing" of the material occurs once more, with an expansion maximum at $t = 107$ fs. The third expansion of the lattice turns out to be irreversible: The expansion and contraction cycles have caused small rearrangements of the atoms which now lead to bond breaking between several of the diamond planes, and the armchair-shaped hexagons of the diamond lattice begin to straighten out. The graphitization starts at the upper left side of the MD supercell and propagates into the remaining three dimensionally bonded material during the following 80 to 100 fs. At $t = 240$ fs, perfect graphite planes have formed. Nevertheless, these planes are still strongly vibrationally excited and relaxation of the high electronic temperature must take place before a permanent final graphite state is produced.

---

[1] Fig. 7.9 as well as Figs. 7.10, 7.15, 7.17, 7.18, 7.21, 7.23 and 7.24 were produced with the help of BALSAC (version 2.12, (C) Copyright K. Hermann 1991-9). The BALSAC software was developed by K. Hermann, Fritz-Haber-Institut Berlin (Germany).

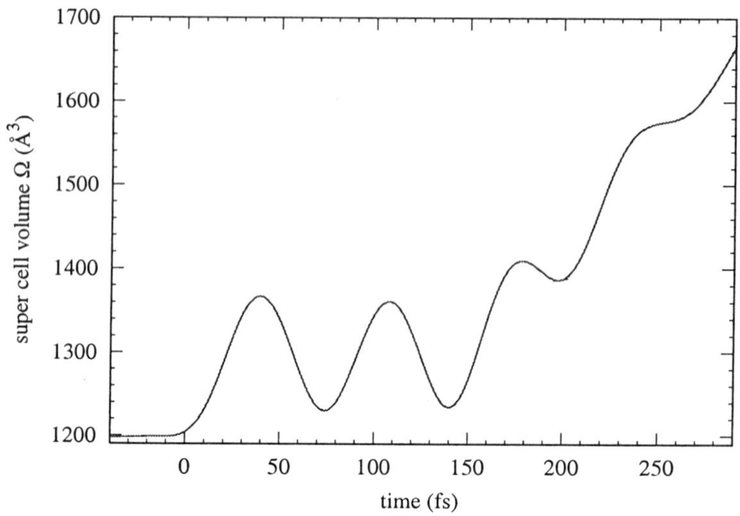

Figure 7.8: Time evolution of the MD super cell volume of a (100) diamond sample during ultrafast graphitization. Time is measured with respect to the maximum of the laser pulse with a duration of $\tau = 20$ fs. The pulse triggers two rapid expansions of the material with subsequent contractions before the graphitization sets in at app. $t = 200$ fs. At this time, strong expansion of the material sets in.

This result, shown in Figs. 7.9 and 7.10 for a (100) diamond MD supercell at constant pressure, has been observed for a large range of laser pulse durations and intensities; in fact, for the entire pulse duration and intensity region between the damage and the ablation thresholds, an ultrafast graphitization is the first relaxation process. The theoretical observation of this laser-induced graphitization has also been confirmed by calculations for a film geometry.

The MD calculation for the bulk material at constant pressure was performed at $P = 10^5$ Pa. The trajectories determined by this method correspond to an isobaric-isoenthalpic ensemble. This means that the entire MD supercell is subjected to an isotropic constant pressure. Note that our volume expansions upon graphitization relate to the considered subregion due to our use of a MD supercell with periodic boundary conditions. While the electronic transition $sp^3 \rightarrow sp^2$ behind the graphitization occurs very fast (in less than 50 fs), the time scale for the following volume and shape changes is characterized by a velocity of typically $v_{cp} \approx 0.03$ Å/fs, which may be compared with the velocity of sound in diamond, $c_D = 0.18$ Å/fs. Thus, although the theoretically observed graphitization occurs on a femtosecond time scale, it involves lattice deformations at a velocity well below the theoretical upper limit defined by the sound velocity.

In Fig. 7.11 we show the densities of states corresponding to the graphitization process of Figs. 7.9 and 7.10. In addition to the total DOS, σ and π

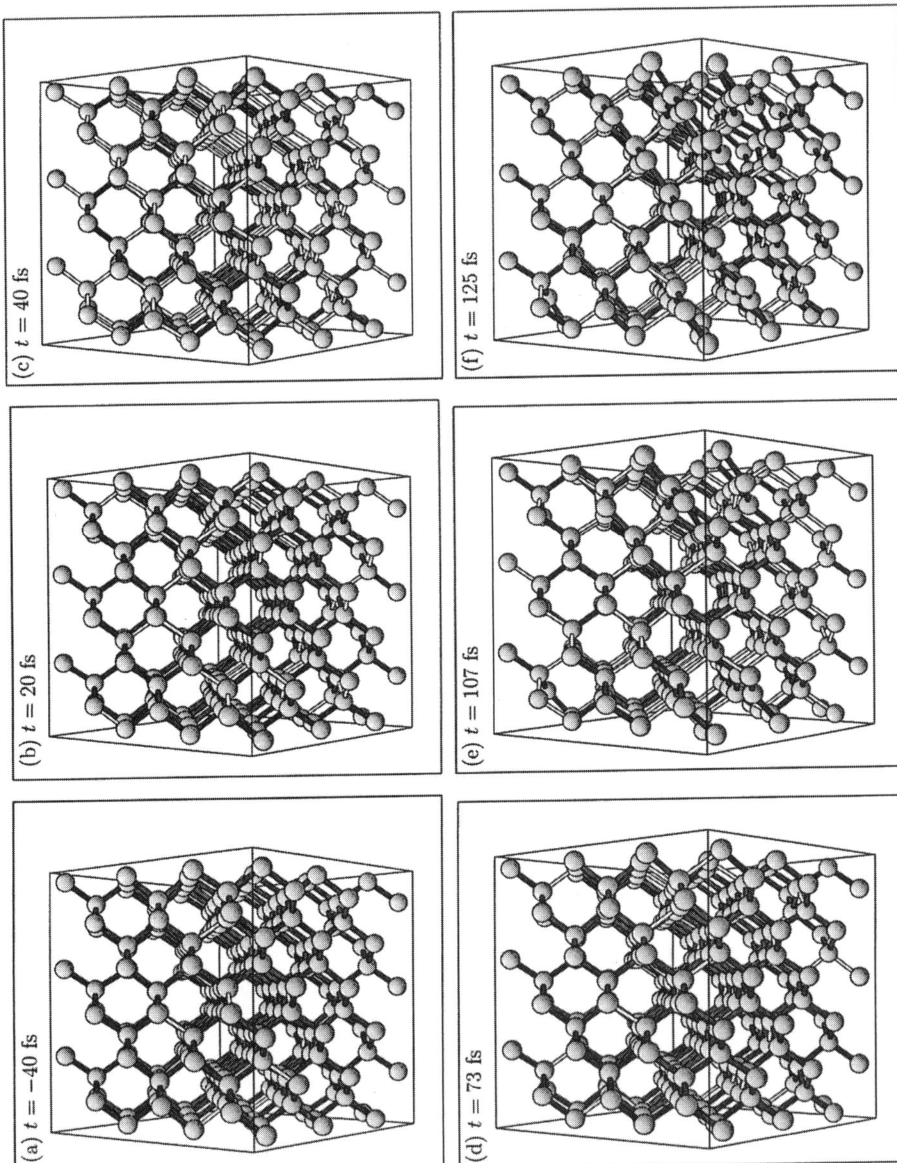

Figure 7.9: Ultrafast graphitization of a (100) diamond crystal (initial stage). The energy $E_0 = 1.0$ eV/atom was absorbed from the laser pulse of $\tau = 20$ fs duration. The MD supercell contains $N = 216$ atoms. Bonds up to a length of $l = 1.6$ Å are drawn in dark grey, weaker bonds of 1.6 Å $< l < 2.0$ Å in light grey. For a description of the transformation see the text.

contributions are shown. The DOS should be compared to the DOS of ideal diamond and graphite shown in Fig. 7.12 (a) and (b).

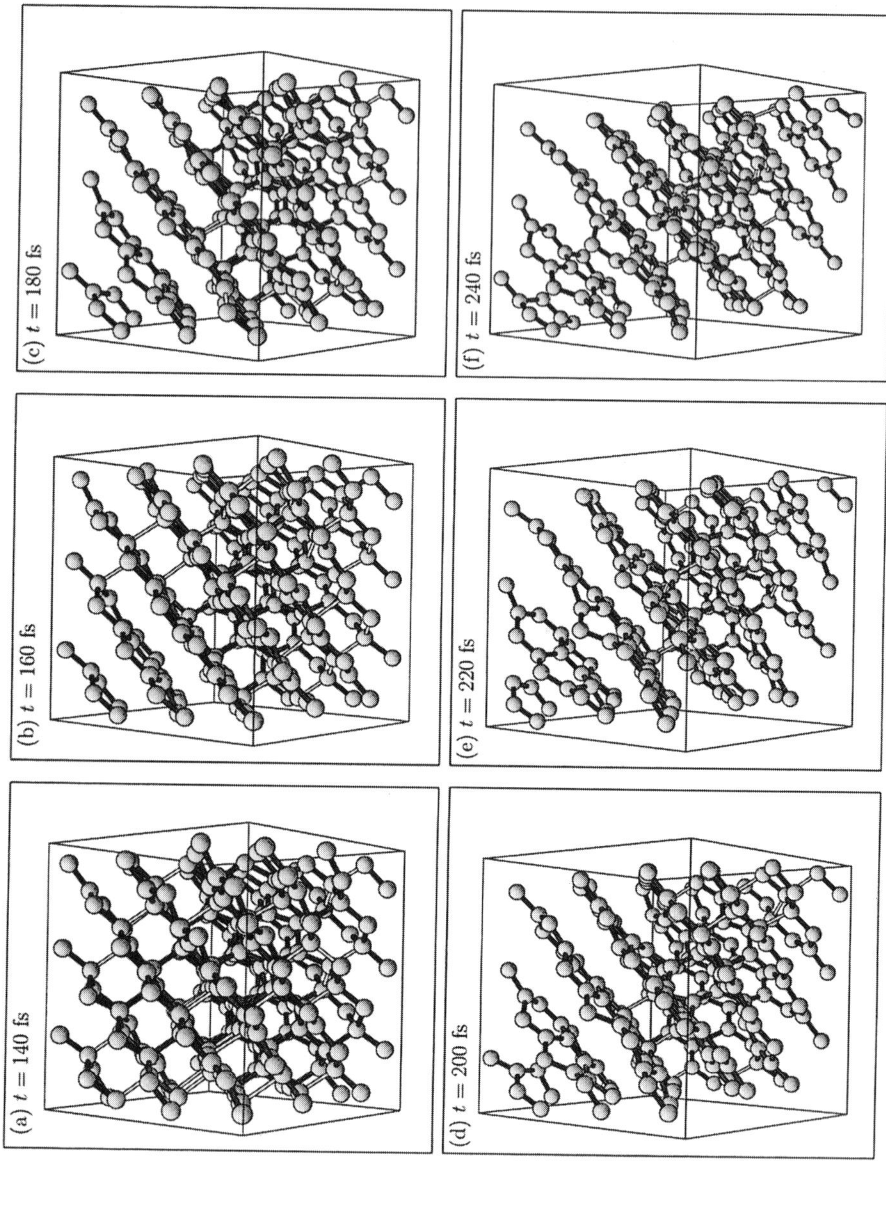

Figure 7.10: Ultrafast graphitization of a (100) diamond crystal (later stage). The energy $E_0 = 1.0$ eV/atom was absorbed from the laser pulse of $\tau = 20$ fs duration. The MD supercell contains $N = 216$ atoms. Bonds up to a length of $l = 1.6$ Å are drawn in dark grey, weaker bonds of 1.6 Å $< l < 2.0$ Å in light grey. For a description of the transformation see the text.

At a time $t = 10$ fs (see Fig. 7.11 (a)), measured from the maximum of the laser pulse of $\tau = 20$ fs duration, the DOS is still that of intact diamond

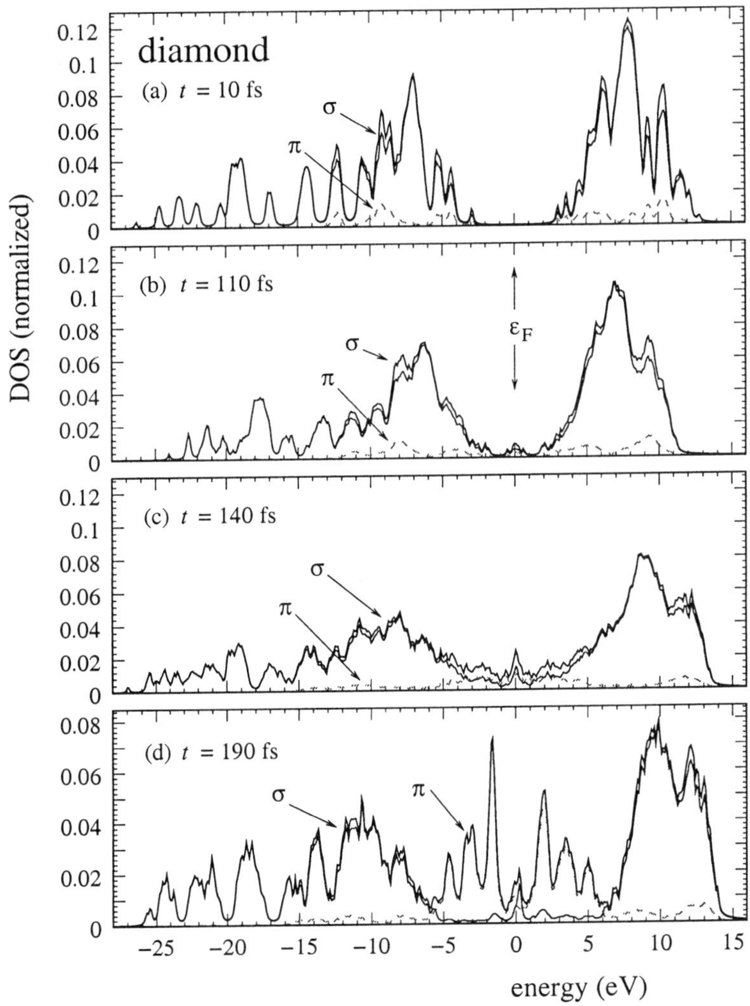

Figure 7.11: Densities of states during a nonequilibrium graphitization of diamond. The time is given with respect to the peak of a laser pulse of $\tau = 20$ fs duration. The densities of states correspond to the structures in Figs. 7.9 and 7.10. The π contribution to the DOS is indicated by a dashed line. Note, the ordinate scale of (d) has been changed with respect to (a)-(c). For a detailed description of the changes to the DOS see the text.

with a very small π contribution and a gap of $E_g = 5.5$ eV. At $t = 120$ fs (see Fig. 7.11 (b)), the structures of diamond DOS are broadening, and we see the emergence of a small number of states in the gap. At $t = 140$ fs (see Fig. 7.11 (c)), the structures of the diamond DOS have broadened even further, and close to the Fermi level, the density is increasing. This increase is to a large part due to π states. At $t = 190$ fs, (see Fig. 7.11 (d)) the features of the graphite DOS have developed. The π states now contribute strongly to the DOS, forming the structures on both sides of the Fermi level. These structures

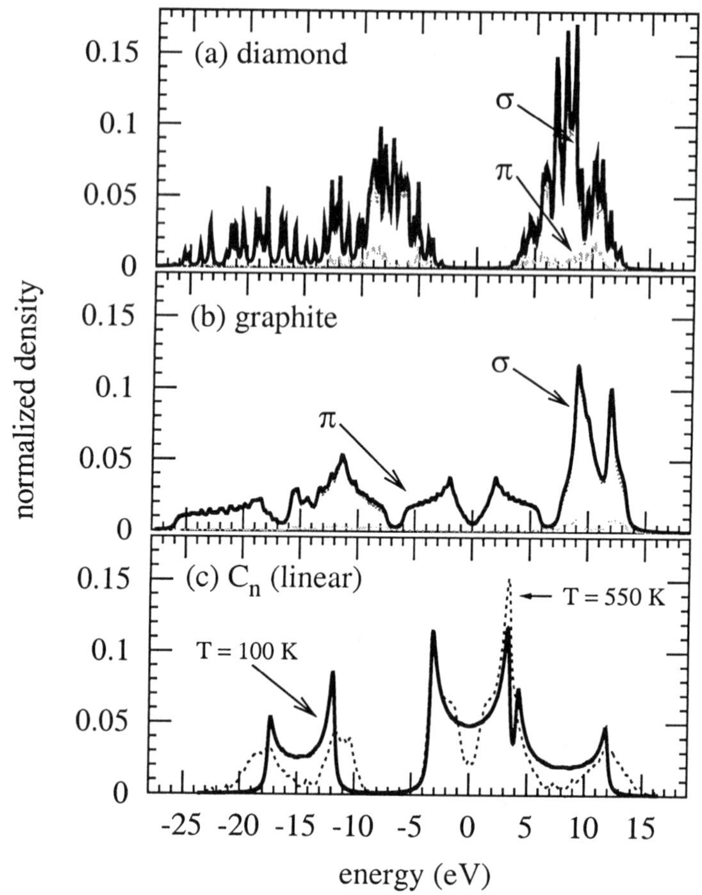

Figure 7.12: Densities of states of (a) diamond, (b) graphite, and (c) linear carbon chains. The DOS of diamond and graphite are computed at $T = 0$ K, while the DOS of C_n chains is given at $T = 100$ K (solid line) and $T = 550$ K (dashed line). For diamond and graphite, the solid lines indicate the total DOS, while σ and π contributions to the DOS are indicated by dotted and dash-dotted lines, respectively. For all graphs, the Fermi level ϵ_F determines the zero of the energy scale. (a) In the case of diamond, the material is characterized by the very large band gap of $E_g = 5.5$ eV. The DOS is dominated by the σ bonds of the sp^3 hybridized valence orbitals. The contribution of π electrons to the DOS is almost negligible. (b) In the case of graphite, the structures on both sides of the Fermi level ϵ_F are almost exclusively formed by the π bonds, while the remainder of the DOS reflects the σ bonds of the sp^2 hybridized orbitals. The fact that the DOS is slightly above zero at the Fermi level is an artifact caused by the broadening of the energy levels with Lorentzians of a width $\gamma = 0.1$ eV. (c) For the linear carbon chains, the higher temperature leads to a significant broadening of the structures. As in graphite, the structures on both sides of the Fermi level ϵ_F are determined by π bonding, while the rest of the DOS reflects σ bonds. As in a linear chain only two of the valence orbitals take part in the sp hybridization, the weight of the π contributions to the DOS is increased compared to graphite.

show relatively sharp spikes when compared to the DOS of graphite at $T = 0$ K. These can be attributed to the strong vibrational excitation of the graphite that has been newly formed from diamond.

4. Ablation mechanisms in graphite

In this section we present results for damage and ablation thresholds of graphite, and we analyze in detail the structural changes occuring above these two distinct thresholds. The thresholds have been calculated for bulk materials at a constant pressure of $P = 10^5$ Pa. Under ablation threshold we understand the deposited energy that is necessary to completely destroy the cohesion of the graphite planes and vaporize the material within 2 ps after the peak of the laser pulse. Even at laser intensities somewhat below the threshold determined in this way, the deposited energy may be sufficient to melt and vaporize the material on a time scale of tens or hundreds of picoseconds which is not accessible to our calculations. Thus, the thresholds for ultrafast ablation shown here can be considered as upper limits for the ablation thresholds found in the experiment by *post mortem* analysis of the sample. The damage threshold is determined by the energy density that causes, again within 2 ps after the pulse, significant structural changes without completely destroying the cohesion of the material.

In Fig. 7.13, we show the ablation threshold for graphite as a function of pulse duration. It was calculated using a MD cell containing $N = 64$ atoms. As in the case of diamond the ablation is defined as an evaporation of the material. Below the ablation threshold interesting structural changes occur and Fig. 7.13 shows a damage threshold that is approximately 1 eV lower in energy than the ablation threshold. Calculation with a much higher number of atoms ($N = 576$) in a film geometry confirmed that in fact a second ablation threshold below the one presented in Fig. 7.13 can be distinguished. It includes a strong vibrational excitation of the graphite planes. These remain intact, but they collide because of the strong oscillation of the atoms perpendicular to the planes and then, ablation of entire planes can occur as a consequence of the momentum transferred in these collisions. This process will be described in detail below.

Both thresholds shown in Fig. 7.13 increase very slightly with pulse duration, but this effect is not very pronounced.

The examination of a graphite film which had absorbed a moderate energy density has produced a surprising result: There is an ablation mechanism which does not lead to a destruction of the graphite planes. The principle of this mechanism is illustrated in Fig. 7.14. The graphite planes are separated in thermal equilibrium by a very large distance of $d = 3.4$ Å and consequently their interaction is given by a very weak van-der-Waals energy of $E \approx 20$ meV. Meanwhile, a photoinduced electron-hole plasma can lead to a very strong vibrational excitation of the graphite layers, and at the turning point of their oscillations the atoms of two different layers can come as close as $d \approx 2$ Å. At this distance, the interaction becomes already quite strong and is on the order of $E \approx 1$ eV. In other words, the laser-induced strong vibrations can lead to

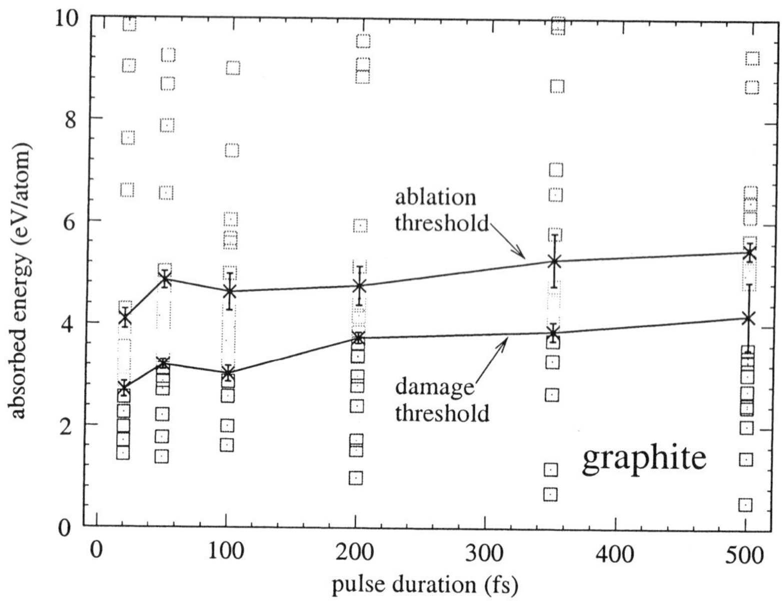

Figure 7.13: Ablation threshold of graphite as a function of laser pulse duration. Each result (□) in the figure corresponds to a trajectory calculated for a $N = 64$ atom MD supercell.

collisions of the planes in which momentum is transferred. Thus, a plane on the surface of a graphite sample which usually has zero total momentum can gain enough momentum to leave the surface.

In Fig. 7.15 we present snapshots of the dynamics of graphite for a $\tau = 20$ fs laser pulse with an absorbed energy of $E_0 = 2.4$ eV/atom. The number of atoms is $N = 576$, and they are in a film geometry, *i. e.* the material has periodical boundary conditions in the horizontal directions, while in the vertical direction it borders on large slabs of vacuum. This film was thermalized to a temperature of $T = 300$ K before the action of the laser pulse. The trajectory of Fig. 7.15 shows the very interesting new ablation mechanism. At a deposited energy density that is far below the ablation threshold we find the ablation of entire graphite planes by the mechanism explained above: The laser pulse leads to a strong vibrational excitation of the graphite layers, with a significant movement of the atoms perpendicular to the graphite planes. This excitation is already strong at $t = 80$ fs after the laser pulse maximum, but for the topmost plane it reaches a maximum at $t = 160$ fs (see Fig. 7.15). At that time the distance between the graphite planes has narrowed to approximately $d = 1.8$ Å, and the first and second planes strongly interact. We can speak of a collision of the two topmost graphite planes, which then leads to a momentum transfer. Consequently, the surface plane that had up to that time zero total momentum

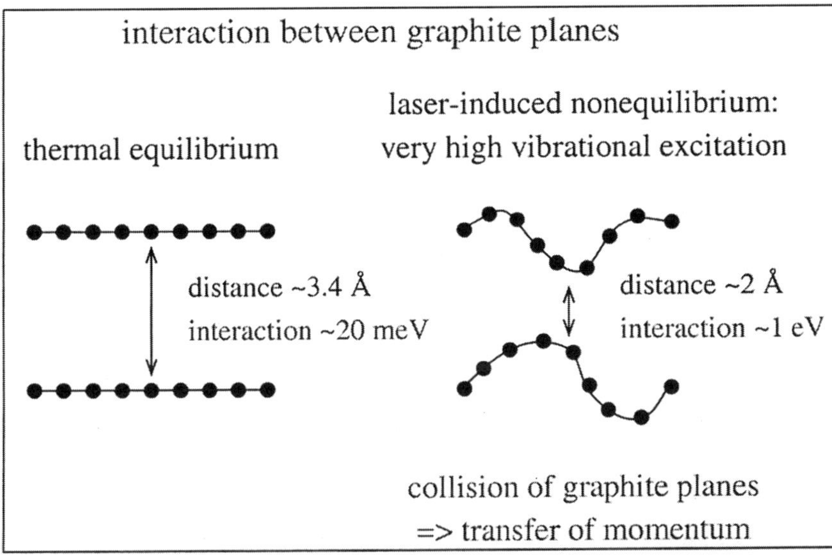

Figure 7.14: Illustration of a new ablation mechanism in graphite.

starts to leave the crystal in positive z direction. In the same way the bottom layer of Fig. 7.15 can be seen to leave the rest of the film in negative z direction. This ablation mechanism in graphite has not been described before. It solely relies on the fact that strong vibrational excitation of the graphite planes can increase the interaction between the planes to such a degree that the momentum transferred in this collision process causes ablation of entire planes. It is to be expected, of course, that in reality we are dealing with graphite planes that show slight defects. These would result in large plane fragments being removed from the surface instead of entire graphite planes as the simulation with periodic boundary conditions in the horizontal directions suggests.

In Fig. 7.16 we show the densities of states during the ablation process presented in Fig. 7.15. The densities of states in Fig. 7.16 (b) and (c) at $\tau = 80$ fs and $\tau = 160$ fs respectively show only slight modifications in comparison to the DOS of the $T = 300$ K graphite film of Fig. 7.16 (a). The violent vibrations present in Fig. 7.15 (b) and (c) are reflected in a broadening of the features of the DOS. It is interesting that in Fig. 7.16 (d) for a time $t = 860$ fs the sharp features of the $T = 300$ K graphite film have been almost recovered. This indicates that the high-amplitude oscillations of the graphite planes are strongly damped.

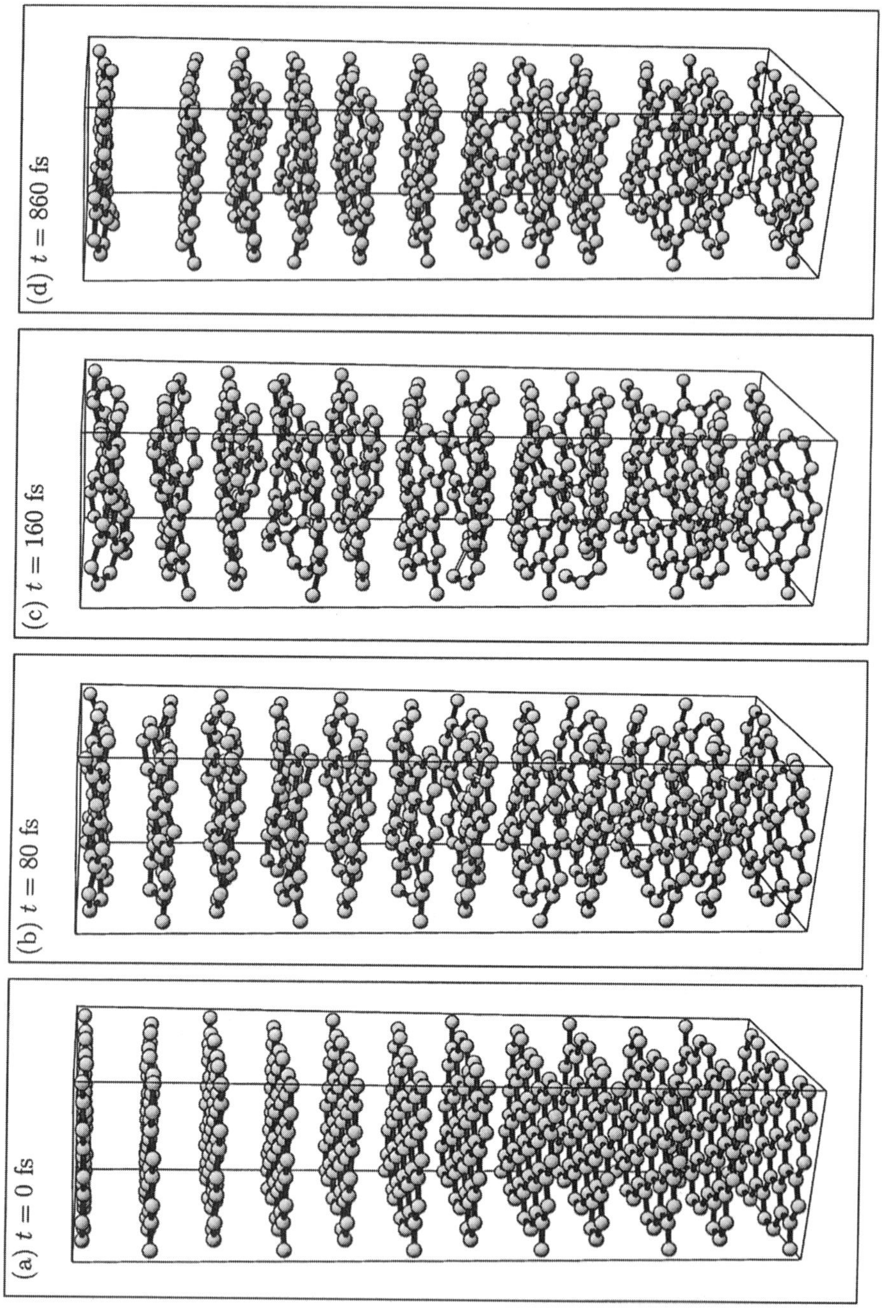

Figure 7.15: Ablation of graphite at an absorbed energy of $E_0 = 2.4$ eV/atom. The pulse duration was $\tau = 20$ fs. This energy density is below the threshold for the destruction of graphite planes. Thus, the laser energy is transferred into a very strong vibration of the graphite planes. At $t = 160$ fs, this vibration leads to a collision of the planes, during which momentum is transferred. As a consequence, the planes at the top and at the bottom are removed from the surface of the film.

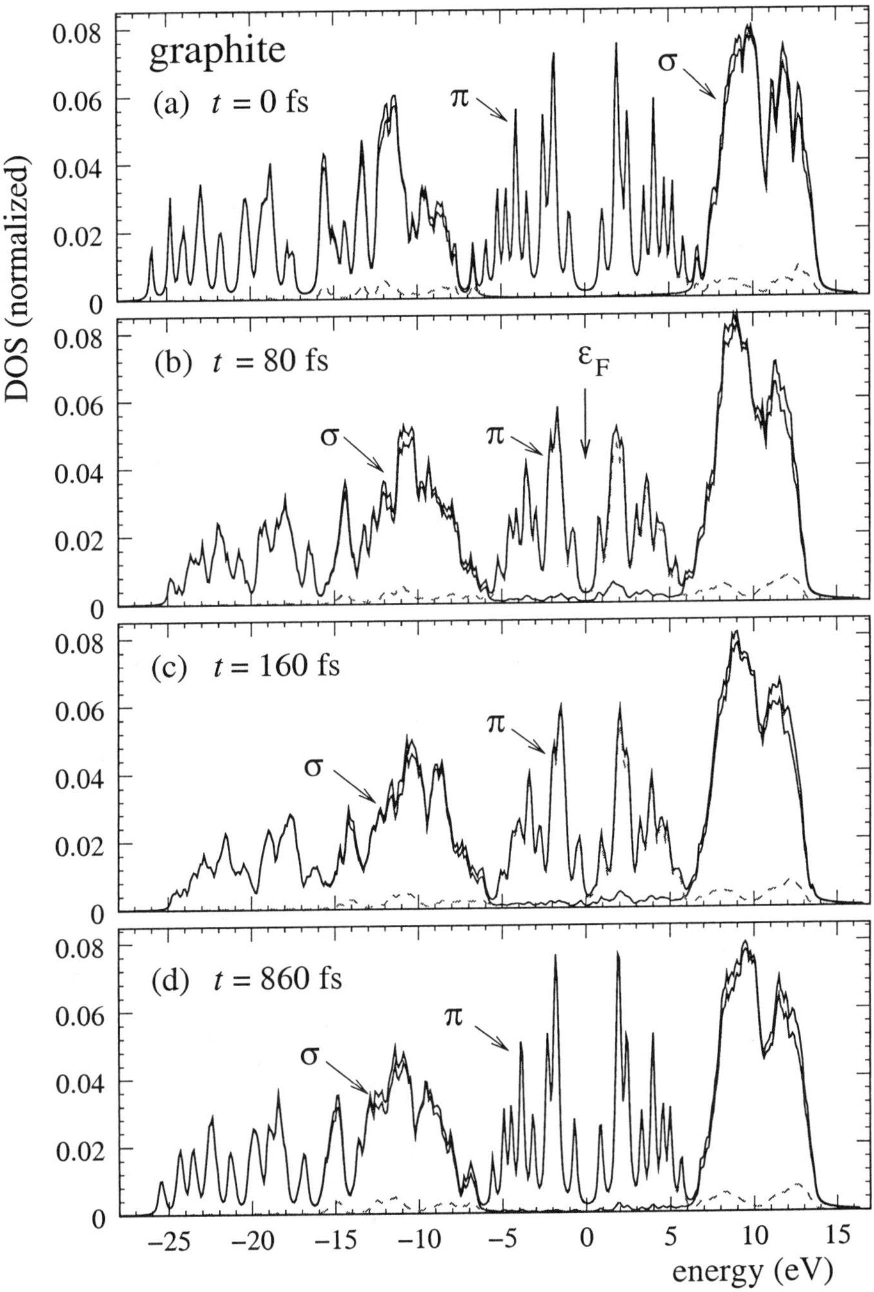

Figure 7.16: Densities of states for the ablation of graphite at an absorbed energy of $E_0 = 2.4$ eV/atom. This figure corresponds to the trajectory of Fig. 7.15. The π contribution to the DOS is indicated by a dashed line. Due to the strong vibrational excitation of the graphite planes at $t = 80$ fs and $t = 160$ fs the π contribution to the DOS is slightly diminished and the corresponding structures in the DOS are broadened. At $t = 860$ fs the DOS is recovering its original shape, corresponding to the falling amplitude of the vibrations in Fig. 7.15 (d).

5. Nonequilibrium melting and ablation of carbon

The processes leading to a melting of graphite are of high fundamental interest. While thermal melting has been studied already [183], the nonequilibrium melting discussed in this section has not been analyzed theoretically before. This study is important for a great number of experiments where nonequilibrium conditions are induced in carbon materials.

In Fig. 7.17 and Fig. 7.18 we show snapshots of a long trajectory of a graphite film in which melting and subsequent evaporation can be observed. The film consists of $N = 576$ atoms. The dynamics shown are for a $\tau = 20$ fs laser pulse with an absorbed energy of $E_0 = 4.0$ eV/atom. This energy density is high enough to break the bonds of the graphite planes. After $t = 40$ fs, we see that especially in the lower part of the film, bonds are formed between the graphite planes. Note, the initial graphite film was thermalized to $T = 300$ K, so that at the time the laser pulse hits the material all atoms are in an individual state of motion. This determines the initial conditions for the amplified oscillations due to the presence of the electron-hole-plasma. Thus, it is not surprising that on a very short time scale there are differences in the motion of different graphite planes. At $t = 80$ fs, we can see already the first carbon monomer being evaporated from the material, confirming that the graphite planes have been excited above their fragmentation threshold. In the following snapshot at $t = 140$ fs, four further monomers are leaving the two surfaces of the film and the graphite planes have been dissolved completely.

In Fig. 7.18, the further time development of the excited graphite sample of Fig. 7.17 is shown. In the snapshots presented here, a strong volume expansion is observed, *e. g.* from 5480 Å^3 at $t = 220$ fs to 6040 Å^3 at $t = 320$ fs. This expansion is reflected in Fig. 7.18 in the decreasing width of the column of the material rather than in an increasing z extension of the column which would have made for an awkward representation in the figures. In all graphs of Fig. 7.18, four monomers that had already moved far away from the surface were omitted. Fig. 7.18 now shows the advancing ablation process. Further carbon monomers are ejected from the material, and between $\Delta t = 420$ fs and $\Delta t = 920$ fs we can observe how the expansion of the material takes place especially rapidly in the two surface regions of the film, leading to low densities and the formation of short carbon chains which then start to leave the sample.

In Fig. 7.19 we show the densities of state that correspond to Fig. 7.17. At $t = 80$ fs (see Fig. 7.19 (b)) the features of the graphite DOS have already significantly broadened, and at the Fermi level the DOS has increased. In Fig. 7.19 (c) for $t = 160$ fs, the DOS has become metallic-like. The sharp spike close to the Fermi level corresponds to features of the DOS of the linear chain (see Fig. 7.12 (c)). These features have developed further in Fig. 7.19 (d)) for $t = 860$ fs.

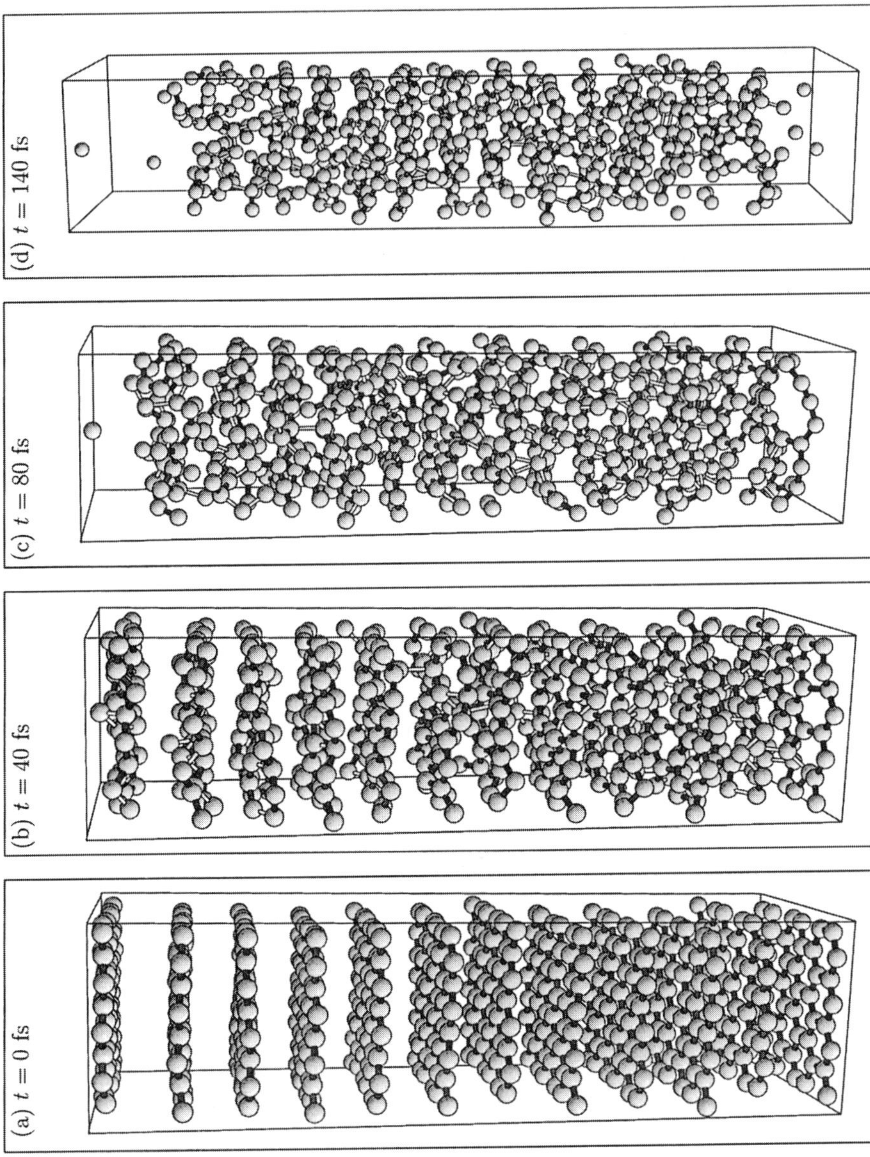

Figure 7.17: Ablation of a graphite film at an absorbed energy of $E_0 = 4.0$ eV/atom. The laser pulse duration was $\tau = 20$ fs. This figure shows the early stages of the ablation process (for the later stages see Fig. 7.18). The graphite planes vibrate strongly (b) and are destroyed (c) in response to the electron-hole plasma. At $t = 140$ fs the material is in violent motion (d), and carbon monomers are emitted.

In Fig. 7.20 we show the pair correlation functions $g(r)$ during the melting of a graphite film. The pair correlation functions of Fig. 7.20 correspond to

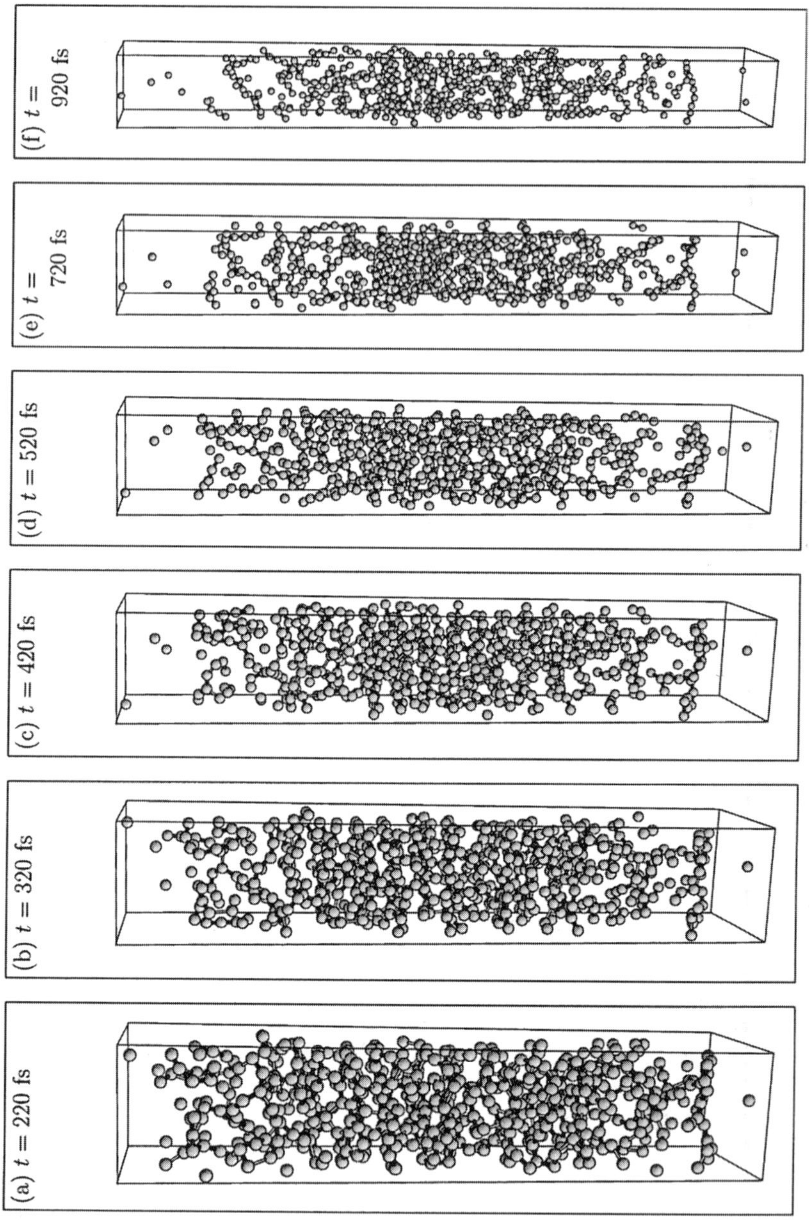

Figure 7.18: Ablation of a graphite film at an absorbed energy of $E_0 = 4.0$ eV/atom. The laser pulse duration was $\tau = 20$ fs. This figure shows the advanced stages of the ablation process (for the early stages see Fig. 7.17). Four emitted carbon monomers far from the surface are not shown. A strong expansion of the material is observed; as the film has been represented with constant height, this is apparent from the narrowing of the columns. The material is in strong, liquid-like motion. In (d)-(f), in addition to the emission of carbon monomers, linear carbon chains are formed and emitted.

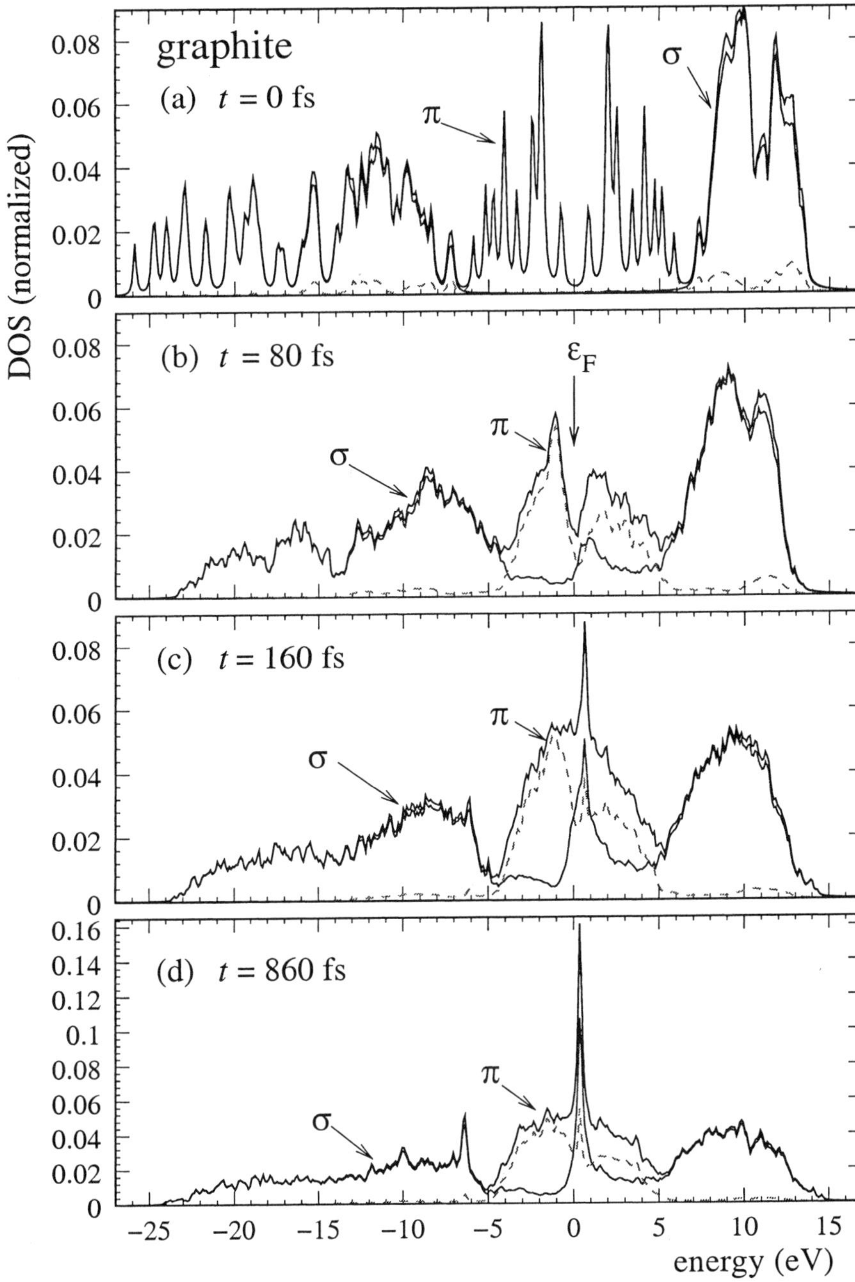

Figure 7.19: Densities of states for the ablation of graphite at an absorbed energy of $E_0 = 4.0$ eV/atom. This figure corresponds to the trajectory of Fig. 7.17. The π contribution to the DOS is indicated by a dashed line. The features of the graphite DOS in (a) at $t = 0$ fs are quickly replaced by a metallic-like DOS (b). The sharp peak at the Fermi level that appears at $t = 160$ fs in (c) is due to contributions of linear carbon chains. Note, the ordinate scale of (d) is different from (a) to (c).

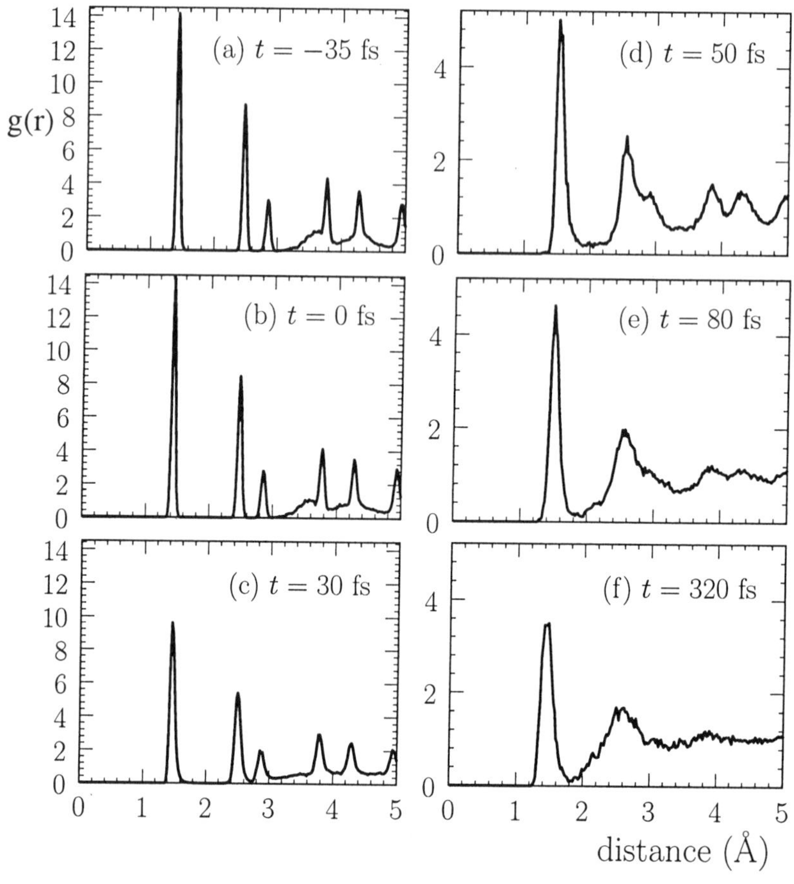

Figure 7.20: Pair correlation function $g(r)$ during the melting of a graphite film at an absorbed energy of $E_0 = 4.0$ eV/atom. The pulse duration was $\tau = 20$ fs. The time of (a) is negative because we measure time with respect to the laser pulse maximum. The pair correlation functions correspond to the same trajectory as Figs. 7.17 to 7.19. Note, the ordinate scale of the left figures is different from that on the right. At $t = 0$ fs, $g(r)$ is not significantly modified with respect to $g(r)$ before the pulse (see (a)). At $t = 30$ fs (see (c)) the structures of $g(r)$ have slightly broadened. Shortly afterwards (see (d) and (e)), the pair correlation function of liquid carbon is formed.

the same trajectory as Figs. 7.17 to 7.19. $g(r)$ at the pulse maximum (see Fig. 7.20 (b)) does not yet differ much from $g(r)$ before the pulse (Fig. 7.20 (a)). But then a rapid broadening of the structures of $g(r)$ occurs (Fig. 7.20 (c) and (d)) and at $t = 80$ fs the pair correlation function of liquid carbon is recognizable.

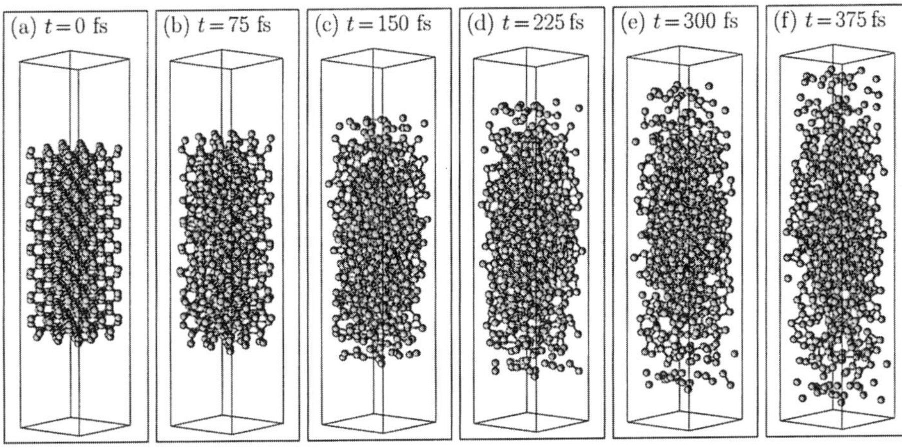

Figure 7.21: Ablation of a silicon film at an absorbed energy of $E_0 = 6.1$ eV/atom. The laser pulse duration was $\tau = 20$ fs.

6. Ablation of silicon

Silicon was among the first materials for which ultrafast structural transformations were observed [163, 164]. Its optical response following laser irradiation was studied in great detail [166]. This prompted early theoretical investigations, namely studies of the softening of phonon modes as a result of a laser induced electron hole plasma [179, 180, 181] and of the melting of the material due to a high electron temperature [184].

We have performed simulations of the laser melting and ablation of silicon in the bulk and in films. Results for the ablation thresholds from calculations in the bulk have been published elsewhere [190]. In Fig. 7.21 we show snapshots of the ablation of a (111) silicon film. The laser pulse of $\tau = 20$ fs duration had an intensity of $E_0 = 6.1$ eV/atom. This absorbed energy is far above the ablation threshold of the material. Thus, already within 75 fs after the incidence of the pulse, the lattice is strongly disordered. A fast expansion can be observed, and material is evaporating on both surfaces of the film.

In order to analyse the structural response of the material after laser irradiation, we determined the pair correlation function for the silicon film trajectory of Fig. 7.21. The results are shown in Fig. 7.22. The sharp features of the crystalline structure of silicon broaden very rapidly during the first 100 fs after the pulse, and at $t = 150$ fs, calculated from the laser pulse maximum, the pair correlation of liquid silicon is almost reached. This confirms that the melting process involves a nonthermal mechanism as thermal melting would take at least an order of magnitude longer.

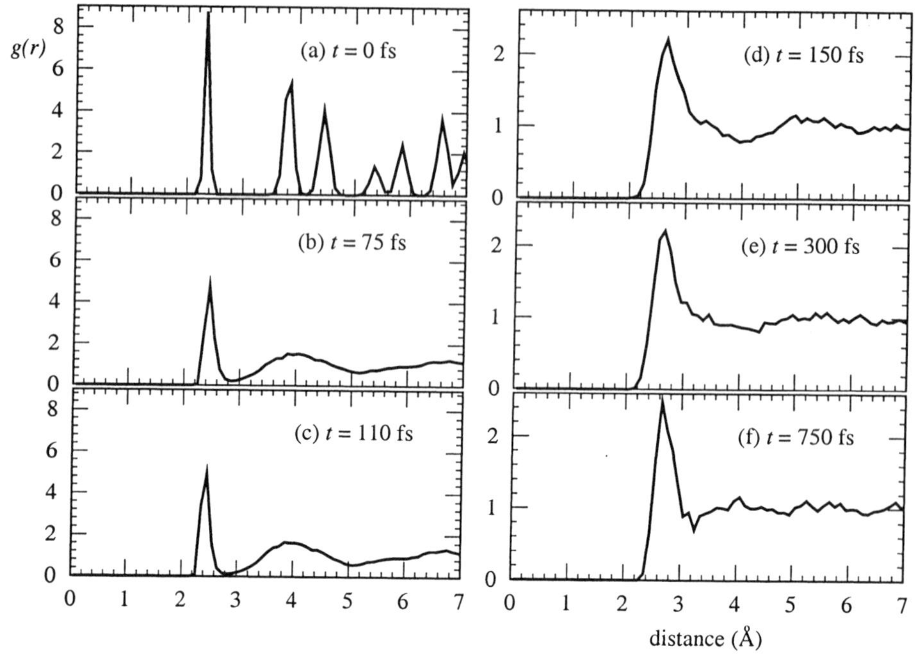

Figure 7.22: Pair correlation function $g(r)$ during the melting of a silicon film at an absorbed energy of $E_0 = 6.1$ eV/atom. The pulse duration was $\tau = 20$ fs. The pair correlation functions correspond to the same trajectory as Fig. 7.21. It was calculated for a slab 20 Å in height from the center of the film in order to exclude the surfaces of the material. Note, the ordinate scale of the left figures is different from that on the right.

7. Laserinduced melting of a C_{60} molecular crystal

The C_{60} molecular crystal is an interesting object of study as it represents a solid carbon material with a density that is app. 20% below that of graphite. Thus, it is a possible starting point for the creation of low density liquid carbon (LDLC) [216]. Laser excitation of fullerites can also lead to polymerization and amorphization [217].

For the simulation of the ultrafast relaxation of a C_{60} molecular crystal we consider a supercell of 240 atoms and periodic boundary conditions.

In Fig. 7.23 we show snapshots of the destruction of a C_{60} molecular crystal induced by intense laser irradiation. The laser pulse is characterized by a duration of $\tau = 20$ fs, and during its action an energy of $E_0 = 3.6$ eV/atom is absorbed by the lattice. A constant external isotropic pressure of $P = 10$ GPa is applied to the system at all times. In the figure we indicate bonds up to a length of 1.6 Å with dark grey, while bonds between 1.6 Å and 2.1 Å in length are shaded in light grey and have a smaller diameter. Bonds longer than 2.1 Å are not shown, as they contribute only negligibly to the binding energy, and thus atoms that are more than 2.1 Å apart can be considered unbound. At $\Delta t = 0$ fs (Fig. 7.23 (a)), i.e. at the pulse maximum, the atoms have not yet had time to react to the change in the PES due to the absorption of nearly half of the laser intensity; they still form the perfect fcc lattice of the solid C_{60} at 300K. The fact that in each of the C_{60} molecules a few atoms seem to be missing while several atoms can be seen that are not directly connected to one of the four C_{60} molecules is a consequence of the periodic boundary conditions. The six atoms on the bottom left of the first panel for example actually form a part of the C_{60} molecule in the background. At $\Delta t = 20$ fs (Fig. 7.23 (b)), the C_{60} molecules have expanded considerably as can be seen from the many light grey, i.e. stretched, bonds that have replace dark grey ones. At $\Delta t = 60$ fs (Fig. 7.23 (c)), the expansion of the C_{60} molecules has proceeded beyond the elastic limit of the C-C bonds, and the original form of the C_{60} molecules appears mainly in the large hollow spaces not yet filled in by the hot carbon atoms. Through the collision processes between the fragments of the C_{60} molecules, the kinetic energy of the system has now increased by 2200 K to $T_l = 2500$ K. At $\Delta t = 160$ fs (Fig. 7.23 (d)), the fragments of the destroyed C_{60} molecules have filled up most of the available space, and the distribution of matter approaches homogeneity. Up to a time delay of approximately $\Delta t = 120$ fs, the volume of the system has decreased very rapidly by 30% because the resistance of the nearly incompressible C_{60} molecules to the external pressure has vanished. The panels for $\Delta t = 260$ fs and $\Delta t = 860$ fs (Fig. 7.23 (e) and (f)) show rapid structural rearrangements of the very hot ($T_l \approx 3000 - 3500$ K) material. The density at $\Delta t = 860$ fs is $\rho = 2.5$ g/cm^3. A very long simulation of several hundred picoseconds will be needed to determine the final structure of the material after cooling through slow processes like heat diffusion. Recently, results on the laser induced melting of fullerites have also been reported on an *ab initio* [218] and on a TB basis [216].

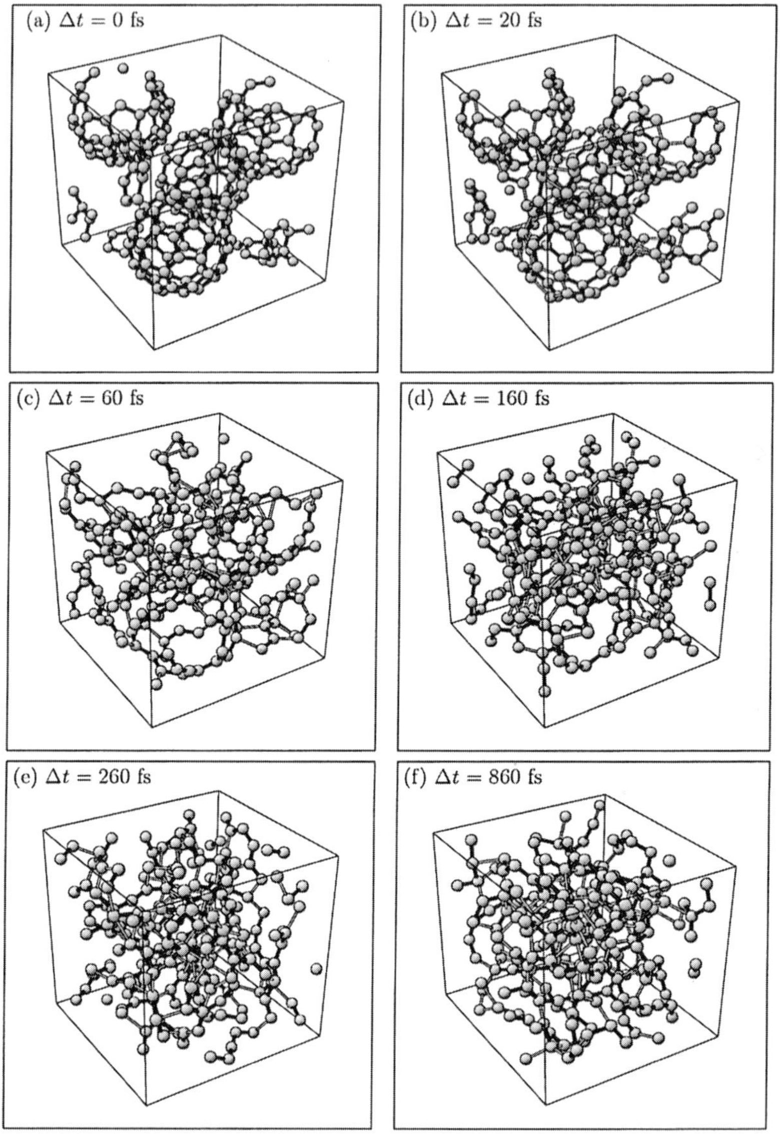

Figure 7.23: Snapshots of a C_{60} molecular crystal during and after irradiation by a ultrashort laser pulse. Pulse duration is $\tau = 20$ fs, an energy of $E_{abs} = 3.6$ eV/atom is absorbed during the laser excitation. A constant external pressure of $p = 10$ GPa acts on the crystal. Time delays Δt are given with respect to the peak of the pulse. Bonds up to a length of 1.6 Å are shown in dark grey, bonds between 1.6 Å and 2.1 Å in length are shown in light grey and are thinner in diameter. For a detailed discussion of the dynamics please refer to the text.

8. Fragmentation of nanotubes

In this section, we will analyze fragmentation mechanisms of carbon nanotubes. These materials are fascinating objects of investigation in its own right as the many different nanotube structures show diverse physical properties.

We present results for the fragmentation of two different sizes of nanotubes. The nanotubes were treated with periodical boundary conditions in the z direction. Thus, they are assumed to be infinite in length. In Fig. 7.24, we present snapshots of the fragmentation of a (10,0) nanotube. An energy of $E_0 = 2.2$ eV/atom was absorbed from a laser pulse of $\tau = 20$ fs duration. This absorbed energy is slightly above the fragmentation threshold and only a few monomers are emitted. At $t = 900$ fs after the pulse, five monomers have left the nanotube.

In Fig. 7.24, we can observe two typical damage mechanisms which have been found in all trajectories of fragmenting nanotubes. The first is the emission of monomers and the creation of a hole in the otherwise undamaged wall of the nanotube. This process takes place on the upper left hand side of the nanotube in Fig. 7.24 (b) and (c). Although the atom that can be seen to leave the nanotube in Fig. 7.24 (c) appears to be removed from the ragged edge of the tube, the mentioned periodical boundary conditions in the z direction mean that this process actually takes place on the side of a perfectly regular, infinite nanotube. From this hole a total of four carbon monomers are emitted. At $t = 900$ fs (see Fig. 7.24 (d)) a neat hole is visible.

The other typical damage creation process is taking place on the right hand side of the nanotube in Fig. 7.24. Starting at $t = 110$ fs (see Fig. 7.24 (a)) we observe the emergence of a tear in the tube. At $t = 160$ fs, one carbon monomer is emitted from this damage spot. Although no more emission of fragments is observed after that, the damage continues its development. At $t = 900$ fs (see Fig. 7.24 (d)) a group of 12 atoms has straightened out into a segment of a graphite plane. Thus, the tear in the side of the tube has caused a relaxation which lowers the surface tension which distinguishes the nanotube from an even graphite plane.

The fragmentation threshold for the (10,0) nanotube was found to be 2.01 ± 0.13 eV/atom, for the (7,0) nanotube 1.83 ± 0.14 eV/atom. Both results are for laser pulses of $\tau = 20$ fs duration. Thus, the nanotube with the smaller diameter is less stable with respect to laser excitation; the diameter of the (10,0) nanotube is $d = 7.8$ Å, of the (7,0) nanotube $d = 5.5$ Å. This can be understood as follows: The π bonds formed by the $2p_z$ orbitals of carbon in a graphite plane are affected strongly when the plane is rolled into a narrow tube. The narrower the nanotube, the lower is the overlap of the p orbitals on the outside of the tube. In other words, the (7,0) nanotube shows a higher surface tension than

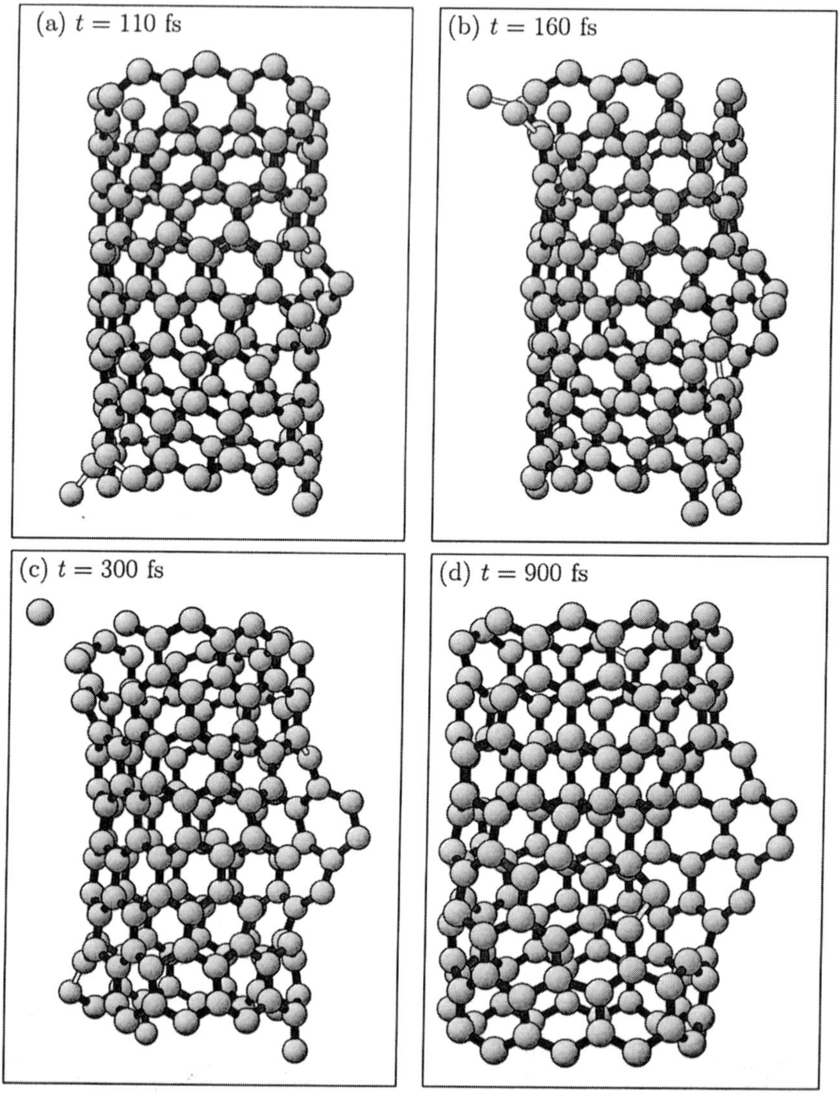

Figure 7.24: Fragmentation of a (10,0) carbon nanotube. In (c), two atoms on the left and one atom on the right of the tube that had moved far from the tube after emission have been removed; in (d), a total of five distant monomers are not shown. Note, periodical boundary conditions in the z direction have been employed, so that the edges on the top and the bottom of the nanotube segment are not present in the calculation. Two important mechanisms of damage development can be observed in this sequence: On the upper left hand side a neat hole is created via the emission of four carbon monomers. On the right hand side a tear appears and leads to a partial unfolding of a graphite plane segment.

the (10,0) nanotube, and this explains the higher vulnerability in the presence of an electron-hole plasma.

The fragmentation thresholds given here are valid for a time up to 2 ps. Thus, they refer to the nonequilibrium fragmentation. It is to be expected that also slightly below the threshold the energy deposited in the cluster can after redistribution lead to a thermal fragmentation. Consequently a threshold including nonequilibrium and equilibrium fragmentation processes would be a little lower than those given here.

9. Summary

An approach to the laser induced structural instabilities of covalent materials has been developed on the basis of a tight binding molecular dyanmics (TBMD) formulation. Writing equations of motion for the occupations of the electronic levels of the system, we can study the response to the shape, intensity and colour of an ultrashort electromagnetic field interacting with the material. The time evolution of the electronic occupations leads to molecular dynamics simulations on time dependent potential energy surfaces. By applying our method in the form of constant pressure MD simulations to bulk systems or alternatively to film geometries where surface effects can be described explicitly, we make sure that we do not artificially suppress relaxation channels. Our findings can, in part, be compared to methods introducing a high electron temperature into Car-Parrinello or TBMD calculations or to methods simulating excited states in molecular dynamics calculations on an empirical pair potential basis.

We presented applications of our method to a variety of materials, focusing our attention on the microscopic processes underlying laser induced structural phase transformations. In diamond, we find that the dominant damage mechanism upon absorption of an energy of roughly 1 eV/atom from an ultrashort laser pulse is a rapid graphitization. Within approximately 200 fs of the peak of an above-threshold pulse, the density of states of the material dramatically changes its character from that of diamond to that of graphite. In the study of the ablation of graphite we find a previously unknown damage mechanism that doesn't involve the breaking of bonds in the graphite planes. Instead, strong vibrational excitation of the intact planes lead to inter-plane collisions and to the removal of entire planes through the transferral of momentum. This mechanism may be relevant in the interpretation of time dependent reflectivity measurements and for understanding the formation of carbon nanotubes. At higher deposited laser energies we describe the nonequilibrium melting of graphite which leads to a carbon phase dominated by linear carbon chains before the material evaporates. In a study of the laser melting of a silicon film we show that the development of the pair correlation function of the liquid state takes less than 200 fs, confirming that the laser induced electron hole plasma leads to a nonthermal melting process. This observation also holds for the other materials we have studied. The interaction of an intense laser pulse with a C_{60} molecular crystal under pressure leads to ultrafast destruction of the C_{60} molecules and the formation of a homogeneous carbon melt. Finally, in the laser fragmentation of a carbon nanotube we can distinguish two damage processes, namely the emission of carbon monomers and the unfolding of the tube after laser induced bond breaking.

Chapter 8

ULTRAFAST LATTICE DYNAMICS IN SOLIDS PROBED BY TIME-RESOLVED X-RAY DIFFRACTION

A.M. Lindenberg
Advanced Light Source, Berkeley National Laboratory, Berkeley, CA 94720, USA
aml@socrates.berkeley.edu

Abstract It is described in this article how the vibrational modes of a solid can be directly observed in real time using time-resolved x-ray scattering techniques. A general technique for performing phonon spectroscopy in the time-domain is developed. These techniques are applied to the study of transient states of matter under highly non-equilibrium conditions, near the transition from an ordered to a disordered state. It is shown how one may extract microscopic information about the coupling between carriers and phonons in a material by directly resolving the transfer of energy between the two subsystems. Finally, it is shown how time-resolved diffraction may be used in conjuction with coherent control techniques to control the evolution of a solid-state material.

1. Introduction

X-rays have been used for the last century to probe the structure of matter on atomic length-scales. With wavelength on the order of a typical atomic spacing, the measurement of how x-rays scatter off materials allows one to map out the positions of individual atoms. This technique has been applied to a wide variety of systems, from simple crystals [158, 157, 219, 220] to complex molecular structures of biological importance [221, 222]. To begin with, we ask the following question: What is the time-scale on which motion occurs within a molecular or solid-state system? Proteins fold on time-scales of milliseconds, although smaller-scale structural rearrangements occur on much faster time-scales [223, 224, 221]. In a simple diatomic molecule, molecular vibrations occur on time-scales of order 100 femtoseconds. In a solid, in contrast, there exist a range of vibrational time-scales. Optical phonon excitations have periods on the order of 100 femtoseconds. There exist as well acoustic modes which correspond to vibrational time-scales over a wide range of frequencies, extending all the way down to q=0, zero frequency modes. It is on such vibrational timescales that phase transitions in solids occur, molecules rearrange, and chemical reactions occur. This article begins by detailing the experimental setup used to directly probe structural dynamics at the Advanced Light Source Synchrotron. It then develops a simple theory for the time-resolved diffracted intensity in the presence of coherent phonon excitations, based on an already existing theoretical description of x-ray scattering from incoherent phonon distributions. It then moves on to discuss experimental results, in particular showing how one can do phonon spectroscopy in the time-domain by probing individual vibrational modes. From the detailed fitting of the results, a variety of fundamental parameters relating to the semiconductor InSb are obtained. Data is also presented studying the high-fluence regime, where the induced lattice displacements are large. Evidence is given for a laser-induced transition to a disordered state, though the nature of this state requires further investigation. The final section is concerned with some aspects of coherent control, in which it is demonstrated that individual modes of a solid can be driven to large amplitude or entirely cancelled out using specially shaped acoustic pulses. Some of the work in this article may also be found in [158, 225].

2. Experimental Setup

2.1 The Advanced Light Source

The experiments are performed at a bending magnet beamline at The Advanced Light Source (ALS) synchrotron at Lawrence Berkeley National Laboratory. The experimental setup is shown in Figure 8.1. The x-rays are first incident on a platinum-coated grazing incidence toroidal focusing mirror which one-to-one images the source size to give an x-ray spot size on order 200 μm in diameter in the experimental chamber. A set of verticle and horizontal apertures collects an angle of 0.3 mrad in the vertical and 3 mrad in the horizontal. The beam is then incident on a rotating chopper, spinning at 100 Hz with 10 small slits cut into it. It transmits x-ray pulses of width $\sim$50 μsec, thereby reducing the heat load on the monochromator crystals by a factor of 50, but still allowing one to run the experiment at a 1 kHz repetition rate. A double-crystal monochromator selects a single wavelength out of the broad spectrum from the source, chosen by setting the crystal angle to satisfy the Bragg equation

$$\lambda = 2d \sin \theta \tag{8.1}$$

A Si (111) crystal gives energy resolution of better than a few eV at 5 keV. Finally, the beam is incident on the sample to be studied and used to probe the dynamical properties of the material in real time through measurement of the time-resolved diffracted intensity.

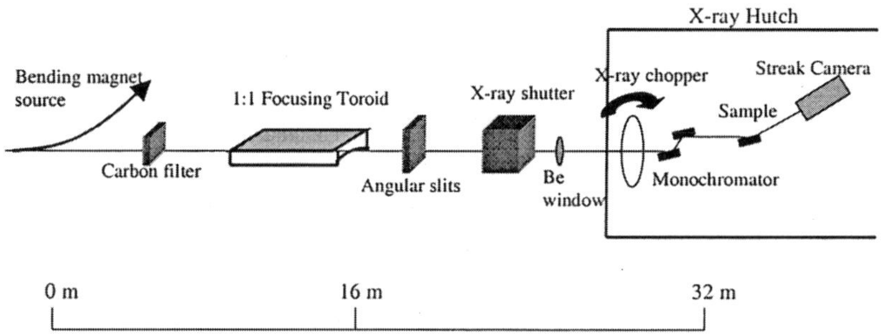

Figure 8.1: Beamline 7.3.3 Schematic

2.2 X-ray Time-Structure

At the ALS, the electrons circulate in the storage ring in bunches, driven by a radio frequency signal at 500 MHz, corresponding to a separation of 2 ns between bunches. The time for an electron to complete a single revolution around the storage ring is 656 ns. In the most typical operating mode, only 276 buckets are filled, with a single bucket inside the resulting 100 ns gap, the so-called "camshaft" pulse. Figure 8.2 shows the measured signal on an avalanche photodiode placed in the x-ray beam. This detector has 10 ns resolution so

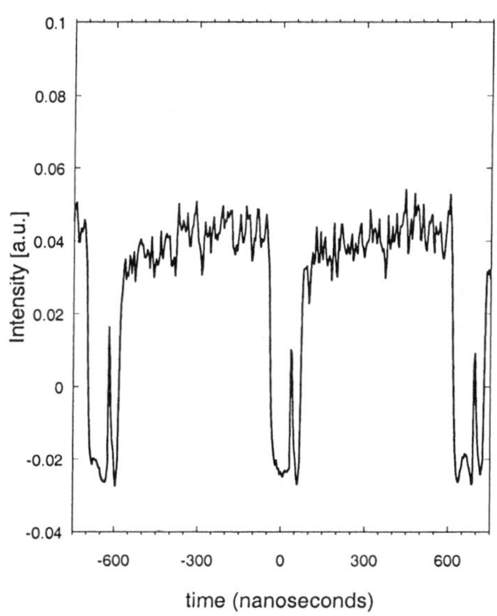

Figure 8.2: X-ray time-structure at the Advanced Light Source, as measured on a fast diode

that the multibunches separated by 2 ns run together into a single dc level. However, the gap and the single camshaft pulse inside the gap are clearly visible. There are on order of 10^4 x-ray photons in a single camshaft pulse after the monochromator in a 0.1% bandwidth.

This time-structure is explicitly taken advantage of in the experiments described hereafter. For ns or slower temporal resolution experiments, one may consider the high repetition rate synchrotron as effectively a continuous source, as most users do. For faster time-resolution, one may work in a pump-probe geometry, scanning a short laser pulse with respect to the camshaft pulse. The time-resolution in such a scheme is limited by the duration of the probe i.e. the duration of a single x-ray pulse. This is of order 50 ps but varies as a function of the number of electrons in a single bunch, because of electron-electron scattering effects [226]. For resolution faster than the intrinsic ALS pulse length, a streak camera detector is employed to resolve changes within the 50 ps x-ray pulse window.

2.3 Laser Synchronization

In any time-resolved pump-probe scheme, synchronization of the excitation pulse to the probe pulse is of utmost importance. For the case of visible light pump-probe geometries, both pump and probe are typically obtained from the same source and are thus absolutely synchronized. In these experiments, since the pump (laser) and probe (x-rays) are derived from independent sources, one needs to be more careful. The synchronization system starts with the oscillator, where the cavity length is controlled in such a way to phase lock its output to the external RF signal which drives the electron bunches. In essence, it is similar many other phase-locked loops, containing a phase sensitive detector (a mixer), a voltage controlled oscillator (the laser itself), and a feedback mechanism. It is required that the oscillator output be phase locked to the synchrotron RF frequency at 500 MHz. The oscillator runs at 83.3 MHz so in practice, it is the 6th harmonic of the oscillator output which is phase locked. This is accomplished by mixing the 500 MHz RF signal with the 6th harmonic of the laser output. Since the laser outputs pulses, there is already a strong 6th harmonic contained in a fast diode looking at a small fraction of the laser output. This signal is passed through a 500 MHz bandpass filter and then sent to the mixer. The output of the mixer is a signal related to the difference in frequency and in phase of the two signals. In particular, a small phase difference results in a small dc output, with sign depending on the phase difference. This output is amplified, filtered, and then sent to a piezoelectric device on which the end mirror of the laser cavity is mounted. Thus the cavity length is controlled in such a way to phase-lock the two signals. Experimental timing jitter is of order 5 ps, much less than the ALS pulse duration.

2.4 Streak Camera

The x-ray streak camera we use forms the heart of the experimental apparatus. It is an x-ray detector with better than two picosecond temporal resolution in averaging mode and subpicosecond when used single shot[227]. A simplified schematic is shown in Figure 8.3. The basic principle of a streak camera is to

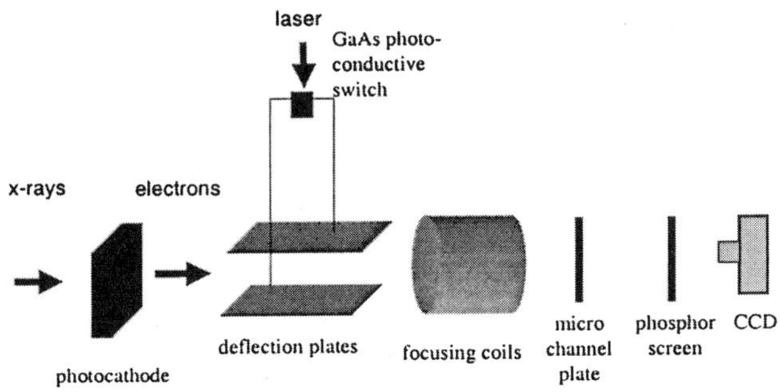

Figure 8.3: Diagram of picosecond x-ray streak camera

transform the desired temporal history of an incident signal into spatial information, much like an oscilloscope. X-rays are first incident on a CsI photocathode which transforms the incident light signal into an electron signal. Any modulation in the incident x-ray signal is directly transformed into a modulation in the extracted electron signal. The optimum CsI thickness is a function of the energy of the x-rays[228] but is on the order of 1000 Å for the experiments described here. The photo-generated electrons are extracted from the cathode region by an electric field of order 30 kV/cm and travel down the streak tube. As they propagate, the electrons are subject to a fast ramping electric field transverse to the propagation direction. Thus electrons arriving early are deflected a small amount while electrons arriving late are deflected a large amount. The end result is a spatially dispersed image (or "streak") at the end of the streak tube, reflecting any temporal modulations in the incident x-rays. This electron streak is magnetically imaged by a solenoidal current loop onto a gateable microchannel plate (MCP)-phosphor screen assembly and then read out by a cooled CCD camera. Running at a 1 kHz repetition rate, one can average 1000 incident x-ray pulses per second. Typically we expose the CCD camera on order 30 seconds. In order to obtain high temporal resolution, it is important to trigger the sweep plates that generate the transverse field at exactly the same time relative to the incident x-ray photons. This is accomplished using a GaAs photoconductive switch. An adjustable voltage bias is applied across an insulating GaAs crystal. Excitation by a femtosecond laser pulse split off from the main excitation pulse, closes the open circuit on a rapid time-scale by generating carriers which then move in the applied field. The two outputs, of opposite polarity, each drive one of the deflection plates. Thus one can turn on the sweep electric field at a precise time relative to the arrival of the x-ray signal on the streak camera photocathode. Displayed in Figure 8.4 are a series of femtosecond uv pulses separated by 20 ps incident on the streak camera photocathode. Measured time

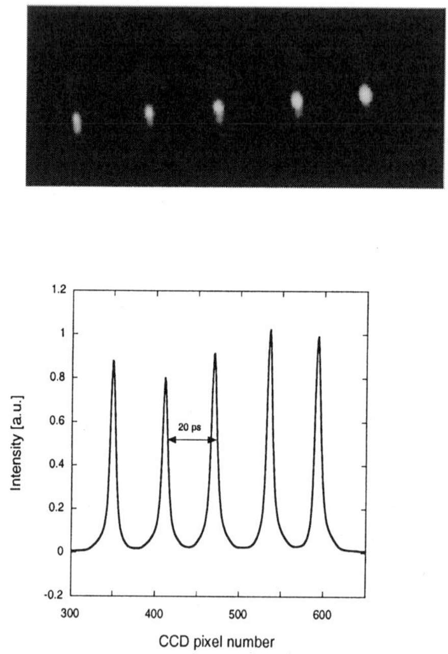

Figure 8.4: Image of UV calibration pulses

resolution is better than 2 ps in averaging mode. For the experiments discussed here, the dominant contribution to timing jitter is probably fluctuations in the laser energy incident on the photoconductive switch. To estimate the magnitude of this effect, consider that the width of the linear region of the sweep is about 200 ps (this is mainly set by the capacitance of the sweep plates and the size of the GaAs sample inside the photoconductive switch). A 1% fluctuation in the laser energy results in an effective jitter of (200 ps)(.01)(.5)=1ps.

3. Theory of Time-Resolved X-Ray Diffraction

In this section, the theory of time-resolved x-ray diffraction is developed, concentrating in particular on the time-resolved scattering of x-rays by phonon excitations. Since one may usually decompose an arbitary atomic motion into the normal modes of the solid, the approach is quite general. Following this development,it is shown in the later sections how these results are observed experimentally.

Standard books describing the theory of x-ray diffraction (for example, [229, 230]) describe in detail how x-rays scatter off of incoherent lattice vibrations. This scattering gives rise to well- known effects like the Debye-Waller factor, corresponding to a reduction in the Bragg peak intensity, or diffuse scattering, a corresponding increase in the wings of the x-ray reflection. In any theoretical description of these effects, one finds, at some point in the calculation, a time-average of the calculated diffracted intensity. When performing measurements on time-scales comparable to vibrational periods, as is done in this work, this approximation breaks down. Consider a longitudinal phonon excitation of wave-vector q, corresponding to a lattice displacement

$$\mathbf{R_n} = \mathbf{r_n} + \mathbf{u_q}\sin(\mathbf{q}\cdot\mathbf{r_n} - \omega\mathbf{t}). \tag{8.2}$$

Any time-dependence due to x-ray propagation effects with the material is neglected i.e. we assume that at each moment in time, the x-rays see some fixed atomic arrangement. The diffracted intensity is then

$$\begin{aligned} I &\propto \left|\sum_{\mathbf{n}} e^{i\mathbf{\Delta k}\cdot\mathbf{R_n}}\right|^2 \\ &= \left|\sum_{\mathbf{n}} e^{i\mathbf{\Delta k}\cdot\mathbf{r_n}} e^{i\mathbf{\Delta k}\cdot\mathbf{u_q}[e^{i(\mathbf{q}\cdot\mathbf{r_n}-\omega t)}+e^{-i(\mathbf{q}\cdot\mathbf{r_n}-\omega t)}]}\right|^2 \\ &\sim \left|\sum_{n} e^{i\mathbf{\Delta k}\cdot\mathbf{r_n}}\left(1 + i\mathbf{\Delta k}\cdot\mathbf{u_q}\left[e^{i\mathbf{q}\cdot\mathbf{r_n}}e^{-i\omega t} + e^{-i\mathbf{q}\cdot\mathbf{r_n}}e^{i\omega t}\right]\right)\right| \\ &= \left|\delta_{\mathbf{\Delta k},\mathbf{G}} + i\mathbf{\Delta k}\cdot\mathbf{u_q}\delta_{\mathbf{\Delta k+q},\mathbf{G}}e^{-i\omega t} + i\mathbf{\Delta k}\cdot\mathbf{u_q}\delta_{\mathbf{\Delta k-q},\mathbf{G}}e^{i\omega t}\right|^2 \end{aligned} \tag{8.3}$$

For each phonon of wave vector q, one thus obtains two sidebands, centered around the main Bragg peak, one peak corresponding to a phonon absorption and one to a phonon emission event. In the case of a thermally excited distribution of phonons, the effect is then to create wings on the main Bragg peak. These are all well known effects in the theory of temperature diffuse scattering [231, 232, 232] But what of the time-dependence? It is clear that in this approximation, there is no time-dependence. Upon evaluation of the above expression, the cross terms vanish since the delta-functions have zero overlap. On the other hand, it is clearly a poor approximation to treat Bragg peaks as delta-functions. Thus to the extent that the rocking curves have finite angular width, one obtains for the diffracted intensity, to first order in the displacement $\mathbf{u_q}$:

$$I \propto \delta^2_{\mathbf{\Delta k},\mathbf{G}} + \cos(\omega t)\left[\delta_{\mathbf{\Delta k},\mathbf{G}}\delta_{\mathbf{\Delta k+q},\mathbf{G}} - \delta_{\mathbf{\Delta k},\mathbf{G}}\delta_{\mathbf{\Delta k-q},\mathbf{G}}\right]\mathbf{\Delta k}\cdot\mathbf{u_q} + \cdots \tag{8.4}$$

where the δ-functions represent the physical x-ray rocking curves, of finite width. As can be seen, sidebands are created angularly displaced from the Bragg peak. These oscillate up and down, out of phase, at the frequency of the modulating phonon. A general expression for the diffracted intensity as a function of time correct to all orders can be shown to be:

$$I \propto \left| \sum_{m=-\infty}^{\infty} \delta_{\mathbf{\Delta k}+m\mathbf{q},G} e^{-im\omega t} J_m(\mathbf{\Delta k} \cdot \mathbf{u_q}) \right|^2 . \tag{8.5}$$

It is thus seen that the time-dependence is a result of coherent interferences between phonon-zero phonon scattering or phonon-phonon scattering events. In particular, the first order terms are due to 1 phonon/zero-phonon coherent interference while the second order terms are due to two-phonon/zero-phonon interferences as well as 1 phonon/1 phonon interferences. Note that these effects are inherently inelastic; the scattered x-ray photon must undergo a small change in energy in order to conserve momentum and energy. This energy shift is directly analogous to the doppler shift light experiences on reflecting off a moving mirror. However, while the incident photon energy is on the order of 5 keV, a typical acoustic or optical phonon energy is of the order of a few meV. We can thus safely neglect the change in energy in this analysis, at least to first order. The above results should be compared to the rapidly developing field of x-ray inelastic scattering[233], (also termed "X-ray Brillouin Scattering") in which one attempts to probe vibrational excitations by resolving these small changes in energy. In such experiments, meV energy resolution is typically obtained using multiple large angle reflections. We show in the following that in fact one can resolve meV acoustic phonons by time-resolving the x-ray diffracted intensity, even with eV-type energy resolution.

4. Wave-vector Matching Considerations

It has been shown that a phonon of wave vector q induces sidebands angularly displaced from each Bragg peak,which oscillate at the corresponding phonon frequency ω. We now derive the frequency of oscillation as a function of detuning angle. Consider first the case of symmetric reflections, for which the lattice planes are parallel to the crystal surface.

4.1 Symmetric Case

Figure 8.5 shows an x-ray photon initially incident at the Bragg angle θ_B. According to the Laue equations, $\mathbf{\Delta k} = \mathbf{G}$, where $\mathbf{G}$ is the reciprocal lattice vector. As shown in the diagram,$|\mathbf{G}| = 2\mathrm{k} \sin\theta$. If we now detune the crystal away from the Bragg angle so that the incidence angle is $\theta_B + \Delta\theta$, then $\mathbf{\Delta k} \neq \mathbf{G}$. Now, the change in the x-momentum is zero. The change in the

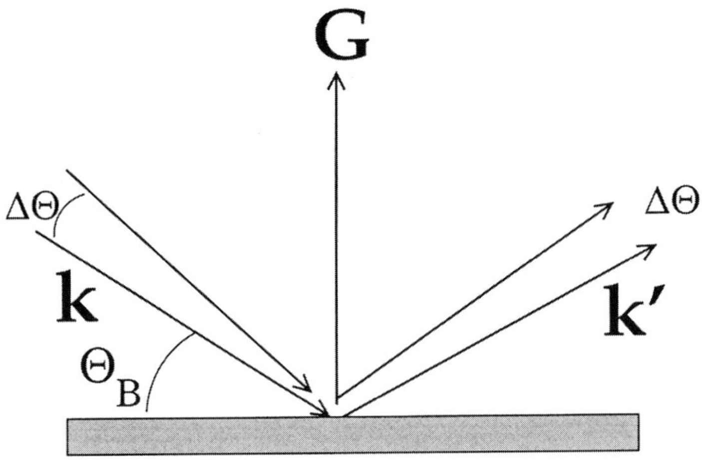

Figure 8.5: Phase matching diagram

y-momentum of the scattered photon is

$$\begin{aligned} \mathbf{\Delta k}_y &= 2k\sin(\theta_B + \Delta\theta) \\ &\approx 2k[\sin(\theta_B) + \cos(\theta_B)\Delta\theta] \\ &= G + 2k\cos\theta_B\Delta\theta \end{aligned} \tag{8.6}$$

The momentum mismatch q is thus

$$\begin{aligned} q &= 2k\cos\theta_B\Delta\theta \\ &= 2k\sin\theta_B\cot\theta_B\Delta\theta \\ &= G\cot\theta_B\Delta\theta \end{aligned} \tag{8.7}$$

Since

$$\omega = qv \tag{8.8}$$

for acoustic phonons, we thus arrive at the following equation for the observed oscillation frequency as a function of deviation angle from the Bragg peak:

$$\omega_{\text{symm}} = vG\cot\theta_B\Delta\theta \tag{8.9}$$

For the case of asymmetric x-ray diffraction (in which the lattice planes for the reflection under study make an angle α with the sample surface), the situation is slightly more complicated. One typically defines a parameter b such that

$$b = \frac{\sin(\alpha + \theta_B)}{\sin(\alpha - \theta_B)}. \tag{8.10}$$

For large b (which is the experimental geometry much of the following data is taken) one may show by an argument similar to the above, that

$$q \sim k\Delta\theta b. \tag{8.11}$$

Thus,

$$\omega_{\text{asymm}} = vkb\Delta\theta \tag{8.12}$$

For example, for b=10, at 100 arcsec off the Bragg peak, one obtains an oscillation period of 8 ps (about 10 times faster than the symmetric case). While the rocking curve width decreases only as the square-root of b, the dispersion increases as b. It is thus easier to probe larger wave-vector phonons in an asymmetric geometry. Interestingly, one may obtain the same equations from an entirely different point of view based on dynamical diffraction theory. Instead of considering a single phonon excitation, we consider the more realistic case in which a superposition of phonon modes are created, forming an acoustic pulse[234] Here, as a simple model of this pulse, we derive an equation for the time-dependent rocking curve for the specific case of a crystal of time-dependent thickness $z = vt$, expanding at the sound velocity. In the notation of Zachariasen [230] we define the variables

$$A = \frac{\pi}{\lambda} |\Psi_H| \frac{vt}{\gamma_0 \gamma_h} \tag{8.13}$$

and

$$y = \frac{\frac{1-b}{2}\Psi_0 + \frac{b}{2}\Delta\theta \sin(2\theta_B)}{\sqrt{|b|}\,|\Psi_H|}. \tag{8.14}$$

Ψ_0 and Ψ_H are the crystal susceptabilities, proportional to the structure factors for the Bragg reflection H. γ_0 and γ_h are the direction cosines of the incident and diffracted wave-vectors, relative to the surface normal. Physically, y is a measure of the deviation from the Bragg angle and A is a dimensional measure of the crystal thickness. It can then be shown that the rocking curve R(y) for a crystal of thickness z has the following dependence [230]:

$$R(y) \propto \frac{1}{y^2 + (y^2 - 1)\cot^2\left[A\sqrt{y^2 - 1}\right]} \tag{8.15}$$

For a thin crystal, it is clear that there exists an interference effect which gives rise to sidebands around the main Bragg peak. Furthermore, as the crystal thickness changes, so do the positions of the sidebands. Consider a deviation $\Delta\theta$ from the Bragg peak at y=0, for the symmetric case (b=-1). In the wings of the peak, for y>>1, but for small times i.e. A<<1, the above equation for the diffracted intensity becomes

$$
\begin{aligned}
R(y) &\sim \frac{1}{y^2 + y^2\cot^2(Ay)} \\
&= \frac{\sin^2(Ay)}{y^2} \propto \sin(2Ay) \\
&= \sin(\frac{2\pi v\Delta\theta\sin(2\theta_B)}{\lambda\sin\theta_B}t)
\end{aligned}
\tag{8.16}
$$

which corresponds to an oscillation at

$$\omega = vG\cot\theta_B \tag{8.17}$$

exactly as found above. For the asymmetric geometry, consider as above the case of $|b| \gg 1$ i.e. the case for which the diffracted beam leaves the crystal at grazing incidence. Then, using the fact that

$$b \sim \frac{\sin(2\theta_b)}{\sin(\theta_b - \alpha)} \tag{8.18}$$

one obtains an oscillation at

$$\omega = \frac{2\pi}{\lambda}vb\Delta\theta \tag{8.19}$$

as found above. Thus the expanding crystal model and the inelastic scattering model give identical results. This should not be suprising as, in a reference frame moving with the expanding crystal (i.e. the propagating acoustic mode), the reflected beam undergoes a doppler shift and thus a slight change in frequency. The next section gives a physical basis for the expanding crystal model.

5. Generation of coherent displacements

Irrespective of the coupling mechanism, the deposition of energy into a solid on a time-scale faster than the material can respond will lead to the generation of a coherent displacement. Thus, without invoking any mechanism for this heating, one may ask the question: How will the the lattice move toward the new equilibrium lattice spacing, from a microscopic viewpoint? This question can be answered by solving the elastic equations of motion of a solid, following the original work of Thomsen et al. [234] Following Thomsen's approach, we treat the solid as a continuum. In the approximation, the elastic equations of motion are

$$\sigma = K\eta + \sigma_{external} \tag{8.20}$$

$$\rho\frac{\partial^2 u}{\partial t^2} = \frac{\partial\sigma}{\partial z}. \tag{8.21}$$

The first equation is Hooke's Law, where σ is the total stress, η is the strain, and $\sigma_{external}$ is the applied external stress due to heating. The second equation is an expression of Newton's Law, where u(z,t) is the displacement field at depth z in the crystal, at time t. For the applied external stress, in the simplest approximation, we use an equation of the form

$$\sigma_{ext} \propto e^{-\frac{z}{\xi}} \tag{8.22}$$

where ξ is the laser penetration depth. This represents a depth-dependent stress applied to the crystal before it has time to respond. Thomsen showed that the final solution for the displacement as a function of time is:

$$u(z,t) = \frac{p\xi^2}{2k}\left(e^{-\frac{|z-vt|}{\xi}} + e^{-\frac{(z+vt)}{\xi}}\right) - \frac{p\xi^2}{k}e^{-\frac{z}{\xi}} \tag{8.23}$$

Thus the lattice strain field is

$$\eta = \frac{\partial u}{\partial z} = \frac{p\xi}{k}e^{-\frac{z}{\xi}}\left(1 - \frac{1}{2}e^{-\frac{vt}{\xi}}\right) - \frac{p\xi}{2k}e^{-\frac{|z-vt|}{\xi}} sgn(z-vt). \tag{8.24}$$

The lattice strain is shown in Figure 8.6 for three different times. It is seen that the resulting lattice motion corresponds to an acoustic pulse propagating into

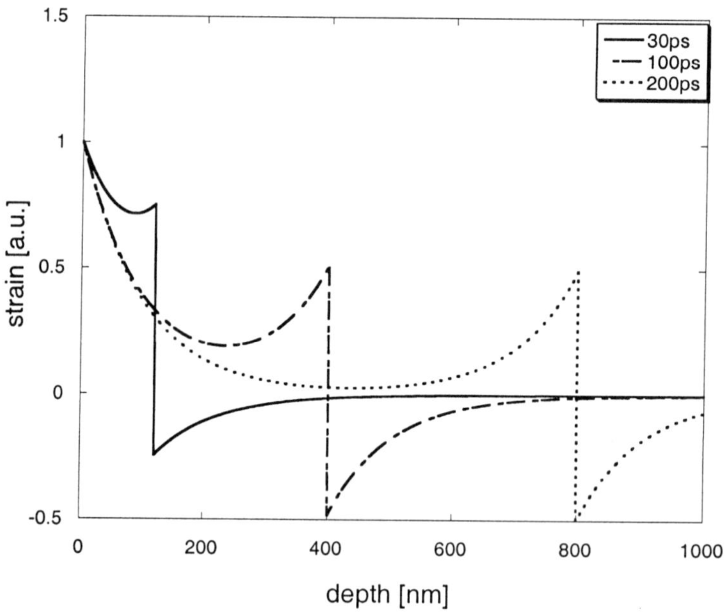

Figure 8.6: Laser-generated coherent acoustic pulses at times t=30,100, and 200 ps

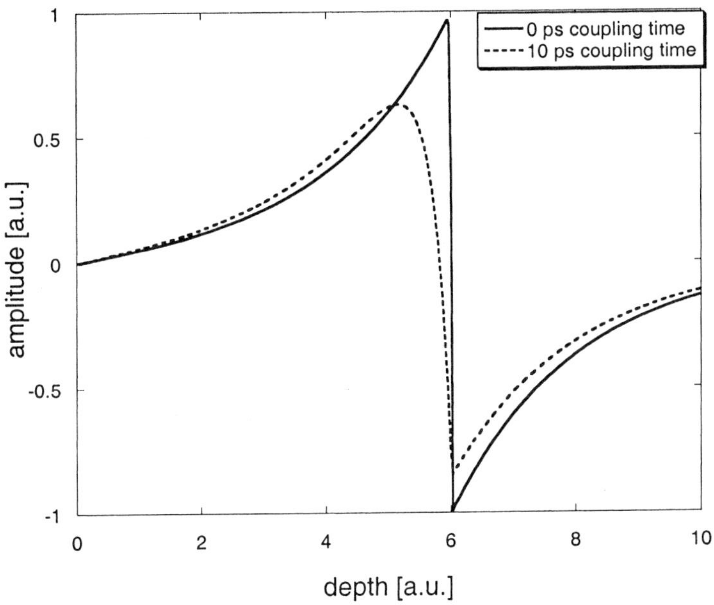

Figure 8.7: Coherent part of the strain for electron-phonon coupling times of 0 and 10 ps

the solid at the velocity of sound. The pulse consists of a region of compression (or negative strain) followed by a region of expansion (positive strain). Superimposed on the coherent motion is an incoherent, time-independent component corresponding to lattice expansion near the surface. The width of the expansive or compressive regions is of order the laser penetration depth ξ. The frequency spectrum of this pulse thus exhibits a maximum at $\nu \sim v_{sound}/\xi$. In other words, the time for the surface layer to thermally expand is just governed by the speed of sound. Rose-Petruck et al [157] first observed such coherent acoustic pulses using time-resolved x-ray diffraction.

6. Extension to Finite Electron-Phonon Coupling Times

In a more realistic model, one needs to take into account the fact that the energy initially placed into the electron-hole system may not be transfered instantaneously to the lattice. Indeed, for the case of polar semiconductors like InSb, it is thought that one important pathway for energy relaxation starts by coupling energy into longitudinal optical phonons. These decay in an acoustic phonon cascade, the net result being that it takes on order 10 ps for this transfer to be complete and the carrier and lattice system to reach a mutual equilibrium. The implication of this is that the excitation of the lattice is not fully impulsive i.e. δ-function like. In particular, phonons with period shorter than this transfer time will not be coherently excited. In contrast, phonons with period longer than this transfer time still see the excitation in the impulsive limit and will be coherently excited. With respect to the shape of the coherent pulse, one would expect that the removal of the high-frequency components should blur the the sharp boundary between the expansive and compressive parts of the strain wave. This is indeed shown to be the case in the following analysis: One would like to find an equation for the displacement field u(z,t) for the case where the external stress is both space and time-dependent, representing the transfer of energy to the atoms on a timescale τ_p, the electron-phonon coupling time. For simplicity, we assume an equation for the external stress of the form

$$\sigma_{ext}(z,t) = a(t)e^{-\frac{z}{\xi}} = \left(1 - e^{-\frac{t}{\tau_p}}\right) e^{-\frac{z}{\xi}} \tag{8.25}$$

This represents a stress which turns on on a time-scale set by τ_p, a phenomenological electron-phonon coupling time. The equation to be solved then becomes:

$$\rho \frac{\partial^2 u}{\partial t^2} = k \frac{\partial^2 u}{\partial z^2} + \frac{\partial \sigma_{ext}}{\partial z} \tag{8.26}$$

with the initial condition $u(z,0) = 0$ and the boundary condition $\sigma_{tot}(0,t) = ku_z(0,t) + \sigma_{ext}(0,t) = 0$. This can be solved by the method of Fourier transforms. For the coherent part, one finds the following result: If $z > vt$ then

$$u(z,t)_{\text{coh}} = \frac{\pi \left[2v\tau_p e^{-\frac{(z+vt)}{v\tau_p}} - (\xi + v\tau_p)e^{-\frac{(z+vt)}{\xi}} + (v\tau_p - \xi)e^{-\frac{(z-vt)}{\xi}} \right]}{4(v\tau_p^2 - \xi^2)} \tag{8.27}$$

If $z < vt$ then

$$u_{\text{coh}} = \frac{\pi \left[2v\tau_p e^{-\frac{(z+vt)}{v\tau_p}} - (\xi + v\tau_p)e^{-\frac{(z+vt)}{\xi}} + 2v\tau_p e^{\frac{(z-vt)}{v\tau_p}} - (\xi + v\tau_p)e^{\frac{(z-vt)}{\xi}} \right]}{4(v^2\tau_p^2 - \xi^2)} \tag{8.28}$$

The coherent part of the strain is shown in Figure 8.7. As can be seen, even a 10 ps electron-phonon coupling time induces a sizeable change to the shape of the acoustic pulse near the boundary between the compressive and expansive regions. Qualitatively, one would expect just such a behavior. In effect, the acoustic pulse starts to propagate as it is built up over the electron-phonon coupling time. As will be seen, this effect is clearly observed in the experimental data.

7. Experimental Results

7.1 Slow (Nanosecond) Time-scale Measurements

To start out with, we show here nanosecond time-scale measurements of the time-resolved diffracted intensity following short-pulse laser excitation of InSb. Using an avalanche photodiode with about 10 ns time-resolution, the x-rays appear as a continuous source, apart from the 100 ns gap. The excitation pulse is timed to arrive during the multi-bunches (separated by 2 ns). One can thus probe the long time-scale behavior by simply recording the ouput of the diode on a fast oscilloscope. Figure 8.8 shows the time-resolved diffracted intensity at different points on the rocking curve, relative to the peak of the unperturbed rocking curve. The laser arrives at $\sim$t=250 ns. As can be seen in this figure and the following ones, the dynamics are quite complicated. From a qualitative standpoint, the results are as expected: On the negative (smaller angle) side of the rocking curve, one observes an increase in the diffracted intensity following excitation. This is a result of the deposition of heat in the material, a resulting lattice expansion, thus causing a shift of the rocking curve to smaller angles. Correspondingly, on the positive side of the rocking curve, one observes an increase following excitation as the rocking curve shifts again to smaller angles. In all cases, the initial increase or decrease occurs on a time-scale limited by the resolution of the detector. Following this initial change, however, the dynamics strongly depend on rocking curve angle. Even the recovery time-scales seem to vary strongly. The return to equilibrium one would expect to correspond to the time for the deposited heat to leave the x-ray probe region (500 nm). But a simple estimate of the time-scale for this diffusive process yields a time-scale of only 10 ns whereas the observed recovery times are, in some cases, more than a microsecond.

7.2 50 ps Resolution Pump-Probe Experiments

Instead of treating the ALS as a CW source, a detector capable of gating and measuring a single x-ray pulse allows one to obtain temporal resolution limited by the pulse length of the probe x-ray pulse. This is accomplished by scanning the delay of the excitation pulse relative to the arrival time of the x-ray pulse on the sample, in a pump-probe scheme. Typical data are shown in Figure 8.9. These results correspond to taking the results of the previous section and zooming in around t=0 one-hundred times. A decrease in the diffracted intensity is observed on a time-scale limited by the ALS pulse duration. Also shown in Figure 8.9 is a simulation using the acoustic strain field derived above in the dynamical diffraction equations. Note that in this pump-probe geometry, the initial drop in the diffracted intensity is convoluted with the ALS pulse width, thus giving rise to a slower than expected observed time-scale. The dynamics are

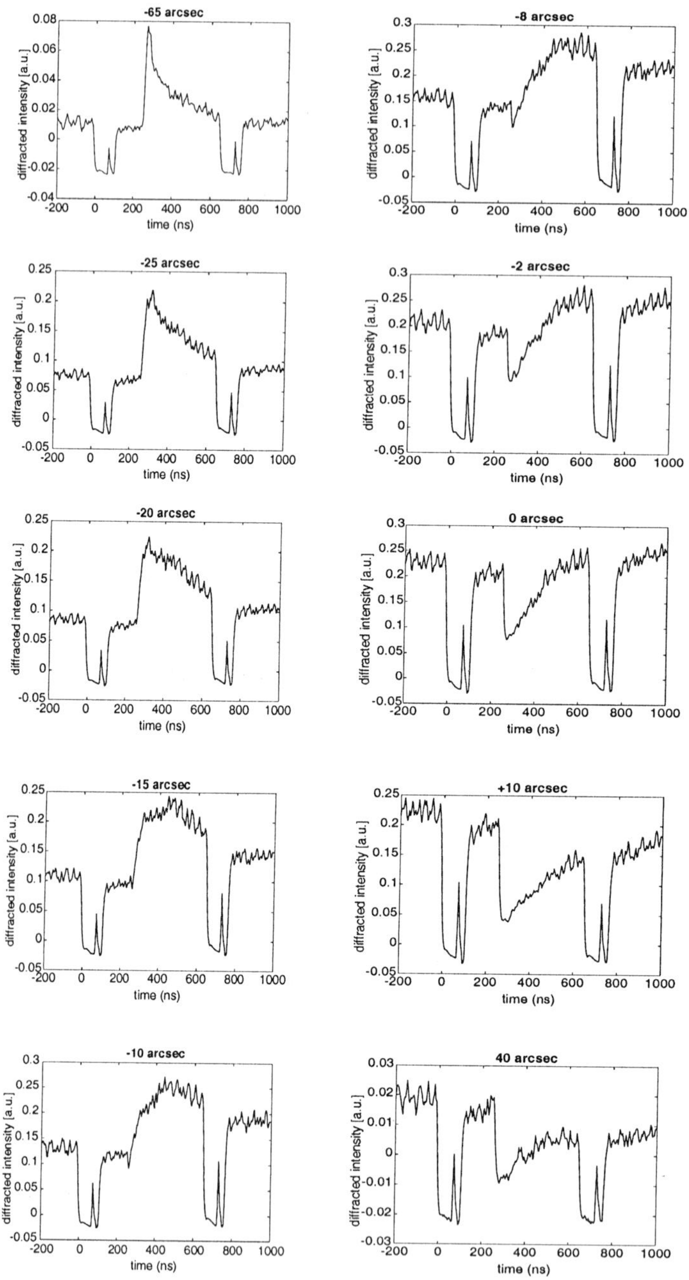

Figure 8.8: Time-resolved diffracted intensity at different rocking curve angles, with nanosecond resolution

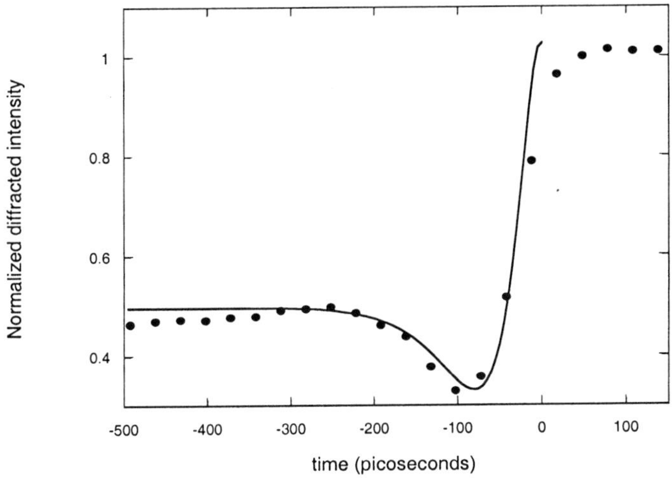

Figure 8.9: Pump-probe measurements of time-resolved diffracted intensity. Solid line is simulation.

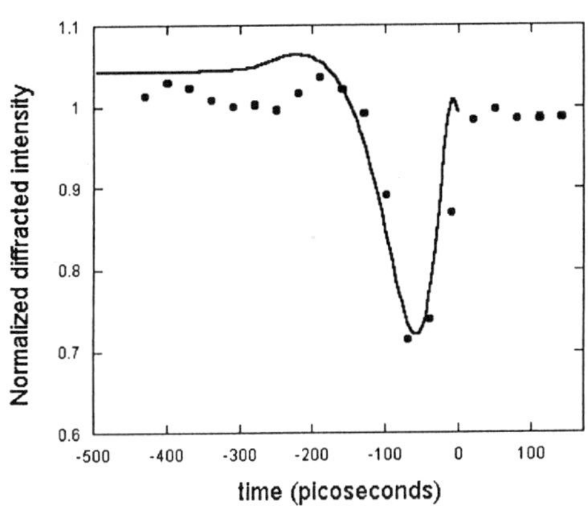

Figure 8.10: Pump-probe measurements of time-resolved diffracted intensity

highly sensitive to the position on the rocking curve peak. Figure 8.10 shows a measurement taken just slightly off the Bragg peak, exhibiting a complete recovery of the diffracted intensity within 200 ps. Although somewhat non-intuitive, these results follow directly from dynamical diffraction simulations using the simple Thomsen model described above. Shown in the figure is a fit to the observed data, in good agreement with the experimental results. That the observed dynamics are highly sensitive to the the angle of the crystal

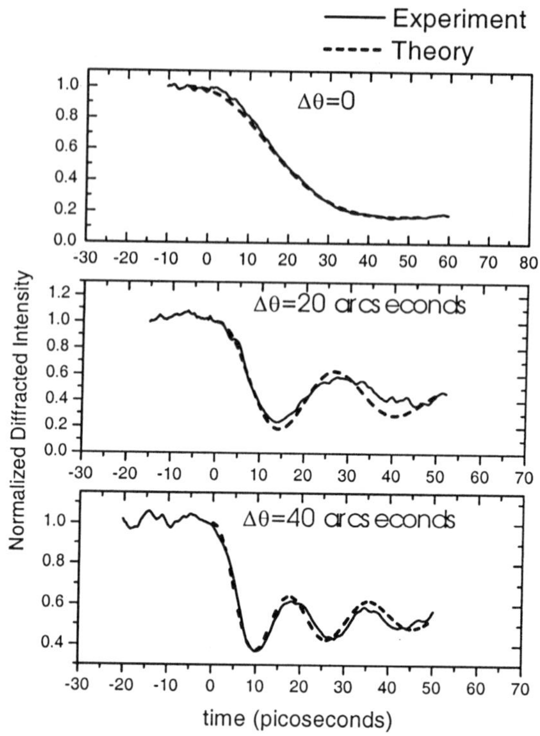

Figure 8.11: Measurements and simulations of time-resolved diffracted intensity at various angles from the Bragg peak

is evidenced by the fact that Figures 8.9 and 8.10 differ by only five arcsec, well within the theoretical rocking curve width ($\sim$ 10 arcsec) and significantly smaller than the experimentally measured width.

7.3 Streak Camera Results

Figure 8.11 shows the measured time-resolved diffracted intensities at a variety of angles near the Bragg peak, using a streak camera with 2 ps resolution. We have now zoomed in on time-zero by three orders of magnitude relative to the initial nanosecond results first presented. In agreement with the theoretical description above, we observe that by varying the angle of the crystal relative to the Bragg peak, different coherent phonon modes are selected out. The figure shown corresponds to phonons with period from 20-40 ps. As is expected, the frequency of oscillation increases as one moves off the rocking peak. By measuring the oscillation frequency as a function of crystal angle, the acoustic

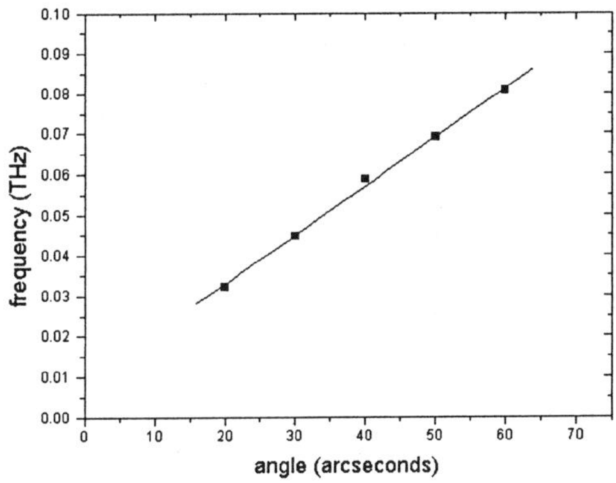

Figure 8.12: Measured acoustic phonon dispersion relation near the zone-center

phonon dispersion relation is mapped out near the Brillouin zone center, as shown in Figure 8.12, with slope proportional to the sound velocity in the crystal. To quantitatively extract a sound velocity, we fit this curve by using the time-dependent strain field derived above. We extract a sound velocity of ~4000 m/s, in agreement with the known value for InSb (3900 m/s) [235] No indication of softening or anharmonicity in the LA phonon mode is observed. Note that the exact probed wave-vector depends on the laser fluence. At higher fluence, the rocking transiently shifts more to smaller angles and one thus probes higher wave vector modes.

8. Extraction of Electron-Phonon Coupling Times

A rough simulation of the above results based on the Thomsen model described above qualitatively matches the experimental results. However, in order to quantitatively fit the experimental results, it is found that one must take into account the physics of how the excited carriers couple to the lattice and in particular, the time-scale for this to occur. Whereas polar coupling of hot carriers to the lattice probably takes of order 10 ps in InSb, there exists a more direct way pathway for energy transfer from initial hot electrons to acoustic phonons through the acoustic deformation potential[236] Assuming that the electrons and phonons maintain separate states of equilibrium characterized by temperature deviations δT_e and δT_p, the combined stress due to these two effects can be written as [237, 238]

$$\sigma \sim \gamma_e C_e \delta T_e + \gamma_p C_p \delta T_p \tag{8.29}$$

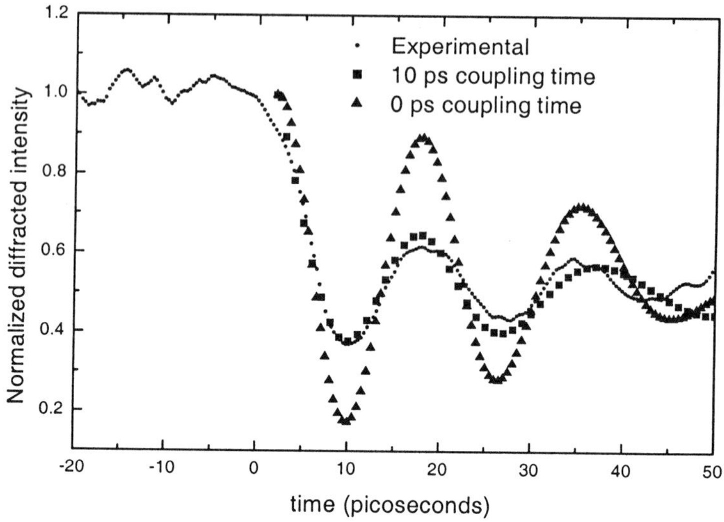

Figure 8.13: Sensitivity to various electron-phonon coupling times

where C_e and C_p are the specific heats of the electrons and the ions, and γ_e and γ_p are the corresponding Gruneisen factors for the electron and phonon subsystems. While the thermal contribution to the stress is dominant after electron-lattice equilibration, the electronic term scales as nT_e^2[238], where n is the carrier density, so that the initial hot, dense carrier distribution ($T_e \geq 1eV$, n$\geq 10^{21}cm^{-3}$) implies a significant electronic contribution to the impulsively driven lattice at short times. In accordance with the above discussion, we find that in order to obtain good agreement between the model and the observed temporal behavior, it necessary to modify the profile of the acoustic pulse to take into account the effect of a finite electron-acoustic phonon coupling time and the two mechanisms for energy transfer. Figure 8.13 demonstrates the sensitivity of the experiment to various electron-phonon coupling times. As is observed, increasing the coupling time reduces the spectral weight at high wave-vectors, and thus the observed amplitude of the oscillations. Figure 8.11 also shows the time-dependent diffracted intensity measured at 0, +20, and +40 arcseconds from the Bragg peak with the simulated (normalized) diffracted intensity at each angle. There are three adjustable parameters in the model: the electron-acoustic phonon coupling time and the amplitudes of the thermal and deformation potential generated stress. A single set of parameters matches all experimental curves with small error. In particular, best fit corresponds to a coupling time of 12 ps (in agreement with previous results [220]), and a thermal strain of 0.17% (just below that of InSb at its melting temperature), together with a nonthermal contribution a factor of two smaller. Inclusion of thermal diffusion following Ref.[234] has negligible effect on the time-scales discussed in this paper; more rapid carrier diffusion on time-scales faster than the electron-phonon coupling time will effectively increase the region in which the laser energy is deposited. However, we obtain good fits to the data using the tabulated laser penetration depth (100 nm) so do not expect this to be a significant effect. The damping of the oscillations over 50 ps is due to the spectral band width set by the Si monochromator crystal and is included in our model by averaging over the Si rocking curve width; this is the time-scale for the finite range of phonons probed at each crystal angle to dephase.

9. High Fluence Results

The observed oscillation amplitudes shown above correspond to maximum induced acoustic displacements of order 1 Å. Lattice dynamical calculations show that at the melting temperature of InSb, the incoherent acoustic phonon root-mean-square amplitude is of this order [239]. A calculation of the temperature rise in the sample at an incident fluence of 15 mJ/cm^2 confirms that the system is driven close to it's melting point (800 K) following excitation. This is in agreement with the Lindemann melting criterion [240], which predicts a transition from an ordered to a disordered state when vibrational amplitudes approach a critical value of 10% of the lattice spacing. We thus observe coherent acoustic phonons corresponding in amplitude to the incoherent phonons one would expect from InSb near its melting point. In the fluence regime discussed above, the laser-induced state is thus characterized by a maximally-strained lattice undergoing large-amplitude oscillations about the equilibrium lattice positions. At a slightly higher laser fluence, 10% below the damage threshold (about 15 mJ/cm^2), the same dispersion effects are observed as in the low fluence regime, but no temporal oscillations occur, as shown in Figure 8.14. On the Bragg peak, a 60% reduction in the diffracted intensity occurs over $\sim$ 20 ps. At +60 arcseconds from the Bragg maximum, a 40% reduction in intensity (smaller due to the larger x-ray penetration depth off the Bragg peak) occurs on a time-scale limited by the streak camera temporal resolution (3 ps). In effect, only the first half-period of the oscillations induced at lower fluences are observed. The data shown in Figure 8.14 are an average over 60,000 laser shots on the same spot. The observed effect is thus completely reversible. The dependence of the oscillation amplitude on the fluence is shown in Figure 8.15. We observe a critical threshold at $\sim 10mJ/cm^2$, above which the lattice does not support coherent modes of oscillation. We interpret these findings as follows. At low fluences, impulsive excitation of the solid induces small-amplitude

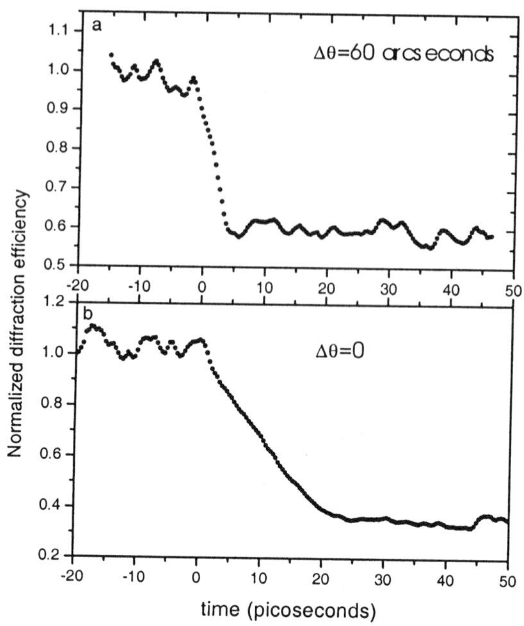

Figure 8.14: Time-resolved diffracted intensity near the damage threshold at 0 and 60 arcsec off the Bragg peak

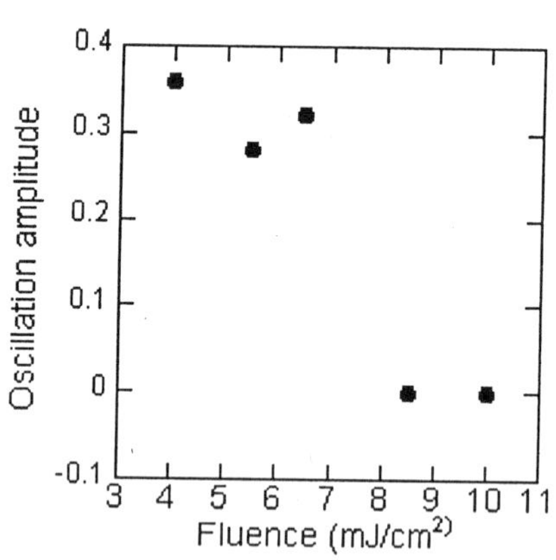

Figure 8.15: Dependence of the coherent phonon amplitude on excitation fluence

coherent vibrational atomic motion. At a critical fluence, corresponding to a critical phonon amplitude, the crystalline lattice is coherently driven into a new, disordered state on a time-scale set by one-half a phonon period, at which point the average atomic displacement is maximum for a given mode. This new state is characterized by the fact that it does not support coherent oscillations. As the crystal angle is varied relative to the Bragg peak, phonons of differing wavevectors $\mathbf{q}$ are probed; equivalently, different length scales are being probed. Thus disorder develops over shorter length-scales (large $\mathbf{q}$) in less time than longer length-scales (small $\mathbf{q}$). One may observe a change in the diffracted intensity on arbitrarily fast time-scales by tuning the crystal angle sufficiently far from the Bragg peak, presumably limited by the time-scale on which the lattice stress develops. Since the observed 3 ps drop in the diffracted intensity occurs faster than the thermal coupling time we conclude that the first step in the observed disordering transition at high laser fluence is the initial excitation of hot carriers which subsequently drive large-amplitude, coherent vibrational motion. More work is needed to understand the nature of this optically-induced transition. Similar work probing this effect has also been carried out in Germanium [219]. In this experiment, no critical threshold was observed, although the oscillations were seen to disappear at high fluence. This effect was attributed to anharmonic damping of the acoustic modes. Note that the probed mode was of much smaller frequency (10 GHz) than those discussed here.

10. Coherent Control

10.1 Introduction

By making use of classical or quantum-mechanical interferences, light may be used to control the temporal evolution of atomic, molecular, and solid-state systems, and to drive them into novel, non-equilibrium states of matter. For example, appropriately timed sequences of femtosecond light pulses maximize the yield of a chemical reaction [241, 242, 243] or selectively excite a vibrational mode in a solid[244, 245]. Until now, the resulting product states of a coherent control experiment have been probed by indirect means, typically using light or wavelength much larger than an interatomic spacing. As a result of the indirectness of visible light scattering techniques, algorithmic feedback techniques [246, 247] have been developed which attempt to converge on an optimal control sequence for a particular experiment. Direct imaging techniques like x-ray diffraction perhaps offer an alternative, in which one learns the correct control parameters by directly probing the products of an experiment.

In this section, time-resolved diffraction experiments with atomic-scale spatial resolution and picosecond temporal resolution are described which attempt to directly observe the coherently-controlled atomic motion as a solid is driven along a vibrational pathway. As a first experiment, a single acoustic mode is driven to large amplitude or cancelled out. It is further shown that by shaping ultrafast acoustic pulses, the x-ray diffraction efficiency of a crystal can be controlled. We note that coherent control of acoustic modes has been demonstrated previously [248, 249] using visible pump-probe techniques. However, this has not been achieved in a bulk material, to our knowledge. This is due to the difficulty of using visible-light techniques to probe non zone-center phonons through low-frequency Brillouin scattering. Recent experiments have thus focused on multiple quantum well structures or semiconductor superlattices, which exhibit zone-folding of the acoustic phonon branch. In contrast, x-rays, with wavelength comparable to atomic spacings, allow one to map out the entire Brillouin zone, in principle up to wave-vectors on order an inverse lattice constant. Two-pulse excitation impulsively excites a particular acoustic mode twice, the second time with some chosen phase relative to the first. In effect, the first femtosecond pulse starts up a coherent motion with all atoms moving in synchrony. The second pulse excites a phase-delayed replica of the first, and the two interfere constructively or destructively. The generated acoustic pulse is the acoustical analog of an excited wave packet, namely, a coherent superposition of eigenstates (phonons) of the crystal Hamiltonian. The final product-state is a system with one selected mode excited to large amplitude or cancelled out. Equivalently, one may view two-pulse excitation as pulse shaping of an acoustic pulse, analogous to the shaping of femtosecond optical pulses used in a variety of coherent control schemes [246, 247].

10.2 Experimental Results

Fig. 8.16a shows the diffracted intensity (normalized to one for $t \leq 0$) for single pulse excitation at an angle 300 arcsec from the (111) Bragg reflection, corresponding to the q=$4 \times 10^5 \text{cm}^{-1}$ mode. Two-pulse excitation creates acoustic pulses of the form shown in the insets to Fig. 8.16. The inset to Fig. 8.16b shows the acoustic pulse generated by two-pulse excitation with a delay of 35 ps, equal to one vibrational period. This profile drives the $4 \times 10^5 \text{cm}^{-1}$ mode to large amplitude and suppresses all nearby modes. In contrast (Fig. 8.16c), for a relative delay of 18 ps, or 1/2 a vibrational period, the probed mode is completely silenced. The insets b and c thus represent the shape of the optimal acoustic pulses for constructive or deconstructive interference of the $4 \times 10^5 \text{cm}^{-1}$ mode. Also shown in Figure 8.16 are simulations of the time-resolved diffracted intensity. These correspond to solutions of the dynamical diffraction equations as a function of time, for a given time-dependent strain field. We extract a maximum amplitude of $\sim$0.3 Å for the coherently enhanced mode. In comparison, for the case of deconstructive interference, the amplitude is zero within our signal to noise ratio. The simulations are similar to those described above. As we are probing relatively low-frequency modes in this section, these results are not sensitive to the time-scale for coupling to the lattice to occur (the electron-phonon coupling time is less than the period of

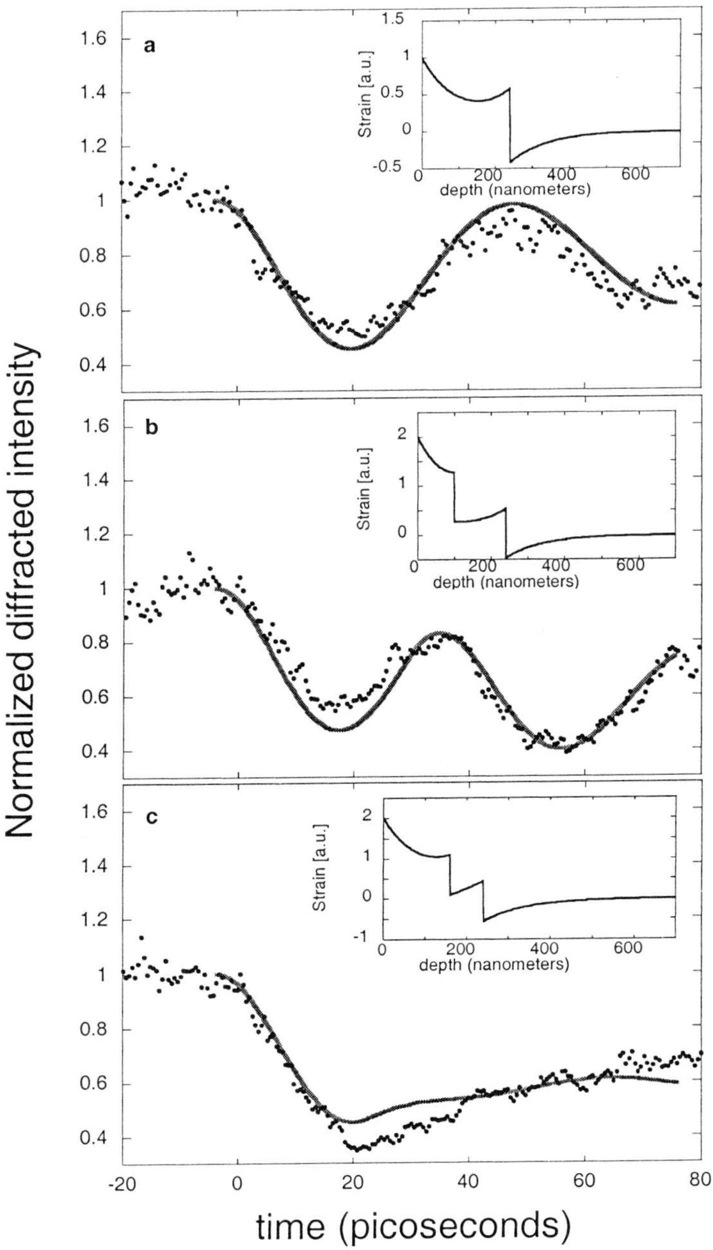

Figure 8.16: Coherent control of the q=5×10^5 cm^{-1} mode

the probed mode). However, we are sensitive to the subsequent decay of the carrier-induced stress as the carriers relax toward the band edge. This is most clearly seen in Figure 8.16c, in which, by canceling the phonon oscillations, one can directly probe the time-scale of this recovery. Note that the oscillation frequency differs slightly between the one pulse and two pulse excitation cases. This is a result of a larger shift of the Bragg peak to smaller angles for the two-pulse case, leading to a slightly larger probed phonon wave-vector for the two-pulse case.

11. Control of the Diffraction Efficiency of a Crystal

A direct result of the ability to control phonons in a crystal is the ability to control the x-ray diffraction efficiency of a crystal. From one perpective, shaped acoustic pulses as shown in Figure 8.16 are being used to define the local amount of order in a crystal, and thus the scattering efficiency of the solid. Such control has a variety of applications, as first noted by reference [250]. See also [123], in which the authors have demonstrated a switch in transmission, using the Bormann effect. In particular, by using optical phonons, which exhibit high-frequency ($\sim$10 THz) oscillations at the Brillouin zone center, a femtosecond x-ray switch becomes possible. In this scheme, one would impulsively excite phonons of a certain wave-vector, and then turn off the oscillations after a half or a single period using a second excitation pulse. Other possiblities for fast switches include the use of semiconductor superlattices which, due to their artifical periodicity, exhibit zone-folding of the acoustic phonon branch.

12. Conclusion

In conclusion, we have described some examples of time-resolved x-ray techniques which enable direct observation of vibrational excitations in solid state, crystalline materials, and give new insight into the coupling of energy into a crystalline lattice. In the future, these techniques will be extended to materials which exhibit only short range order. The scattering in this case gives rise to diffraction rings and a full range of time-resolved powder diffraction techniques may be applied. For example, nanocrystals exhibit perfect crystalline order over length scales of a few nm and serve as model systems for the study of phase transitions and associated dynamics. In the extreme case, the study of liquids (for example, water) is an example of systems with short range order extending over angstroms. Such systems, which exhibit intrinsic structural rearangements on fast time-scales is another example of a system for which a wide range of time-resolved x-ray techniques might be applied.

Chapter 9

SEEING SOUND: MEASURING ACOUSTIC PULSE PROPAGATION WITH X-RAYS

M.F. DeCamp, D.A. Reis, and P.H. Bucksbaum
FOCUS Center and Physics Department, University of Michigan
mattydee@umich.edu,dreis@umich.edu,phb@umich.edu

Abstract Time-resolved x-ray diffraction has become a very powerful tool in the study of structural dynamics of solids. The wave nature of x-rays allows the detection of milliÅngstrom changes in crystalline structure. Coupling the sensitivity of x-ray scattering with ultrafast pump-probe techniques can provide unprecedented studies of ultrafast dynamics in solids. This chapter demonstrates how time-resolved x-ray diffraction can be used to view transient strains in crystalline solids. Several different x-ray scattering geometries are explored.

1. Introduction

Picosecond ultrasonics has been studied extensively using optical scattering techniques [251, 234, 252, 253, 254, 255, 256]. Using ultrafast optical techniques, optical pulses can create and measure large lattice distortions in opaque materials. However, the wavelength of optical radiation is much larger than the crystalline lattice spacing restricting the observation of lattice motion to observing average crystalline changes. X-rays overcome this restriction by scattering from the core electrons making direct observation of lattice motion a possibility.

Pump-probe experiments utilize the pulsed nature of the probe radiation to generate 'snapshots' of the pump induced dynamics. Like the shutter speed on a camera, the time resolution of the snapshot is limited by the duration of the probe radiation. By delaying the probe pulse with respect to the pump radiation, a series of pictures can be strung together to make a 'movie' of the pump induced dynamics. Typical X-ray pulses generated at synchrotron sources have a pulse length $\sim$100ps, which is sufficiently fast for the study of laser induced strain.

For x-ray diffraction experiments, the probe depth is limited by the x-ray extinction depth, which for most strong Bragg reflections is $\sim 1\ \mu m$. Early studies using time-resolved x-ray Bragg diffraction were limited to the generation of large amplitude lattice strain and rapid loss of translational order due to intense laser excitation[257, 258, 157, 158, 219, 259, 171, 124, 169, 143, 170, 260]. By utilizing synchrotron radiation and novel diffraction geometries it is possible to extend these early studies to observe the propagation, evolution, and interaction of ultrafast strain pulses [124, 123, 261]. In this chapter we describe the experimental setup and recent results of measuring acoustic pulses via time-resolved x-ray diffraction.

2. Ultrafast Strain Generation

In solids, the absorption of an ultrafast optical pulse will generate a coherent strain pulse. The acoustic pulse constitutes a coherent phonon wavepacket with the central phonon frequency equal to the sound speed divided by the optical penetration depth. For many semiconductors this is in the 10's of GHz. In this work we will discuss 2 models of laser induced strain, a thermo-elastic model and strain generated by ultrafast carrier diffusion. Other methods of laser generated strain are presented by Akhmanov and Gusev[256].

2.1 Thermo-elastic model

In optically opaque materials, optical radiation is absorbed exponentially at the crystal surface, typically $\leq 1\ \mu m$. In many semiconductors the electron-phonon coupling is very fast ($\sim$1ps) such that the surface temperature of the crystal is approximated to increase instantaneously. The rapid change in the surface crystal equilibrium causes the crystalline lattice to elastically respond. Thomsen *et al.*[234] presented an analytic solution which describes the generation and propagation of this laser induced coherent strain. If the illuminated area is large compared to the square of the optical absorption depth, ζ, the strain will be uniaxial. The amount of energy deposited per unit volume at a distance z into the crystal bulk can be written as:

$$W(z) = (1 - R)\frac{\mathcal{F}}{\zeta}e^{-z/\zeta} \tag{9.1}$$

where R is the surface reflectivity and $\mathcal{F}$ is the incident optical fluence. In the absence of thermal or carrier diffusion, the generated thermal gradient will follow the spatial profile of the energy deposition:

$$\Delta T(z) = \frac{W(z)}{C} \tag{9.2}$$

where C is the specific heat per unit volume. The generated temperature differential produces a thermal stress in the $\hat{z}$ direction:

$$-3B\beta\Delta T(z) \tag{9.3}$$

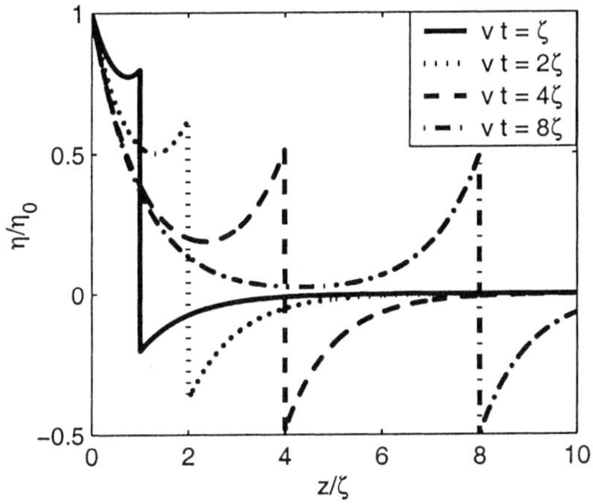

Figure 9.1: Calculated strain profiles at four different time delays using the Thomsen model.

where B is the bulk modulus of the material and β is the linear expansion coefficient.

Since the stress is uniaxial the only nonzero component of the strain tensor is $\eta_{33} = \frac{\partial u_3}{\partial z}$. The thermo-elastic equations of motion can then be simplified:

$$\rho \frac{\partial^2 u_3}{\partial t^2} = \frac{\partial \sigma_{33}}{\partial z} \tag{9.4}$$

where σ_{33} is the component of stress tensor in the $\hat{z}$ direction given by:

$$\sigma_{33} = v^2 \rho \eta_{33} - 3B\beta\Delta T(z) \tag{9.5}$$

where v is the longitudinal sound velocity in the material. The harmonic response of the material is represented by the first term in equation 9.5 . Thomsen *et al.* solved equation 9.4 with the assumption that initially the strain is zero everywhere and that the stress is zero at the crystal surface ($z = 0$):

$$\eta_{33}(z,t) = (1-R)\frac{F\beta}{\zeta C}\frac{v^2\rho}{3B}\left[e^{-z/\zeta}\left(1 - \frac{1}{2}e^{-vt/\zeta}\right) - \frac{1}{2}e^{-|z-vt|/\zeta}\mathrm{sgn}(z - vt)\right] \tag{9.6}$$

Equation 9.6 represents a lattice strain with two separate components; a static thermal strain and a coherent strain pulse. By calculating the strain as a function of crystal depth for a few time intervals, this partitioning of acoustic energy becomes clear (see figure 9.1).

2.2 Plasma Diffusion

When electron or thermal diffusion is faster than the associated attenuation, the simple thermo-elastic model cannot describe the laser induced strain accurately. For a dense electron-hole plasma at the crystal surface, the evolution of the plasma will be governed by plasma diffusion. This plasma changes the lattice equilibrium and thus provides the driving force for the laser induced strain. When the effects of the diffusion cannot be neglected the propagation of the strain field is not limited by the sound speed, but rather the speed of the electron-hole diffusion.

The initial electron-hole plasma density is directly related to the optical penetration depth:

$$n(z,t) = \frac{n_0 e^{-z/\zeta}}{\zeta} \tag{9.7}$$

where n_0 is the number of photons per unit area absorbed at the crystal surface from the pump pulse. The carrier density is governed by the following equation:

$$\frac{dn}{dt} = D_p \frac{d^2 n}{dz^2} - An^3 \tag{9.8}$$

where D_p is the diffusion constant and A is the Auger recombination rate. We assume that the temperature rise due to the initial electron plasma is equal to:

$$T(z, \Delta t = 0) = n(z, \Delta t = 0)\frac{E_p - E_g}{C_l} \tag{9.9}$$

Where E_p and E_g is the photon energy and the electronic band gap respectively, and C_l is the crystalline heat capacity. The energy release by Auger recombination leads to further lattice heating. The evolution of the temperature profile is governed by carrier and thermal diffusion and the Auger recombination rate.

$$\frac{dT}{dt} = D_t \frac{d^2 T}{dz^2} + An^3 \frac{E_g}{C_l} \tag{9.10}$$

Where D_t is the thermal diffusion constant. To calculate the strain profile, we use the fact that the equilibrium lattice condition depends on the temperature and the plasma density:

$$\eta_e(z,t) = \alpha_t T(z,t) + \alpha_p n(z,t) \tag{9.11}$$

where η_e is the deviation from the equilibrium lattice condition and $\alpha_{t,p}$ are the coupling constants for the temperature and plasma components respectively. The elastic response of the material causes a strain pulse to develop with η_e as the driving force. Optical scattering experiments have shown that these models are qualitatively correct for many materials[262, 163, 164, 166, 263, 251, 234, 255, 264]. It is impossible using optical scattering techniques, however, to get a measurement of the absolute lattice strain as well as the absolute partitioning of energy. Using x-rays to probe the lattice directly may provide information on the initial dynamics of the strain generation.

3. The X-ray Source

The experiments described here were performed at the MHATT-CAT (sector 7) undulator beamline at the Advanced Photon Source (APS). Third generation synchrotron sources, such as the APS, can provide very intense pulses of monochromatic x-rays which facilitate the accurate measurement of lattice dynamics.

3.1 Bunch Timing

The earliest synchrotron x-ray sources were 'parasitic' to large particle accelerators. Through the decades, these sources have been refined to create the most intense sources of hard x-rays on earth. Synchrotron sources produce intense x-ray radiation by the acceleration of electron pulses. The bunch duration depends on the details of the accelerator and the amount of charge in the electron bunch are typically 50-100ps for machines such as the APS[265]. RF accelerating cavities replenish the energy lost due to synchrotron radiation losses. A stable orbit requires the electron arrive at the accelerator in phase with resect to the RF. For experiments that are sensitive to bunch timing, a signal is relayed to the experiment hall which is phase locked to the RF accelerating cavities ($\sim$352MHz). Due to electronic noise, there is a timing jitter of $\sim$20ps between the relayed RF signal and the electron bunch train.[1] The standard fill pattern[2] of the APS at the time of these experiments was structured such that there are 22 individual electron bunches ('singlets') and a 'superbunch' made of 6 electron bunches each separated by a single RF cycle (figure 9.2) .[3] Each electron bunch is separated by 52 RF cycles ($\sim$150ns). Separating each pulse train a gap of 300ns is incorporated to provide a method of counting individual revolutions of the storage ring. A signal locked to the superbunch ('P0'=272kHz) provides synchronization to the individual bunches for time-resolved experiments. The effective x-ray pulse length becomes the electron bunch length convolved with the timing jitter associated with the RF reference. For the current parameters of the APS, the effective x-ray pulse length is $\sim$100ps.

For the experiments described here, x-ray radiation is produced in an undulator insertion device. An undulator consists of a periodic series of magnets which causes the radiating electron pulse to oscillate with a specific frequency [266]. The resultant emitted radiation is extremely intense and semi-coherent. However the spectral bandwidth of the emitted radiation is too large to ac-

[1]Recent beamline upgrades have decreased the timing jitter to under 2ps

[2]editor's note: standard fill pattern at the time of the experiment. The superbunch no longer exists. See fig. 6.4 for another fill pattern

[3]The superbunch has now been replaced with a singlet during the normal operations of the APS.

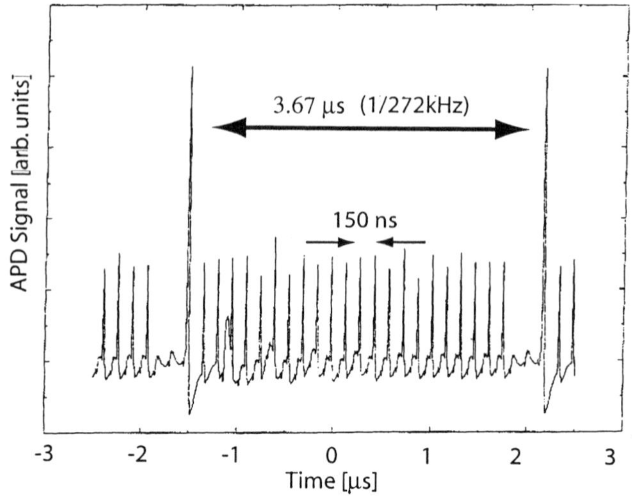

Figure 9.2: Synchrotron pulse train

curately probe lattice changes. To ensure the incident x-rays are sufficiently monochromatic, the introduction of x-ray optics is necessary.

Immediately after the undulator a set of water cooled slits spatially restricts the x-ray beam to 400x400μm^2. This spatial restriction limits the average power, spatial divergence, and the spectral bandwidth of the emitted x-ray pulse. To limit the x-ray spectral bandwidth even further, a cryogenically cooled double crystal Si [111] x-ray monochrometer is used. The bandwidth exiting the monochrometer is limited by the acceptance of the diffraction peak ($\sim 10^{-4}$ fractional energy bandwidth). At 10k eV, there are $\sim 10^6$ monochromatic x-ray photons per pulse available for the experiment.

The experiments are performed ~55 meters from the undulator source. At this distance the x-ray beam has diverged significantly to a spot size of 1x4mm^2. A second pair of slits is used to limit the x-ray spot size to $\sim$ 100x200μm^2. At the experiment, the crystal of interest is supported by a 4 circle goniometer. The goniometer allows full control of the angular orientation of the crystal as well as control of the angular position of the detector with 0.3mdeg resolution.

To utilize the 100ps temporal resolution that the x-ray pulses provide, synchrotron sources require x-ray detectors that can differentiate between individual x-ray bunches. A fast silicon Avalanche Photodiode (APD) is such a device[267]. The rise time of the APD is on order of 3-5ns (figure 9.3). This is fast enough to temporally resolve the singlet spacing of the emitted x-ray bunches (figure 9.2). Because the APD's can distinguish between individual bunches, the time-resolution is not limited by the electronic response of the

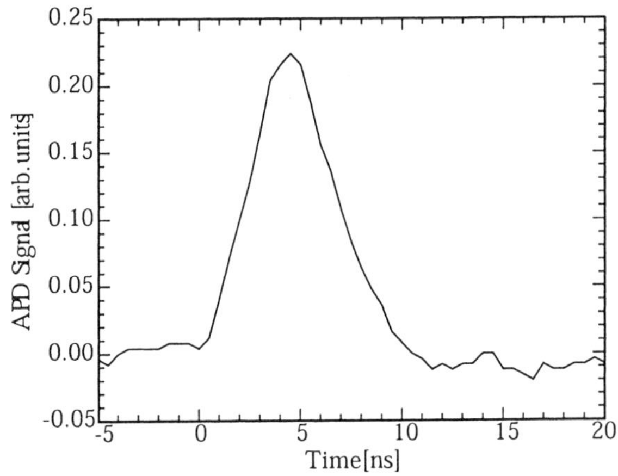

Figure 9.3: Temporal response of the APD.

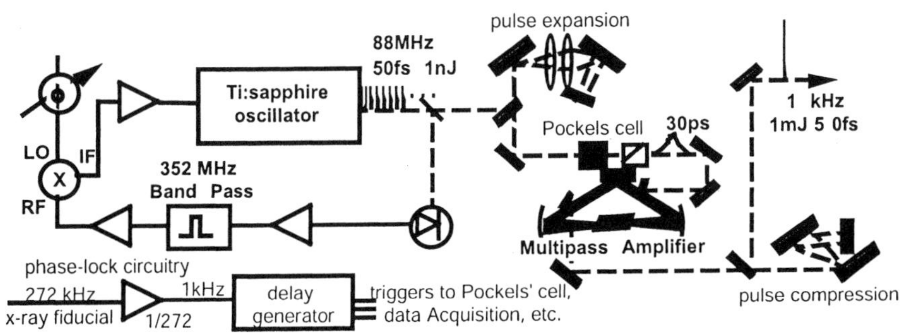

Figure 9.4: A schematic for the amplified laser system

APD but rather by the x-ray pulse width. Electronic gating is used to measure the intensity of a given x-ray pulse.

4. Ultrafast Laser

The generation of coherent strain pulses on a picosecond timescale requires a laser system that is capable of producing pulse widths shorter than the ratio of optical penetration depth and the speed of sound[234]. To generate large strains ($> 10^{-3}$), optical fluences in excess of 10 $\frac{mJ}{cm^2}$ are desirable. Commercial laser systems are now available which provide both stable and high average power ultrafast optical pulses.

4.1 Ti:sapphire oscillator

Passively modelocked titanium doped sapphire (Ti:sapphire) readily produce optical pulses ten's of femtoseconds in duration. The large gain profile of the

Ti:sapphire (greater than 300nm) allows the generation of sub-100fs optical pulses using a relatively simple design[268]. The oscillator produces sufficient bandwidth to support a 25fs optical pulse, with an average pulse energy of 5nJ. Locking the repetition rate of the Ti:sapphire oscillator to the electron bunches of the synchrotron is required to study laser induce dynamics with synchrotron radiation. The RF accelerating cavities of the APS oscillate at 352MHz. Since the repetition rate of the Ti:sapphire oscillator is determined by laser cavity length, the cavity length is adjusted with very high precision to lock the laser to the synchrotron. Active feedback is used to control the cavity length such that the oscillator frequency is the fourth sub-harmonic of the RF cavities (88MHz). To control the cavity length of the laser, a cavity mirror is mounted on a small piezoelectric crystal.

At the output of the oscillator a fast photodiode samples the pulsing frequency (see figure 9.4). The output of the photodiode is filtered with a 352MHz band-pass filter and then mixed with a RF reference that is synchronized to the APS accelerating cavities. The mixer creates a DC error signal which represents the phase difference between the RF reference and the oscillator. The error signal sent to a feedback loop which adjusts the voltage on the piezo in the attempt to zero the error signal. The feedback loop has sufficient bandwidth to respond to noise sources of up to several kHz and to lock the oscillator to the RF to better than 4 ps.

To generate fluences of greater than 10 $\frac{mJ}{cm^2}$ amplification is required. Since the optical pulse has a very high peak amplitude, amplifying the ultrafast pulse directly out of the oscillator will damage the amplifier before sufficient gain has been achieved. Introducing optical elements can reduce the risk of optical damage while increasing the potential gain of the amplifier. The method called chirped pulse amplification (CPA) is used to amplify the oscillator pulses by several orders of magnitude[269, 268].

4.2 Chirped Pulse Amplification

CPA 'stretches' a short pulse into a long pulse, thereby reducing the peak intensity. Transmitting a transform limited optical pulse through a dispersive medium, will cause the different colors to travel different distances thus 'stretching' or 'chirping' the pulse. At the output of the stretcher, the 25fs optical pulse is stretched to $\sim$25ps reducing the peak intensity by almost 3 orders of magnitude. A general diagram of the CPA setup is given in figure 9.4.

To generate a uniaxial strain pulse, the laser spot size must be very large compared to the optical penetration depth. Typically the laser spotsize is a $\sim 1mm^2$. To generate large strains the laser pulse energy must be on order of a milliJoule. To amplify the oscillator pulses sufficiently, a laser is needed which can supply at least that amount of energy to the amplifier. A 1kHz

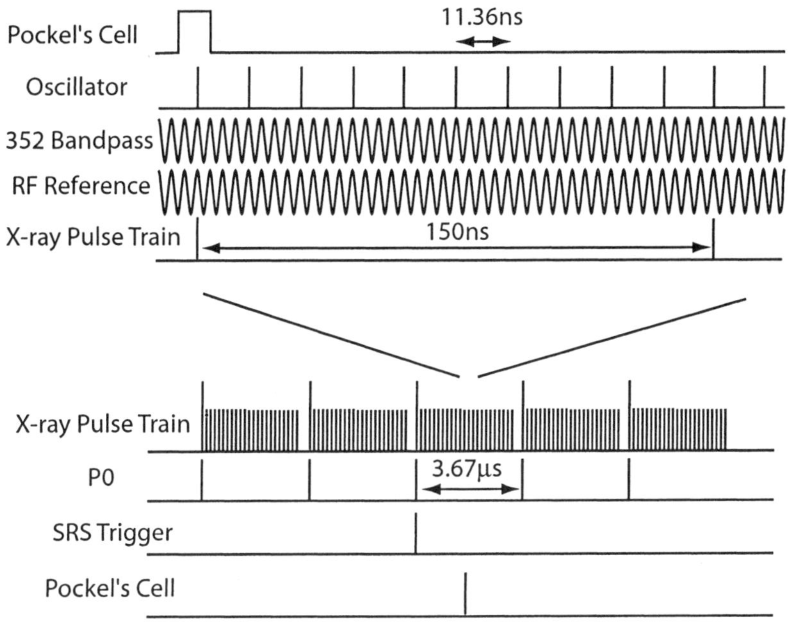

Figure 9.5: Flow diagram of the Pump-probe setup

Q-switched Nd:YLF laser is used to provide 12mJ to the amplifier crystal. To match the pump rate, the 88MHz pulse train emitted from the Ti:sapphire oscillator is reduced to 1kHz using an optical switch (Pockels cell). A digital delay generator controls the Pockels cell firing sequence as well as the firing time of the Nd:YLF laser to maximize the gain in the amplifier.

After a particular pulse has been selected, the chirped pulse is amplified using an 8 pass amplifier. After 8 passes, the optical pulse energy has increased $\sim$60dB while still maintaining the input spectral bandwidth. To reconstruct the short pulse, the fourier components of the optical pulse must be 'compressed'. This is accomplished using another dispersive medium. The final compressed and amplified pulse is $\sim$70fs, $\sim$0.5mJ. With modest focusing the optical fluence can easily exceed 1000 $\frac{mJ}{cm^2}$.

4.3 Laser/X-ray Timing

Ensuring that the ultrafast laser pulse and x-ray pulse arrive at a given time delay is imperative in performing time-resolved diffraction studies of laser driven processes. By adding phase to the reference RF, the relative delay of the laser and the x-ray source can be controlled. The feedback loop recognizes that the phase is not correct and adjusts the laser cavity accordingly. This in effect will change the relative timing between the optical pulse and the x-ray pulse by the amount of phase that was added to the reference RF. The resolution of the digital phase shifter is limited to 19ps, which is well within the x-ray pulse length. The maximum delay that the phase shifter can impart is 4.75ns.

This method of pulse delaying, while effective in the study of relatively short phenomena, does not allow the continuous monitoring of long lived crystalline strains. To study the evolution of acoustic pulses requires picosecond precision over many microseconds of delay. Traditional optical scattering experiments require 100's of meters of optical path to achieve microsecond delays. However, because these experiments utilize two independent light sources, digital electronics can be used to adjust the relative delays of the x-ray and laser pulses to arbitrary delays. A digital delay generator generates an arbitrary delay line for the optical pulse. When used in conjunction with the digital phase shifter, delays of -1ms to 1ms (limited only by the laser repetition rate) are achievable with 19ps resolution. By following a specific x-ray pulse with the boxcar integrator a dynamic range of ~80dB in the timing delay is possible.

5. Time-resolved x-ray Bragg diffraction

A phonon, of wavevector $\pm q$, will generate sidebands on a Bragg diffraction pattern (rocking curve). The positions of the sidebands are determined by momentum conservation: $\mathbf{k} + \mathbf{G} \pm q = \mathbf{k}_H$. Using Bragg's law the location of the sidebands may be calculated.

$$q = \Delta\theta |\mathbf{G}| \tan \theta_B \tag{9.12}$$

To resolve small phonon wavevectors ($\sim 1\mu m^{-1}$), Bragg rocking curves must have an angular extent very close theoretical width (Darwin width). Crystals with diamond or zinc blende structure can be grown with very few crystalline defects so that the experimental x-ray rocking curves are limited only by the incident x-ray bandwidth.

In these experiments, an 840 nm 70 fs laser pulse illuminates a (111) InSb or a (001) Ge single crystal. The optical penetration depth (ζ) for 840 nm light is ~100nm and ~220nm for InSb and Ge respectively. This very shallow penetration depth allows the generation of large surface strains with sizeable phonon wavevectors, $|q| \sim \frac{1}{\zeta}$. The x-rays are able to sample the laser induced strain directly in these two systems since the x-ray extinction depth (~1 μm) is significantly larger than the optical penetration depth.

InSb is an ideal crystal for the study of acoustic phonon generation due to its small optical penetration depth and relatively slow sound speed (~3400m/s in the 111 direction). Using moderate laser intensities, very large strains can be produced ($> 10^{-3}$).

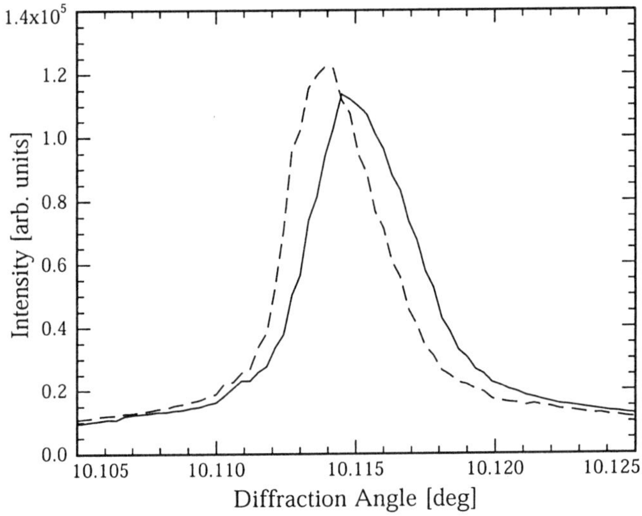

Figure 9.6: Rocking curve of the heated (dashed) and the unstrained 111 symmetric reflection in InSb measured with an ionization chamber[261].

Upon modest laser excitation (fluence $\sim$10 $\frac{mJ}{cm^2}$), the incident laser energy causes a significant increase in the average surface temperature[4]. The rise in surface temperature causes the average spacing of the crystal lattice to increase, thus shifting the diffraction pattern (figure 9.6). Using the Bragg equation, the lattice parameter is inferred to have expanded by $\sim$0.007% ($\sim$0.26mÅ). Assuming a linear expansion coefficient of $4.7 \times 10^{-6} K^{-1}$ [270], the average crystal temperature is increased by $\sim$ 15°K upon laser exposure.

The strain is significantly larger at the time of laser excitation. Sidebands on either side of the diffraction peak are visible 100 ps after laser excitation (figure 9.7). In the kinematic diffraction limit, the existence of sidebands on the diffraction peak indicates that there are three distinct regions inside the crystal; unstrained, compression (positive sidebands), and rarefaction (negative sidebands). In less than 1ns the shape of the diffraction pattern has almost returned to the static case, though the diffraction peak has shifted $\sim$3mdeg to the expansion side indicating a transient strain has propagated out of the detecting region leaving a residual heated layer at the surface. The static thermal layer at the crystal surface is $\sim$ 60°K warmer than the pre-laser excited crystal.

[4]The laser fluence is not enough to cause surface melting, though over many repeated heatings surface scarring is visible. Spectroscopic data indicate that the repeated heating causes a simple oxidation on the crystal surface, which leads to the observed surface scarring which does not effect the x-ray diffraction patterns.

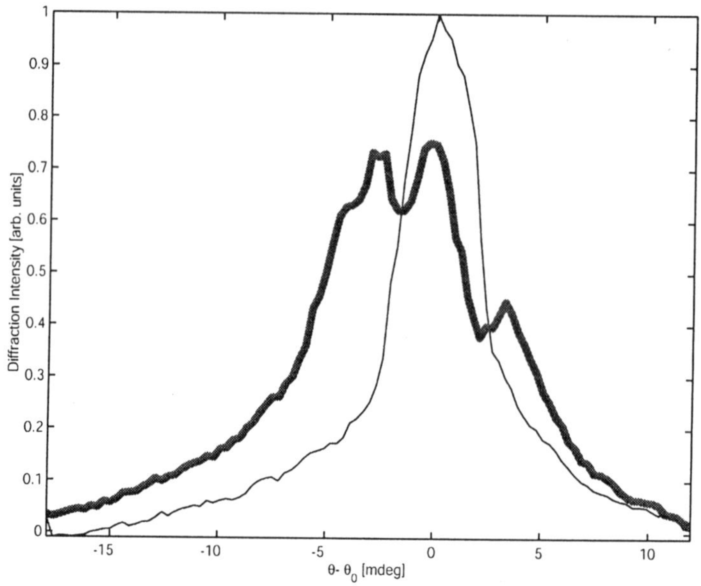

Figure 9.7: The diffraction patterns of the 111 InSb at time delays of -100ps and 100 ps [261].

Recording individual rocking curves at different relative time delays generates a movie of the strain propagation (figure 9.8). At zero time delay sidebands are immediately apparent on the main Bragg peak. The positive sidebands remain for ~400ps indicating the compression layer only exists within the x-ray extinction depth for that amount of time. The rarefaction sidebands, however, are apparent for ~500ps. These two clues indicate that the initially generated strain is comprised of two components: a propagating strain pulse with the compression layer leading the rarefaction layer and a static surface strain.

On either side of the rocking curve, temporal oscillations are apparent on the sidebands. The frequency of these oscillations changes with crystal angle. A Fourier transform at each diffraction angle reveals that the central frequency of the oscillations changes linearly with offset from the Bragg condition (figure 9.8b). Since the deviation from the Bragg condition is a measurement of phonon momentum, the time-resolved x-ray diffraction is a graphical representation of the acoustic dispersion relation, $\omega = ck$.

Data taken with the symmetric 004 reflection in single crystal Ge shows qualitatively similar features to the InSb data (see figure 9.9). However, there are two distinct differences between the InSb and the Ge cases. The most apparent difference is the global increase in diffraction efficiency at positive time delays. The natural Darwin width (and thus the angular acceptance) of the Ge reflection is much smaller than the incident x-ray bandwidth. By distorting the crystal, the x-ray reflectivity rises due to the increased coupling of the non-

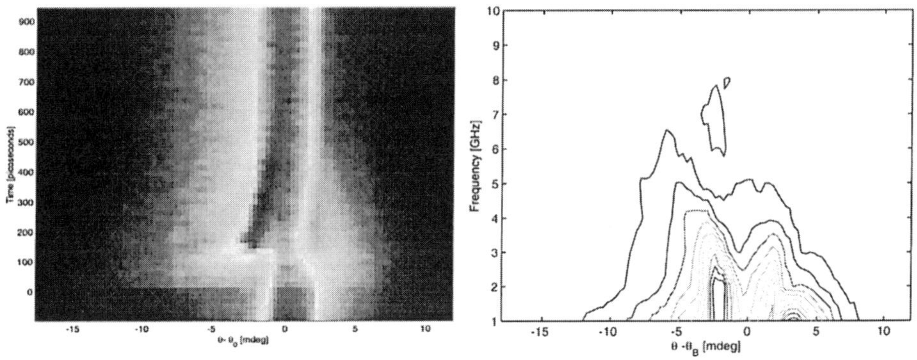

Figure 9.8: (a)Time-Resolved Bragg diffraction curves of the symmetric 111 reflection in laser strained InSb. Adapted from [124] (b) Fourier transform of the Time-Resolved Bragg Diffraction. The solid line has been added for emphasis.

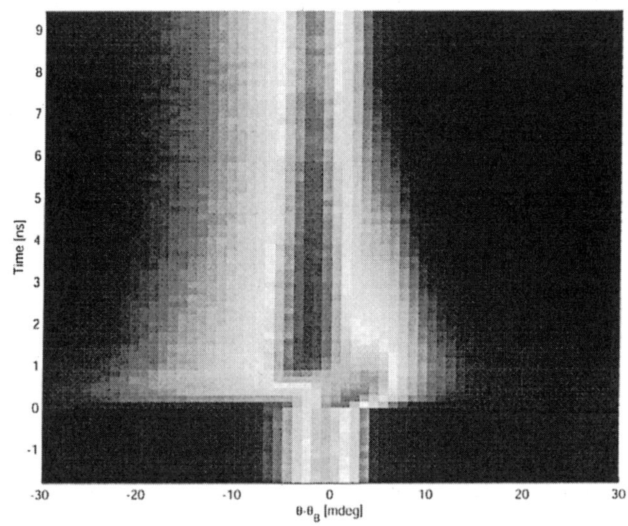

Figure 9.9: Time-Resolved Bragg diffraction curves of the symmetric 400 reflection in laser strained Ge. Adapted from [261].

zero x-ray bandwidth. The second difference is the lack of time dependent modulations on either side of the diffraction peak. This is due to the faster sound speed (~4800m/s) and the larger Bragg angle which makes the time-dependent oscillations faster than time-resolution of the experiment (i.e. the x-ray pulse width).

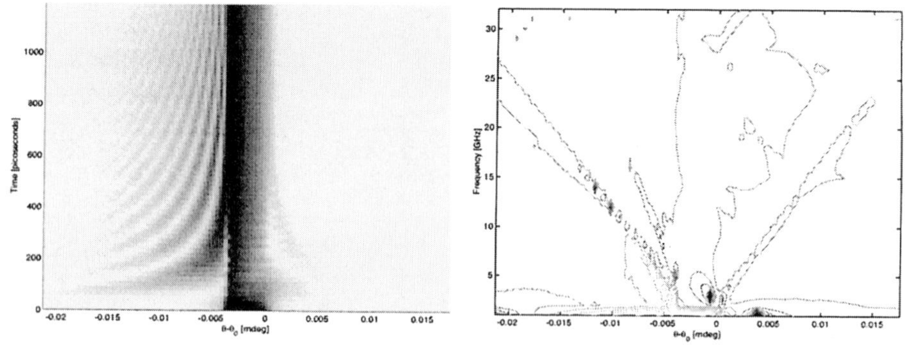

Figure 9.10: (a) Time-resolved x-ray diffraction of the symmetric 111 reflection in InSb calculated using the Thomsen model of strain generation. Adapted from [124].(b) Fourier transform of the time-resolved diffraction calculation. The contours are on a log scale.

5.1 Dynamical diffraction calculations

To determine the strain generation mechanism, comparisons between the data and diffraction simulations are necessary. Because we are using thick crystals, the diffraction patterns must be calculated using dynamical diffraction theory. We use the method put forward by Wie *et al.* [271] to calculate the diffraction patterns due to a strained crystal.

Figure 9.10a shows the calculated time-resolved x-ray Bragg diffraction picture for the symmetric 111 reflection in InSb assuming that the strain mechanism is determined by the Thomsen model and that x-ray probe pulse is perfectly monochromatic and infinitely short. The time dependent oscillations observed in the data are accurately predicted in the simulation. The frequency of the oscillations change linearly as a function of angle representing the acoustic dispersion relation (figure 9.10b). Simulations for single crystal Ge shows qualitatively similar results.

The incorporation of the experimental parameters (i.e. non-zero x-ray spectral bandwidth and non-zero x-ray pulsewidth) is taken into account by performing a two-dimensional convolution on the simulation (see figure 9.11). In the case of InSb the time dependent oscillations as well as the lifetime of the sidebands is correctly predicted. In the case of Ge, the increase in diffraction efficiency and the existence of an interference fringe of the compression side is correctly predicted. Although these general features are maintained, there are distinct differences between the theory and the experiments.

The experimental diffraction patterns and the calculated diffraction patterns highlights the difference between the thermo-elastic model of strain and the experiment . Figure 9.12 shows the calculation and data at a laser time delay of 100ps for the symmetric 111 reflection in InSb. It is clear that the Thomsen model underestimates the strain in both the compression and rarefaction side-

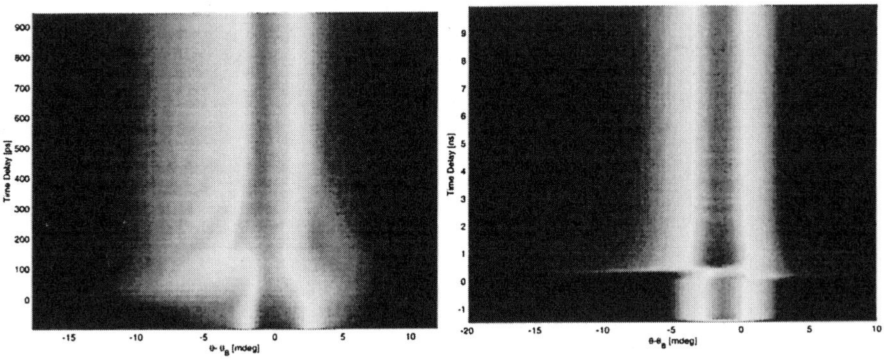

Figure 9.11: (a)Time-resolved Bragg diffraction of InSb using the Thomsen model of strain incorporating 'real world' constraints. Adapted from [124]. (b)'Real world' simulation of Time-resolved Bragg diffraction using the 400 reflection in Ge.

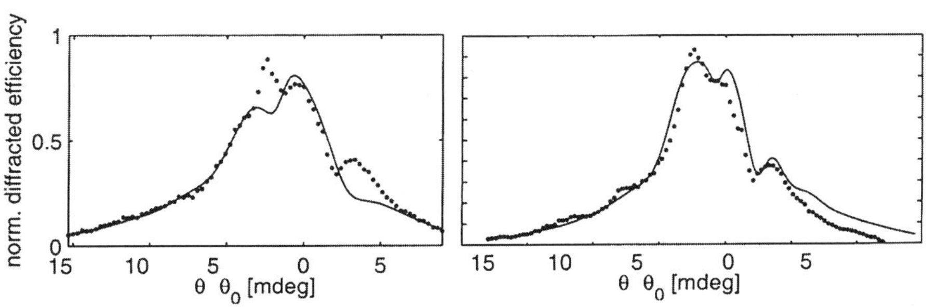

Figure 9.12: Left: InSb data and simulation using the Thomsen model at a time delay of 100ps. Adapted from [124]. Right:InSb data and simulation incorporating the best fit partitioning of energy.

bands. Yet, at time delays >500ps the Thomsen model accurately predicts the residual thermal layer. This comparison indicates that the relative partitioning of the two components of the thermo-elastic model (the static heated layer and the strain pulse) is not accurate. Increasing the relative partitioning of phonon to heating by a factor of two appears to minimize the error between the simulation and the experiment (figure 9.12) [124].

In the case of Ge, simulations using the thermo-elastic model do not accurately represent the data. The compression sideband is visible long after thermo-elastic model predicts. Cavalleri *et al.* [219, 259] deduced that ultrafast carrier diffusion modified the depth where the strain is generated in Ge . It was claimed that the strain in Ge was initially produced up to 1 μm in depth, 5 times the optical penetration depth. The simulation displayed in figure 9.13 shows that the data is consistent with a 2 μm acoustic pulse and thus the carrier diffusion model.

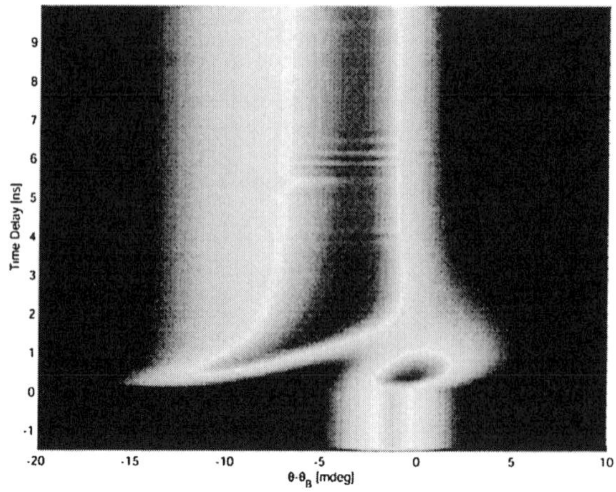

Figure 9.13: Simulation of the time-resolved strain in Ge assuming the strain was generated in a $2\,\mu m$ depth.

5.2 Acoustic Pulse Evolution

The previous experiments were limited to the study of acoustic pulse propagation within the x-ray extinction depth. For allowed Bragg reflections, this length scale is $\sim$1 micron. A quasi-forbidden reflection, however, can probe depths greater than the x-ray extinction depth of the strong Bragg reflection. Though the diffraction efficiency is very weak for these quasi-forbidden reflections, their sensitivity to structural changes makes them an attractive choice for doing time-resolved measurements.

As a proof of principle experiment, we used the quasi-forbidden 222 reflection in InSb to study the acoustic pulse propagating into the bulk. In this geometry the x-ray probe depth was $\sim 10\,\mu m$, thus increasing the detection depth by an order of magnitude. The time-resolved x-ray diffraction data for the quasi-forbidden 222 reflection in InSb is shown in figure 9.14. At zero time delay the main diffraction peak reduces in intensity very quickly at the same time weak sidebands are visible[124].

5.3 Acoustic Reflections

Since acoustic pulses are relatively large wavelength, the acoustic wavepacket should not disperse significantly over propagation distances as long as many meters[252, 253, 254]. Yet the diffraction efficiency of a quasi-forbidden reflection is very small making precise measurements of the strain over long times very difficult. If the acoustic pulse can reenter the x-ray probe region, the effective detection distance inside the crystal can be extended. Using acoustic

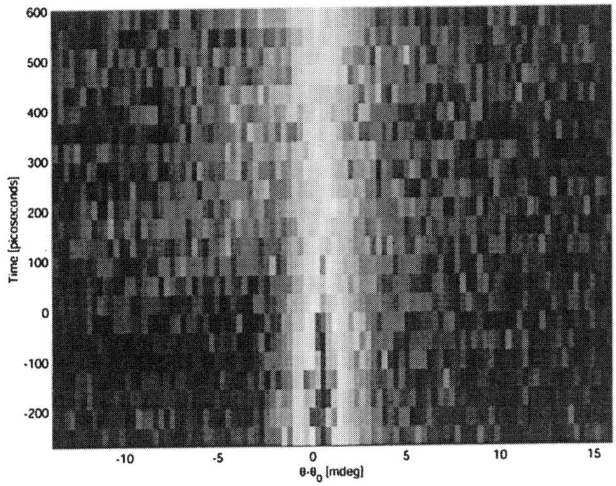

Figure 9.14: Time-Resolved Bragg diffraction of the quasi-forbidden 222 reflection in laser strained InSb. Adapted from [124]

reflections we can achieve this goal, thus strong x-ray Bragg diffraction peaks can then be used to study acoustic pulse evolution over long times.

The impedance mismatch at a polished crystal/air interface allows a nearly perfect reflection of an acoustic pulse. At the reflection, the acoustic pulse will experience a π phase shift since the impedance of the air at the crystal surface is approximately infinite[272, 252]. The resulting strain at the crystal surface is double the amplitude of the acoustic pulse.

Measuring the diffraction efficiency of the symmetric 004 reflection in a laser excited 280 μm thick piece of single crystal 2 side polished 001 Ge, it is seen that at regular intervals the x-ray intensity increases dramatically for a period ~3ns (figure 9.15). This characteristic period is the time for the acoustic pulse to echo through the 280 μm crystal. The time dependent background is due to the static heating of the crystal lattice which diffuses into the crystal bulk.

5.3.1 Acoustic Dispersion. Figure 9.16 shows the time-resolved x-ray diffraction measurement of the first and tenth returns of the acoustic pulse [261]. Each revival corresponds to the acoustic pulse travelling an additional ~550 μm. Upon the first acoustic revival, the compression sideband appears to have increased due to the doubling of the strain at the crystal surface. At the same time, the rarefaction sideband is diminished due to the acoustic pulse interfering with itself. There is also an interference fringe which crosses the diffraction peak in half a nanosecond.

During later revivals significant changes are apparent in the diffraction picture. The large momentum components seem to have disappeared leaving only

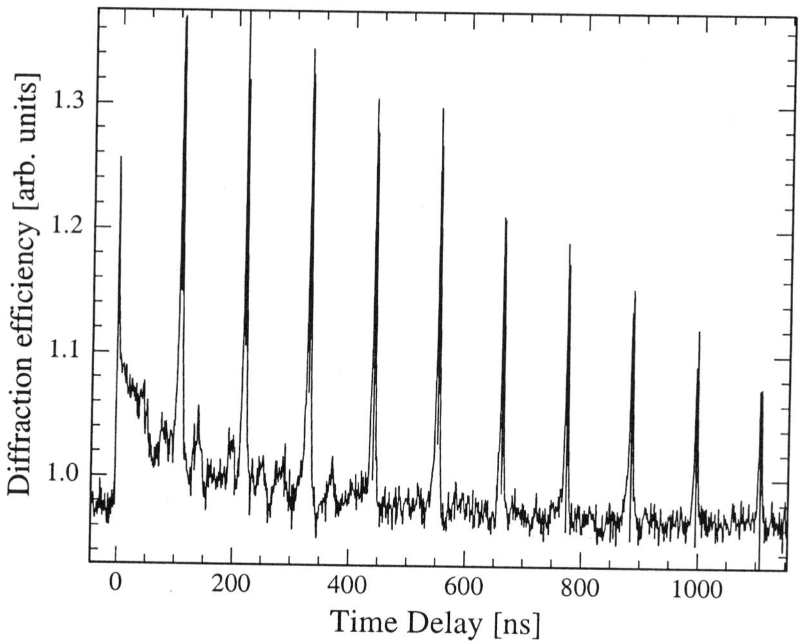

Figure 9.15: Diffracted x-ray intensity as a function of time delay on the rarefaction side of the x-ray rocking curve. Adapted from [261]

the low wavevector components of the acoustic pulse. In the tenth revival the interference fringe crosses the entire diffraction peak in about 3ns, indicating that the acoustic pulse has broadened by over 10 μm. The interference fringe indicate that the rarefaction components of the strain travel about 0.5% faster than the compression components. This dispersion is much larger than expected by simply adding higher order terms in the acoustic dispersion relation [252, 253, 254].

To simulate the experiment it is assumed that the electron-hole diffusion dominates the strain generation, thus the strain pulse has a spatial extent of 2 μm. There is very good qualitative agreement between the simulation and the data of the first acoustic return. To accurately simulate the later returns, the acoustic pulse is modified by adding a spatial dispersion. The exact cause of this dispersion is not yet known, though there are a few possible explanations.

First, the crystal faces may not be perfectly polished. If there are crystal surface variations that are on order of the acoustic wavelength, the acoustic pulse will not experience a perfect reflection due to wave diffraction effects. Because of the nature of diffraction, the high wavevector components will diffract much faster than the low wavevectors components leading to a spatial broadening of the acoustic pulse. The second possible reason is the non-uniform frequency

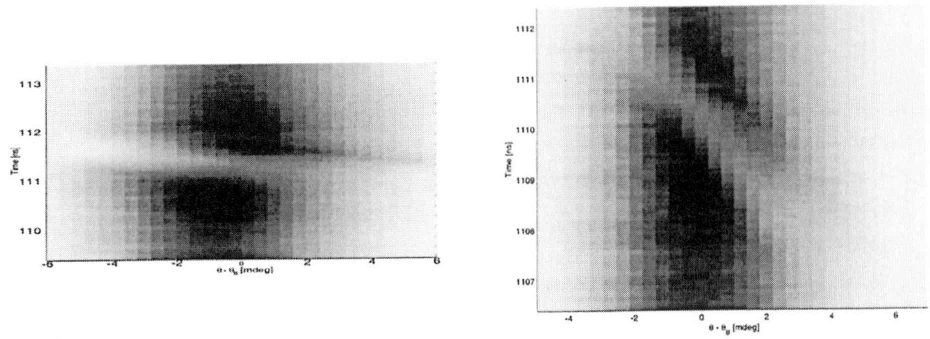

Figure 9.16: Time-Resolved diffraction at an acoustic reflection (a)First revival, (b) tenth revival. Adapted from [261]

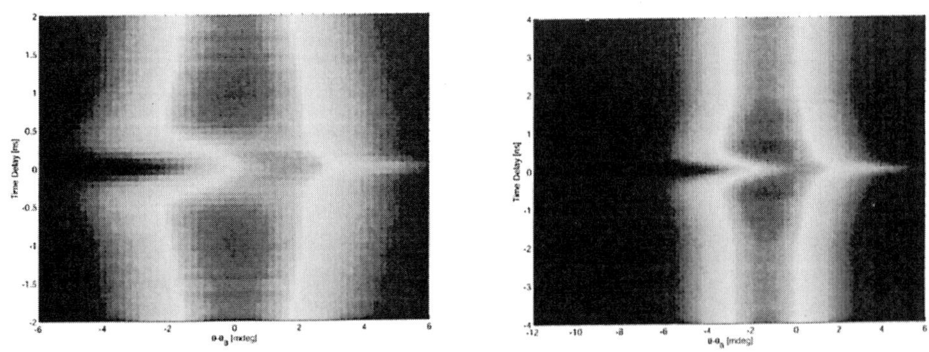

Figure 9.17: Simulation of an acoustic pulse reflection from a Bragg reflection surface. (a) Static heated layer included (b) heated layer not included

attenuation of the crystals. The attenuation length is different for different frequencies, which will lead to a dispersed acoustic pulse.

6. Time-resolved Laue diffraction

Although the x-ray Bragg diffraction is an effective tool for studying time dependent strains, the relatively small penetration depth of the diffracting x-rays does not allow the continuous monitoring of a propagating strain in thick crystals. Laue geometry can circumvent this problem due to the fact that the x-rays diffract through the entire crystal bulk.

Traditional Laue diffraction is typically limited to very thin crystals (<10 μm) or very large x-ray energies ($>$20 keV). Though thick crystals can be probed with high energy x-ray photons, the generation and detection of these photons makes time-resolved experiments difficult (APDs are not efficient photon detectors at energies $>$ 20keV). Nevertheless, coherent acoustic phonons with MHz frequencies have been studied using 30keV x-rays in the Laue geometry[273]. There is a classical diffraction effect which extends the probe depth of x-rays, x-ray anomalous transmission (Borrmann effect) [73, 274, 275].

At the Laue condition, there are two linearly independent solutions to the x-ray wave equation which we will label α and β. The α solution, anomalous x-ray transmission, is a transverse standing wave where the nodes of the x-ray field lie in regions of relatively low electron density. Because the x-ray attenuation is dominated by photo-electric absorption, the α solution is allowed to propagate with little or no attenuation, thereby 'channelling' x-ray energy through the crystal. The β solution is a transverse standing wave where the nodes of the x-ray field lie in regions of high electron density, i.e. the lattice planes. This solution experiences a state of enhanced absorption, and decays at twice the rate of the normal x-ray absorption depth. At the output of the crystal, there are two exterior diffracted waves, '0' and 'H', can be written as a superposition of α and β.

$$I_0 = |a\vec{E}_\alpha e^{i\vec{k}_\alpha\cdot\vec{z}} + b\vec{E}_\beta e^{i\vec{k}_\beta\cdot\vec{z}}|^2 \tag{9.13}$$

$$I_H = |c\vec{E}_\alpha e^{i\vec{k}_\alpha\cdot\vec{z}} - d\vec{E}_\beta e^{i\vec{k}_\beta\cdot\vec{z}}|^2 \tag{9.14}$$

where $I_{0,H}$ are the diffracted beam intensities, $\vec{E}_{\alpha,\beta}$ are the complex wave fields, $\vec{k}_{\alpha,\beta}$ are the complex wave vectors for the two solutions, and a, b, c, d are coefficients which are dependent on the crystal orientation.

In thick crystals, due to the enhanced absorption of the β solution, the diffracted x-ray intensity is dominated by the α solution. In thin crystals, the β solutions can influence the diffraction efficiency significantly. Because the α and β solutions experience different electron densities, and thus indices of refraction, the two solutions travel at different phase velocities through the crystal bulk. As the crystal thickness changes, the two solutions can interfere causing a beating in the two exterior wave solution (the Pendellösung effect).

Because x-ray anomalous transmission allows x-ray to propagate through very thick crystals, crystalographers have studied lattice dislocations and crystal defects in crystals that are much thicker than the x-ray absorption depth [275]. Even small thermal gradients were shown to completely disrupt the x-ray channelling [90]. Early time-resolved studies using the Borrmann effect were limited to microsecond time resolution [276, 277]. The experiments described here extend these time-resolved studies to the picosecond time scales.

6.1 Pump-Probe X-ray Anomalous Transmission

A Ge 001 crystal is oriented to diffract from the $20\overline{2}$ diffraction plane making the crystal planes asymmetric with respect to the crystal surface. An ultrafast optical pulse generates an acoustic disturbance on the output face of the crystal. In this geometry, the generated longitudinal acoustic pulse will have a compo-

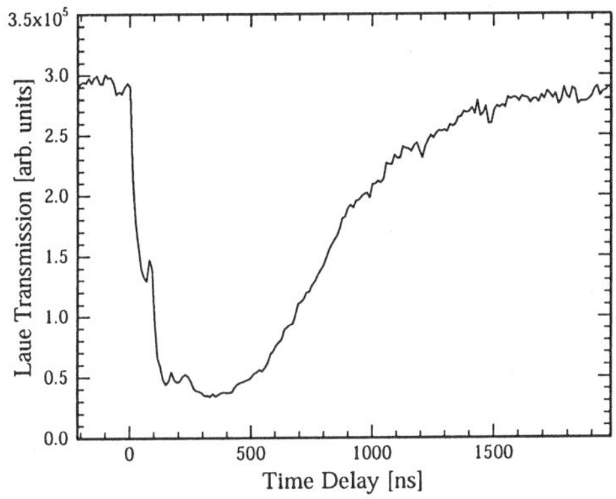

Figure 9.18: $20\overline{2}$ diffraction efficiency under intense laser excitation.

nent of its wavevector along the reciprocal lattice vector making detection via x-ray diffraction a possibility.

Under intense laser excitation ($>10\,\frac{mJ}{cm^2}$) the diffraction efficiency of the asymmetric reflection goes to zero for an extended period of time (figure 9.18). This loss of diffraction efficiency is due to the substantial thermal gradient induced by the laser [261]. Reducing the laser fluence, reduces the generated thermal gradient allowing x-rays to propagate though the entire crystal bulk. However, as the laser fluence is reduced the diffraction efficiency of the deflected beam recovers while the forward beam does not.

In figure 9.19 the peak diffraction intensity of both the forward and deflected beams of the Ge crystal is shown as a function of time delay. Immediately after the laser absorption, the intensity of the two diffracted beams changes rapidly [123]. The forward beam decreases while the deflected beam increases by the same amount, indicating a coherent transfer of energy. Over 60% of the x-ray intensity is coherently transferred between the two diffracted beams. The time dependence of this phenomenon is comparable to the x-ray pulse width indicating the switching mechanism is at least as fast as 100ps.

Following the rapid population transfer, oscillations in the diffracted intensity are visible. These oscillations are due to the coupling of the two interior wave solutions by the propagating strain pulse. This redistribution of the α, β basis causes the onset of Pendellösung oscillations as the acoustic pulse travels through the crystal.

6.2 Multiple crystal model

Localized static strains, crystalline defects, and continuous wave acoustic excitation can couple the two interior wave solutions (α and β) causing a change in the relative amplitude of the forward-diffracted and deflected-diffracted beams [275, 110, 125]. Since the sound speed is much less than the group velocity

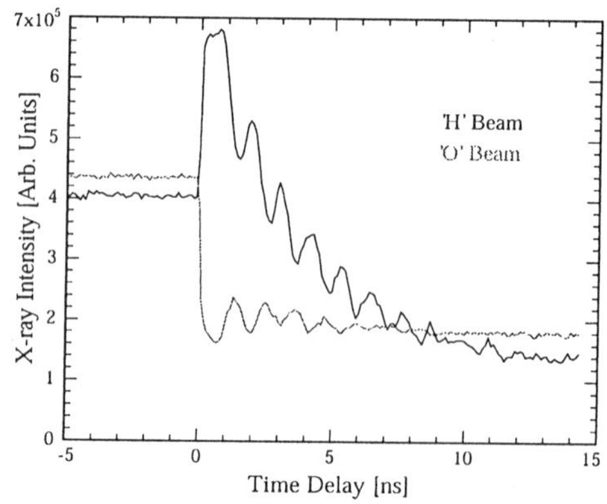

Figure 9.19: The peak diffraction efficiency of the asymmetric $20\overline{2}$ Laue reflection as a function of Laser/X-ray time delay.

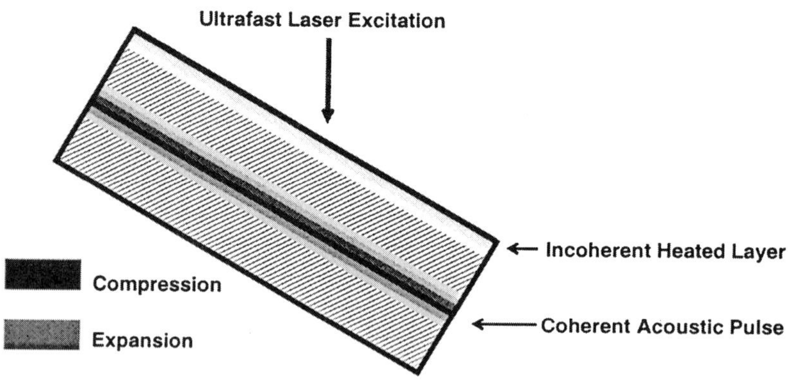

Figure 9.20: A schematic representation of the two crystal model.

of the x-rays, the acoustic pulse will appear to be a static strained layer. If the spatial dimension of the acoustic pulse is much less than that of a Pendellösung depth, this situation can be described fairly well as two perfect unstrained crystals separated by a very thin strained interface (see figure 9.20).

Physically we will treat the strained interface is treated as a simple rotation of the α, β basis [123]:

$$\begin{bmatrix} \cos\Theta & -\sin\Theta \\ \sin\Theta & \cos\Theta \end{bmatrix} \begin{pmatrix} \alpha \\ \beta \end{pmatrix} \tag{9.15}$$

After the x-rays interact with the strained region, the transmitted x-ray fields evolve as governed by the theory of dynamical diffraction. To achieve a full transfer of energy between the two exterior solutions, assuming that the β solution was complete absorbed a strain layer must propagate one quarter of a Pendellösung length (for Ge this would be $\sim 2\,\mu m$).

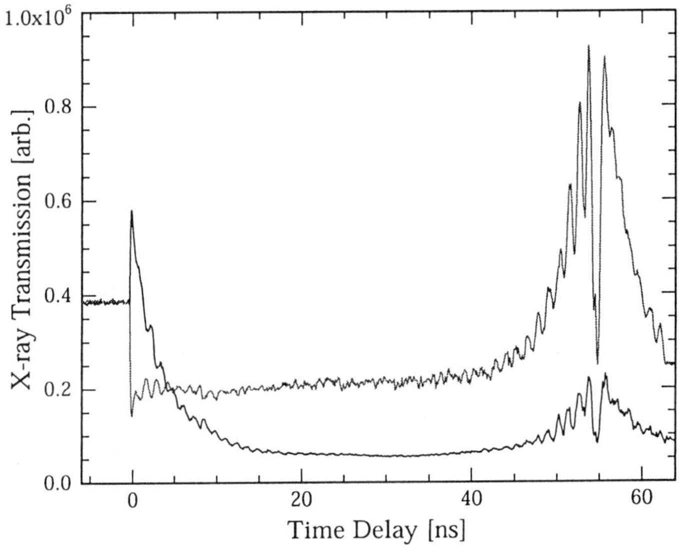

Figure 9.21: X-ray transmission as a function of acoustic pulse time delay. The acoustic pulse is now approaching the input face of the Ge crystal. Adapted from [123].

At acoustic velocities, a laser generated surface strain in Ge would take about 400ps to travel the necessary 2 μm to duplicate the observed results. The data indicate that the laser generated strain was moving into the crystal bulk at supersonic speeds. From the data taken in the Bragg geometry, we know that the strain generation is dominated by ultrafast carrier diffusion in Ge. It has been determined that ultrafast carrier diffusion can provide the necessary mechanism to produce the observed results[261].

6.3 Acoustic Reflections

As the acoustic pulse approaches the opposite crystal face of the crystal, a 'Borrmann revival' is observed. During the revival Pendellösung oscillations in the diffracted intensity are seen. As the acoustic pulse approaches the input face the oscillations in the two beams are in phase (figure 9.21). At the time of collision with the input face of the crystal, the intensity of the forward beam is about 2.5 times that of the static crystal case and greater than the sum of the two diffracted beams of an unstrained crystal [123].

This dramatic increase in the diffraction efficiency can be explained by the two crystal model. In the static crystal case the sum of the output intensities can be no larger than one half of the original input intensity due to the strong absorption of the β solution. In the best case scenario the amount of increase in one beam due to the rotation of the α, β basis on the output face of the crystal is only a factor of 2. However, if the basis rotation occurs on the input face, the

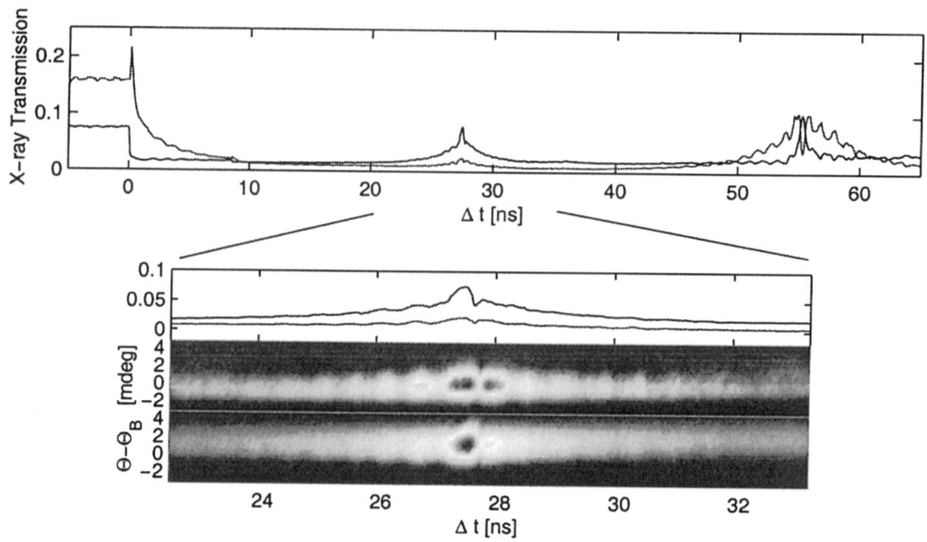

Figure 9.22: Time-resolved diffraction of the $20\overline{2}$ reflection in Ge. The angular blow up is the collision of two counter-propagating acoustic pulses. Adapted from [123].

entire initial β population can be transferred to the α solution, increasing the output of one beam by upto a factor of 4.

6.4 Acoustic Collisions

Although the x-rays diffract through the entire bulk of the crystal, the observational power of the asymmetric Laue reflection is still limited by the absorption depth of the β solution. In Ge, this limitation prevents the direct observation of the strain pulse at depths deeper than $\sim$25 μm. In static crystals, buried interfaces or lattice dislocations have been shown to repopulate the β solution after many absorption depths[275]. A similar effect can be created by launching a second acoustic pulse, counter-propagating with the first[123]. This second acoustic pulse can act like a transient crystalline disturbance which can allow the observation of a transient strain deep within the crystal bulk.

In figure 9.22 two counter propagating acoustic pulses are generated simultaneously from opposite faces. As expected, immediately after the initial lattice expansion an ultrafast energy transfer has taken place which decays quickly as the acoustic pulses travel into the center of the crystal. At a time delay of $\sim$ 27.5ns a Borrmann revival is seen. This time delay corresponds to the acoustic pulses travelling half way through the crystal bulk[123].

Like before, the Borrmann revival has oscillations associated with the increase in x-ray transmission, however in this case the oscillation period is twice

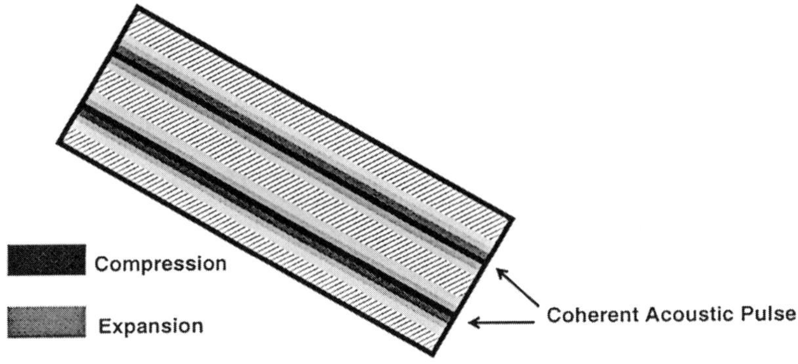

Figure 9.23: The three crystal model

as frequent as compared to the single acoustic pulse excitation. The reason for the frequency doubling is because there are now three distinct crystal regions separated by moving lattice disturbances (figure 9.23). The three crystals are a single thin crystal contracting (expanding) at twice the speed of sound, sandwiched between two thick crystals expanding (contracting) at the speed of sound. Since the thin crystal is changing its thickness by twice the sound speed the Pendellösung oscillations oscillate at twice the frequency.

7. Summary and Acknowledgements

This work demonstrates some of the strengths of time-resolved x-ray diffraction for the study of ultrafast lattice dynamics. The techniques developed here can be extended to the study of more complicated and faster crystal dynamics. Future studies may include looking at higher frequency phonons (including optical phonons), exploring acoustic reflections from buried crystalline interfaces, and studying laser induced dynamics in more complicated systems (liquids, gasses, etc.).

We would like to thank all of our collaborators and contributors including the staff at the MHATT-CAT insertion device beamline at the Advanced Photon Source. This work was conducted at the MHATT-CAT insertion device beamline at the Advanced Photon Source and was supported in part by the U.S. Department of Energy, Grants No. DE-FG02-99ER45743 and No. DE-FG02-00ER15031 and from the NSF FOCUS physics frontier center. Use of the Advanced Photon Source was supported by the US Department of Energy Basic Energy Sciences, Office of Energy Research under Contract No. W-31-109-Eng-38.

Chapter 10

TIME-DEPENDENT DYNAMICAL DIFFRACTION THEORY FOR PHONON-TYPE DISTORTIONS

P. Sondhauss, J.S. Wark
p.sondhauss1@physics.ox.ac.uk,j.wark1@physics.ox.ac.uk

Abstract A rigorous derivation of the dynamical theory of X-ray diffraction from crystals with "optical phonon"-type distortions is presented. The result is a compact, surprisingly simple equation with a strong formal similarity to the well known Takagi-Taupin equation, with the latter included as a special case. Time-dependence is explicitly retained, and thus the analysis is applicable to situations where the crystal is modified on time-scales comparable with that for the X-rays to traverse an extinction depth, as well as being valid for the analysis of diffraction from disturbances that travel at close to luminal velocities, such as polaritons. A comparison is made between the influence of coherent acoustic and optical phonons on the diffraction of X-rays. Numerical and perturbative analytical solutions of the generalised Takagi-Taupin equation are presented in the presence of such phonons.

1. Introduction

A good deal of research in the new and exciting field of ultrafast X-ray science investigates dynamic processes in nearly perfect crystals. The phenomena that have been, or are under study include phase transitions, heat conduction, electron-phonon coupling, and coupled photon-phonon modes (polaritons). The periodic structure makes X-rays a very powerful diagnostic, since diffraction phenomena are very sensitive to any deviation from perfect periodicity. With the *dynamical theory of X-ray diffraction* a well developed theory exists to describe and interpret the measured diffraction patterns. Unfortunately the standard Takagi-Taupin equation, the central equation to describe diffraction from distorted crystals, cannot deal with all types of distortions. It can take into account strain – i.e. modifications to the the local lattice parameter – and thus can be used to model diffraction from crystals which include low frequency acoustic phonons. However, only a few phenomena in crystal dynamics can be described simply on this basis – thermal expansion and sonic waves being probably the most prominent examples. A whole class of dynamic distortions, which we are going to call *"optical phonon"-type* or *non-uniform distortions*, which not just modify the local lattice parameter but also the structure, is not covered by the standard Takagi-Taupin theory. These distortions are also of great interest, since they can occur for example in connection with phase transitions (keyword: "soft modes"), and coherent optical phonons (which modify the structure factor in a periodic manner) which can be directly generated and, to some extent controlled, by short-pulse optical lasers. In this section we are going to present a theory, very similar to the established Takagi-Taupin theory, which overcomes this limitation. It turns out that the central equation in this theory, which we call the *generalised Takagi-Taupin equation* [278], is not particularly more complex than its "classical" counterpart.

If the crystal is modified on very short timescales, as it might be by high frequency coherent optical phonons or polaritons for example, the time dependence of the diffraction process itself may become important [279, 280, 281]. We thus explictly keep time-dependence in the treatment presented here.

In contrast to ch. 3, here use is made of Maxwell's equations in their macroscopic form, i.e. electromagnetic fields in the crystal are given by the dielectric displacement $\mathbf{D}$ instead of the vector potential $\mathbf{A}$ (see app. 5.1). Furthermore, note that in this chapter SI units are used instead of Gauss' cgs-units.

2. The generalised Takagi-Taupin equation

In classical Takagi-Taupin theory the assumption is made that the atoms within a single unit cell are displaced uniformly, i.e. the atoms 1 to M within an M-atom unit cell are displaced by the same vector $\mathbf{u}_1 = \mathbf{u}_2 = \cdots = \mathbf{u}_M \stackrel{\text{def}}{=} \mathbf{u}$. This allows the susceptibility field of the distorted crystal $\chi(\mathbf{r}, t)$ to be related to the susceptibility of the perfect crystal $\chi^{(\mathrm{p})}(\mathbf{r})$:

$$\chi(\mathbf{r}, t) = \chi^{(\mathrm{p})}\big(\mathbf{r} - \mathbf{u}(\mathbf{r}, t)\big) \tag{10.1}$$

For this reason the classical Takagi-Taupin equation cannot be applied directly to non-uniform distortions, by which we mean those where, in principle, each atom within a single unit cell can experience a different displacement (e.g. optical phonons). That is to say in general $\mathbf{u}_1 \neq \mathbf{u}_2 \neq \cdots \neq \mathbf{u}_M$.

For uniform distortions, the crystal merely experiences strain. For small strains in standard Takagi-Taupin theory we make the assumption that the strain does not significantly alter the structure factor of the unit cell, and the main effect of the strain is to alter the local Bragg angle. Thus, in the standard theory, we need only define the structure factor for a unit cell, and we do not concern ourselves with the make-up of the contributions to the structure factor arising from the individual atoms within the basis. Clearly such an approach cannot work for the non-uniform distortions described above, which each atom within a unit cell can experience a different displacement. In this case we must take into account separately the contribution to the overall susceptibility from each atom within the cell, and split up the susceptibility into a sum of M "primitive" susceptibilities $\chi_N^{(\mathrm{p})}$, each corresponding to the susceptibility of an imaginary crystal with only one atom per unit cell. In the case of a non-uniform displacement the susceptibility can therefore be expressed by

$$\chi(\mathbf{r}, t) = \sum_{N=1}^{M} \chi_N^{(\mathrm{p})}\left(\mathbf{r} - \mathbf{r}_N - \mathbf{u}_N(\mathbf{r}, t)\right) \quad , \tag{10.2}$$

with $\mathbf{r}_N$ being the atomic coordinates in a local unit-cell coordinate system. In what follows, and unless stated otherwise, we will assume that $\mathbf{u}_N$ is a function of $\mathbf{r}$ and t.

The perfect crystal susceptibilities $\chi_N^{(\mathrm{p})}(\mathbf{r})$ can be expanded as a spatial Fourier series which leads to

$$\chi(\mathbf{r}, t) = \sum_{h} \chi'_h(\mathbf{r}, t) e^{-i\mathbf{G}_h \cdot \mathbf{r}} \quad , \tag{10.3}$$

where $\mathbf{G}_h$ is the reciprocal lattice vector and

$$\chi'_h(\mathbf{r}, t) \stackrel{\mathrm{def}}{=} \sum_{N=1}^{M} \chi_{Nh} e^{i\mathbf{G}_h \cdot (\mathbf{r}_N + \mathbf{u}_N)} \quad , \tag{10.4}$$

where χ_{Nh} stands for the Fourier coefficients of the primitive susceptibilities, $\chi_{Nh} \stackrel{\mathrm{def}}{=} V_{\mathrm{uc}}^{-1} \int_{\mathrm{uc}} dV\, \chi_N^{(\mathrm{p})}(\mathbf{r}) \exp(i\mathbf{G}_h \cdot \mathbf{r})$. Note that the susceptibility coefficient $\chi'_h(\mathbf{r}, t)$ here is a function of both space and time. Unlike in classical Takagi-Taupin it is not necessary to make the assumption that the spatial variation of the displacement fields $\mathbf{u}_N(\mathbf{r}, t)$ from cell to cell is small. As the displacement of every atom in a unit-cell is considered separately one doesn't need this assumption to ensure that the variation of the displacement vector within a single unit cell is small. Evidently when we make the assumption of small spatial frequencies and further assume $\mathbf{u}_1 = \mathbf{u}_2 = \cdots = \mathbf{u}_M$ we recover results similar to those of Takagi and Taupin.

The wave equation for the X-ray dielectric displacement field $\mathbf{D}$ in a crystalline medium

$$\Delta \mathbf{D} - \frac{1}{c^2} \frac{\partial^2 \mathbf{D}}{\partial t^2} = -\ \mathrm{curl}\ \mathrm{curl}(\chi \mathbf{D}) \tag{10.5}$$

can be solved as is usual by applying the Bloch-type ansatz

$$\mathbf{D}(\mathbf{r}, t) = \sum_{h} \mathbf{D}'_h(\mathbf{r}, t) e^{i(\omega t - \mathbf{k}_h \cdot \mathbf{r})} \quad , \tag{10.6}$$

with the speed of light in vacuum, c, the frequency of the X-rays, ω, the wave vector of the incident wave (in the crystal), $\mathbf{k}_0$, the wave vector of the diffracted wave, $\mathbf{k}_h = \mathbf{k}_0 + \mathbf{G}_h$, and

$$\mathbf{D}'_h(\mathbf{r}, t) \stackrel{\text{def}}{=} \sum_{N=1}^{M} \mathbf{D}_{Nh} e^{i\mathbf{G}_h \cdot \mathbf{u}_N} \quad . \tag{10.7}$$

The X-ray field amplitudes $\mathbf{D}'_h$ here are, like the Fourier coefficients of the susceptibility, a sum over atomic terms. We now substitute equations (10.3) and (10.7) into the wave equation (10.5). The subsequent development and approximations are very similar to those leading to the classical Takagi-Taupin equation [83, 74] but are still worth showing to some level of detail. For the sake of clarity we study this development for each of the three terms in equation (10.5) in turn:

1 *The* $\Delta\mathbf{D}$*-term:* If we assume that the displacement fields $\mathbf{u}_N(\mathbf{r}, t)$ and field amplitudes $\mathbf{D}_{Nh}(\mathbf{r}, t)$ at any time vary slowly in space then second order derivatives of $\mathbf{u}_N$ and $\mathbf{D}_{Nh}$ can be neglected. Under this assumption one obtains

$$\Delta\mathbf{D} = \sum_h \sum_{N=1}^{M} \left[-2i(\mathbf{k}'_{Nh} \cdot \nabla) - k'^2_{Nh}\right] \mathbf{D}_{Nh}\, e^{i(\omega t - \mathbf{k}_h \cdot \mathbf{r} + \mathbf{G}_h \cdot \mathbf{u}_N)} \quad , \tag{10.8}$$

where $\mathbf{k}'_{Nh}$ stands for the *local wave vector* (equivalent to $\mathbf{k}_{h^\circ}$ in ch. 3)

$$\mathbf{k}'_{Nh} \stackrel{\text{def}}{=} \mathbf{k}_h - \nabla(\mathbf{G}_h \cdot \mathbf{u}_N) \quad . \tag{10.9}$$

Slow spatial variation doesn't necessarily mean that this approximation is limited to distortions with small spatial frequencies. For example the displacement fields $\mathbf{u}_N(\mathbf{r}, t)$ of high frequency phonons close to the Brillouin zone boundary have a spatial frequency in the order of an inverse lattice constant but still can be considered slowly varying, provided that their oscillation amplitudes are small compared to the lattice constant.

2 *The* $\partial^2\mathbf{D}/\partial t^2$*-term:* Typical temporal frequencies of diffractable X-rays lie around 10^{19} rad/s, thus far beyond the frequencies of phonons or any other kind of distortion. Therfore terms of the order of $\ddot{\mathbf{D}}_{Nh}$, $\ddot{\mathbf{u}}_N$, $\dot{\mathbf{u}}_N^2$ and $\dot{u}_{Ni}\dot{D}_{Nhj}$ may be neglected. With this approximation the result reads:

$$\frac{\partial^2\mathbf{D}}{\partial t^2} = \sum_h \sum_{N=1}^{M} \left[2i\omega\dot{\mathbf{D}}_{Nh} - 2\omega(\mathbf{G}_h \cdot \dot{\mathbf{u}}_N)\mathbf{D}_{Nh} - \omega^2\mathbf{D}_{Nh}\right] e^{i(\omega t - \mathbf{k}_h \cdot \mathbf{r} + \mathbf{G}_h \cdot \mathbf{u}_N)} \tag{10.10}$$

We would like to draw the reader's attention to the central term of the right hand side of equation (10.10), which contains the time derivative of the displacement field. The temporal evolution of the X-ray field amplitudes $\mathbf{D}_{Nh}$ takes place on timescales of order the *extinction depth traversal time*

$$\tau = \frac{\Lambda_{\text{ext}}}{c\gamma_0} \qquad (10.11)$$

[279, 280, 281], where Λ_{ext} stands for the extinction depth and γ_0 for the direction cosine $\cos\psi_0$, with ψ_0 being the angle between $\mathbf{k}_0$ and the inner surface normal $\mathbf{n}$. One might initially expect that this time would be short compared with the evolution of the atomic displacement fields, and thus the central term can be neglected. However, for weak X-ray reflections extinction depths can be large (tens or hundreds of microns), and thus the relevant timescales for the evolution of the X-ray field may approach or exceed 100 fs – i.e. the period of optical phonons. Thus there may well be situations in which this term must be retained. We keep it in the analysis presented here.

3 *The* curl curl$(\chi\mathbf{D})$*-term:* As the electric susceptibility, χ, for X-rays is very small ($10^{-6}\ldots10^{-5}$), first order derivatives of $\mathbf{u}_N$ or $\mathbf{D}_{Nh}$ and higher can be neglected here.

With such assumptions we find:

$$\text{curl curl}(\chi\mathbf{D}) = k_h^2 \sum_{h,l} \chi'_{h-l}\mathbf{D}'_{l[h]} e^{i(\omega t - \mathbf{k}_h\cdot\mathbf{r})} \qquad (10.12)$$

with

$$\mathbf{D}'_{l[h]} \stackrel{\text{def}}{=} \mathbf{D}'_l - \tfrac{\mathbf{k}_h}{k_h}\left(\tfrac{\mathbf{k}_h}{k_h}\cdot\mathbf{D}'_l\right) \quad . \qquad (10.13)$$

This appears identical to the corresponding term in the classical Takagi-Taupin equation, but one has to bear in mind that χ'_{h-l} and $\mathbf{D}'_l$ are standing for the sums defined in equations (10.4) and (10.7).

Putting the above three parts together, and neglecting terms of the order of $\partial_i(\mathbf{G}_h\cdot\mathbf{u}_N)\,\partial_j\mathbf{D}_{Nh}$ we obtain the *generalised time dependent Takagi-Taupin equation*:

$$\boxed{2i(\mathbf{k}_h\cdot\nabla)\mathbf{D}'_h + k^2\frac{2i}{\omega}\frac{\partial\mathbf{D}'_h}{\partial t} + (k_h^2 - k^2)\mathbf{D}'_h = k_h^2\sum_l \chi'_{h-l}\mathbf{D}'_{l[h]}} \qquad (10.14)$$

with $k \stackrel{\text{def}}{=} \omega/c$ being the spatial frequency of the X-rays in vacuum. Note that all sums over N have disappeared owing to the definition of $\mathbf{D}'_h$ in equation (10.7). A discussion of this equation follows in sec. 3.

3. Comparison with classical Takagi-Taupin theory

Equation (10.14) looks remarkably similar to the classical time dependent Takagi-Taupin equation [279, 280]. In fact the classical equation must be included as the special case for $\mathbf{u}_1 = \mathbf{u}_2 = \cdots = \mathbf{u}_M \stackrel{\text{def}}{=} \mathbf{u}$. This will be demonstrated here, in order to prove consistency.

If the displacements $\mathbf{u}_N$ for all atoms N in a unit-cell are the same, the phase factor $\exp(i\mathbf{G}_h \cdot \mathbf{u}_N)$ is identical for all terms in the sums in equations (10.4) and (10.7), which can thus be written:

$$\chi'_h(\mathbf{r}, t) = \chi_h e^{i\mathbf{G}_h \cdot \mathbf{u}} \quad , \quad \mathbf{D}'_h(\mathbf{r}, t) = \mathbf{D}_h e^{i\mathbf{G}_h \cdot \mathbf{u}} \tag{10.15}$$

with $\mathbf{D}_h$ being the X-ray field amplitude as it is usually defined in classical Takagi-Taupin theory. Inserting these expressions into the generalised time dependent Takagi-Taupin equation (10.14) one obtains:

$$\begin{aligned} &2i(\mathbf{k}_h \cdot \nabla)\mathbf{D}_h - 2(\mathbf{k}_h \cdot \nabla)(\mathbf{G}_h \cdot \mathbf{u})\, \mathbf{D}_h + k^2 \frac{2i}{\omega}\frac{\partial \mathbf{D}_h}{\partial t} \\ &-k^2 \frac{2}{\omega}(\mathbf{G}_h \cdot \dot{\mathbf{u}})\, \mathbf{D}_h + (k_h^2 - k^2)\mathbf{D}_h = k_h^2 \sum_l \chi_{h-l} \mathbf{D}_{l[h]} \quad . \end{aligned} \tag{10.16}$$

The above equation is identical to the time-dependent equation as derived previously, [279, 280, 281] though care should be taken in comparison, as using the present notation the previous work presented equations for $\mathbf{D}'_h(\mathbf{r}, t)$.

Comparing equations (10.14) and (10.16) it turns out that the generalised Takagi-Taupin equation is perhaps slightly less complex and more intuitive, as the distortion affects only the susceptibility, the parameter which defines the X-ray diffraction properties of the crystal.

Given that we have demonstrated that equation (10.14) describes X-ray diffraction from crystals with both uniform and non-uniform distortions, it is of interest to discuss the differences between the diffraction in the presence of the various distortion types. The only term where distortions leave their mark in equation (10.14) is the susceptibility coefficient χ'_{h-l}. In the case of uniform displacements, it has been shown in equation (10.15), that this just adds the phase $\mathbf{G}_h \cdot \mathbf{u}$ to the complex Fourier coefficient χ_h of the perfect crystal. Thus a uniform displacement field leads to a pure phase modulation of the susceptibility coefficient (it is this phase modulation that in the formulism presented here corresponds to the deviation from the local Bragg angle that strain introduces into the standard Takagi-Taupin equations). A non-uniform displacement, however, will not only change the phase but also the modulus (s. fig. 10.1).

4. Coherent phonons and selection rules

An important application of the generalised Takagi-Taupin equation is to study the effects that coherent acoustic and optical phonons have on the diffraction of X-rays. In fact this has been the main motivation for its development.

A general expression for the displacement of single mode acoustic or optical phonons is given by

$$\mathbf{u}_N = \tfrac{1}{2}\mathbf{p}_N \, e^{i(\Omega t - \mathbf{q}\cdot\mathbf{r})} + \text{c.c.} \quad , \qquad N = 1, \ldots, M \tag{10.17}$$

where $\mathbf{p}_N$ is the amplitude vector of the Nth atom in the unit cell – which is proportional to the eigen-vector of the dynamical matrix of the particular mode – Ω the phonon frequency and $\mathbf{q}$ the phonon wave vector. The amplitude vector $\mathbf{p}_N$ is in general complex and thus can contain phase shifts between the different spatial components which is the case for elliptically polarised phonons.

The modulated susceptibility coefficient χ_h' can thus be written

$$\chi_h' = \sum_{N=1}^{M} \chi_{Nh} \, e^{i\mathbf{G}_h\cdot\mathbf{r}_N} \, e^{i\mathbf{G}_h\cdot\{\mathbf{p}_N \exp(i[\Omega t - \mathbf{q}\cdot\mathbf{r}]) + \text{c.c.}\}/2} \tag{10.18}$$

For small amplitudes p_N the second exponential can be expanded as a Taylor series where terms of second order and higher are negligible, leading to

$$\chi_h' = \chi_h + \tfrac{1}{2}\delta\chi_h^+ e^{i(\Omega t - \mathbf{q}\cdot\mathbf{r})} + \tfrac{1}{2}\delta\chi_h^- e^{-i(\Omega t - \mathbf{q}\cdot\mathbf{r})} \quad , \tag{10.19}$$

where

$$\delta\chi_h^+ \stackrel{\text{def}}{=} \sum_{N=1}^{M} i\mathbf{G}_h \cdot \mathbf{p}_N \, \chi_{Nh} \, e^{i\mathbf{G}_h\cdot\mathbf{r}_N} \tag{10.20}$$

and

$$\delta\chi_h^- \stackrel{\text{def}}{=} \sum_{N=1}^{M} i\mathbf{G}_h \cdot \mathbf{p}_N^* \, \chi_{Nh} \, e^{i\mathbf{G}_h\cdot\mathbf{r}_N} \quad . \tag{10.21}$$

In the case of linearly polarised phonons, where the amplitude vector $\mathbf{p}_N$ is real, expression (10.19) becomes

$$\chi_h'^{\,\text{lin}} = \chi_h + \delta\chi_h \cos(\Omega t - \mathbf{q}\cdot\mathbf{r}) \tag{10.22}$$

From this expression it is evident that linearly polarised optical and acoustic phonons do not affect the susceptibility coefficient χ_h' if the direction of polarisation is parallel to the diffracting planes, i.e. $\mathbf{G}_h \cdot \mathbf{p}_N = 0$. In this case the phonons won't be "visible" in the rocking curve. For elliptically polarised phonons this is obviously much less likely to happen, since both $\delta\chi_h^+$ and $\delta\chi_h^-$

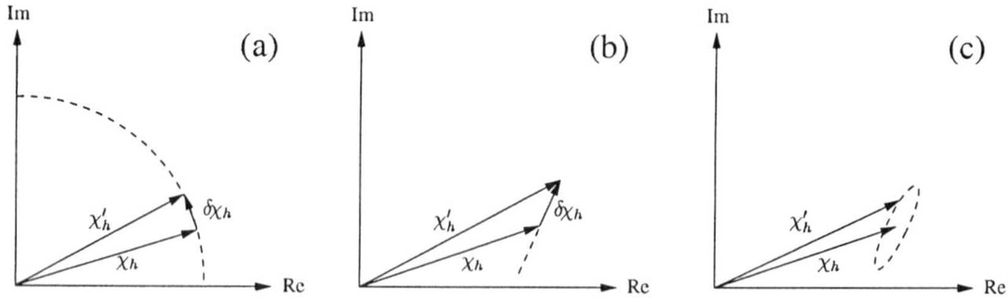

Figure 10.1: Effect of low-frequency acoustic (a), linearly polarised optical (b) and elliptically polarised optical (c) phonons of a single-mode on the susceptibility coefficient χ'_h

would have to vanish. In addition expressions (10.20) to (10.22) reveal that a low frequency acoustic phonon, where $\mathbf{p}_1 = \mathbf{p}_2 = \cdots = \mathbf{p}_M \stackrel{\text{def}}{=} \mathbf{p}$ and

$$\delta\chi_h^{\text{ac}} = i\chi_h \, \mathbf{G}_h \cdot \mathbf{p} \quad , \tag{10.23}$$

is incapable of lifting the ban of a forbidden reflection, because if $\chi_h = 0$ also $\delta\chi_h^{\text{ac}}$ vanishes. In contrast, an optical phonon can, in principle, lift the ban, since $\delta\chi_h^{\pm}$ is not necessarily zero if χ_h vanishes (s. fig. 10.1).

Finally if one substitutes the expression for χ'_h from equation (10.19) into the Fourier series (10.3) of the susceptibility $\chi(\mathbf{r},t)$ one obtains

$$\chi(\mathbf{r},t) = \chi^{(\text{p})}(\mathbf{r}) + \tfrac{1}{2}\sum_h \delta\chi_h^{+} e^{i\Omega t - i(\mathbf{G}_h+\mathbf{q})\cdot\mathbf{r}} + \tfrac{1}{2}\sum_h \delta\chi_h^{-} e^{-i\Omega t - i(\mathbf{G}_h-\mathbf{q})\cdot\mathbf{r}} \quad . \tag{10.24}$$

Besides the perfect crystal susceptibility $\chi^{(\text{p})}(\mathbf{r})$, which has of course the periodicity of the lattice, there are two smaller terms, with slightly modified periodicities. The Laue condition for these terms reads

$$\mathbf{k} - \mathbf{k}_0 = \mathbf{G}_h \pm \mathbf{q} \quad . \tag{10.25}$$

Thus on either side of the rocking curve peak ($\mathbf{k} - \mathbf{k}_0 = \mathbf{G}_h$, $\theta = \theta_B$, θ_B: Bragg angle) smaller diffraction maxima appear at

$$\Delta\theta \stackrel{\text{def}}{=} \theta - \theta_B = \pm\frac{q}{G_h}(\cos\kappa\tan\theta_B + \sin\kappa) - \frac{\chi_{r0}(1-\gamma)}{2\sin(2\theta_B)} \quad , \tag{10.26}$$

where q stands for the spatial frequency $|\mathbf{q}|$, G_h for the modulus of the reciprocal lattice vector, κ for the angle between $\mathbf{q}$ and $\mathbf{G}_h$ (s. fig. 10.2) and γ for the asymmetry factor γ_h/γ_0 ($\gamma_h \stackrel{\text{def}}{=} \cos\psi_h$, $\psi_h \stackrel{\text{def}}{=} \angle(\mathbf{k}_h, \mathbf{n})$). The last term in equation (10.26) represents the shift of the Bragg angle due to refraction.

The periodic structures represented by terms 2 and 3 in equation (10.24) are not static. Obviously the structure with reciprocal lattice vector $\mathbf{G}_h + \mathbf{q}$ moves

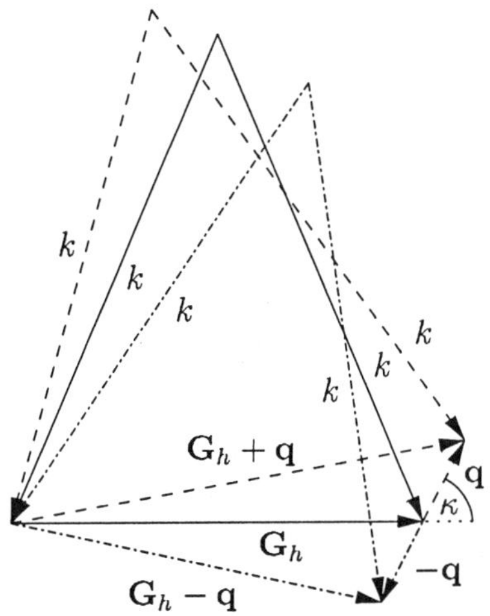

Figure 10.2: Laue condition for a crystal containing single-mode phonons with wave vector $\mathbf{q}$)

with the speed $\mathbf{v}^+ = \Omega\ (\mathbf{G}_h + \mathbf{q})/|\mathbf{G}_h + \mathbf{q}|^2$, and the other with $\mathbf{G}_h - \mathbf{q}$ with $\mathbf{v}^- = -\Omega\ (\mathbf{G}_h - \mathbf{q})/|\mathbf{G}_h - \mathbf{q}|^2$. This becomes important if the speed of sound is so high that relativistic Doppler effects have to be taken into account, as may be the case for e.g. polaritons. More about this in section sec. 5.

5. Perturbative analytical solution of the generalised Takagi-Taupin equation

In many cases of distortion we will need to solve the generalised Takagi-Taupin equation numerically. However, in this section we present a novel perturbative method to analytically solve the Takagi-Taupin equation in the presence of a small periodic distortion. For such a small distortion the susceptibility coefficient χ'_h can be separated into the susceptibility coefficient of the unperturbed crystal χ_h and a small perturbation term, e.g. $\delta\chi_h^+ \exp(i[\Omega t - \mathbf{q} \cdot \mathbf{r}])/2 + \delta\chi_h^- \exp(-i[\Omega t - \mathbf{q} \cdot \mathbf{r}])/2$ in the case of single mode phonons (s. eqn. (10.19)). Also, one can expect that a small perturbation only has a small effect on the X-ray field amplitudes $\mathbf{D}'_h$, thus it is useful to separate them as well: $\mathbf{D}'_h = \mathbf{D}_h^{(\mathrm{p})} + \delta\mathbf{D}'_h$, where $\mathbf{D}_h^{(\mathrm{p})}$ stands for the field amplitude in the perfect unperturbed crystal and $\delta\mathbf{D}'_h$ for a small perturbation (we return to these assumptions later in the section). Inserting these expressions into the generalised Takagi-Taupin equation (10.14), making use of the fact that $\mathbf{D}_h^{(\mathrm{p})}$ is the perfect crystal solution and neglecting terms of the order $\delta\chi_h\delta\mathbf{D}'_h$ one obtains the following non-homogeneous, linear differential equation system for the perturbation terms $\delta\mathbf{D}'_h$:

$$\frac{2i}{k}(\mathbf{s}_h\cdot\nabla)\delta\mathbf{D}'_h+\frac{2i}{\omega}\frac{\partial(\delta\mathbf{D}'_h)}{\partial t}+\frac{k_h^2-k^2}{k^2}\delta\mathbf{D}'_h-\sum_l\chi_{h-l}\delta\mathbf{D}'_{l[h]}$$
$$=\tfrac{1}{2}\sum_l\delta\chi^+_{h-l}\mathbf{D}^{(\mathrm{p})}_{l[h]}e^{i(\Omega t-\mathbf{q}\cdot\mathbf{r})}+\tfrac{1}{2}\sum_l\delta\chi^-_{h-l}\mathbf{D}^{(\mathrm{p})}_{l[h]}e^{-i(\Omega t-\mathbf{q}\cdot\mathbf{r})} \quad (10.27)$$

What has been gained by this? In contrast to the Takagi-Taupin equation itself this differential equation system has constant coefficients. The only space and time dependent term is the inhomogeneity. Therefore the solution of this system is very straight forward, as demonstrated here for the two-wave case, where only a single diffracted wave field $\mathbf{D}'_h$ exists apart from the incident $\mathbf{D}'_0$. The system of equations (10.27) can then be reduced to

$$\frac{2i}{k}\frac{\partial(\delta D'_0)}{\partial s_0}+\frac{2i}{\omega}\frac{\partial(\delta D'_0)}{\partial t}-C\chi_{\bar{h}}\delta D'_h$$
$$=\tfrac{1}{2}\delta\chi^+_{\bar{h}}CD^{(\mathrm{p})}_h e^{i(\Omega t-\mathbf{q}\cdot\mathbf{r})}+\tfrac{1}{2}\delta\chi^-_{\bar{h}}CD^{(\mathrm{p})}_h e^{-i(\Omega t-\mathbf{q}\cdot\mathbf{r})}$$
$$\frac{2i}{k}\frac{\partial(\delta D'_h)}{\partial s_h}+\frac{2i}{\omega}\frac{\partial(\delta D'_h)}{\partial t}-C\chi_h\delta D'_0+2\beta_h\delta D'_h$$
$$=\tfrac{1}{2}\delta\chi^+_hCD^{(\mathrm{p})}_0 e^{i(\Omega t-\mathbf{q}\cdot\mathbf{r})}+\tfrac{1}{2}\delta\chi^-_hCD^{(\mathrm{p})}_0 e^{-i(\Omega t-\mathbf{q}\cdot\mathbf{r})} \quad (10.28)$$

with s_0 and s_h being the spatial coordinates whose axes are parallel to the wave vectors $\mathbf{k}_0$ and $\mathbf{k}_h$, C the polarisation factor which is 1 for σ-polarised and $\cos(2\theta_B)$ for π-polarised X-rays, and finally

$$\beta_h \stackrel{\mathrm{def}}{=} \frac{k_h^2-k^2}{2k^2}-\frac{\chi_0}{2} \quad . \quad (10.29)$$

With the ansatz [1]

$$\delta D'_h = d_h^+ e^{i(\Omega t-\mathbf{q}\cdot\mathbf{r})}+d_h^- e^{-i(\Omega t-\mathbf{q}\cdot\mathbf{r})} \quad (10.30)$$

the differential equation system can be written as an algebraic system for $d_h^\pm$ with the solution:

$$d_h^\pm=\frac{C}{2}\,\frac{\chi_h\delta\chi^\pm_{\bar{h}}CD^{(\mathrm{p})}_h\pm A_0\delta\chi^\pm_hD^{(\mathrm{p})}_0}{\pm A_0(\pm A_h+2\beta_h)-\chi_h\chi_{\bar{h}}C^2} \quad , \quad (10.31)$$

with

$$A_0\stackrel{\mathrm{def}}{=}\tfrac{2}{k}\,\mathbf{s}_0\cdot\mathbf{q}-2\tfrac{\Omega}{\omega} \quad , \quad A_h\stackrel{\mathrm{def}}{=}\tfrac{2}{k}\,\mathbf{s}_h\cdot\mathbf{q}-2\tfrac{\Omega}{\omega} \quad . \quad (10.32)$$

[1] The index h stands here also for 0, i.e. the incident wave-field.

The total solution

$$D_h(\mathbf{r},t) = D_h^{(p)} + d_h^+ e^{i(\Omega t - \mathbf{q}\cdot\mathbf{r})} + d_h^- e^{-i(\Omega t - \mathbf{q}\cdot\mathbf{r})} \tag{10.33}$$

is the unperturbed rocking curve given by $D_h^{(p)}$ plus satellites on either side at $|\mathbf{k}-\mathbf{k}_0| = |\mathbf{G}_h \pm \mathbf{q}|$ with amplitudes $d_h^\pm$ oscillating at the phonon frequency Ω.

For the exact angular position of these satellites refraction effects have to be taken into account. Therefore in equation (10.29) the wave-vectors $\mathbf{k}_0$ and $\mathbf{k}_h$ must be expressed by the vacuum wave-vectors $\mathbf{k}_0^{(a)}$ and $\mathbf{k}_h^{(a)}$, i.e.

$$\mathbf{k}_0 = \mathbf{k}_0^{(a)} + \frac{k\chi_0}{2\gamma_0}\mathbf{e}_z \quad , \quad \mathbf{k}_h = \mathbf{k}_h^{(a)} + \frac{k\chi_0}{2\gamma_h}\mathbf{e}_z \quad . \tag{10.34}$$

This leads to

$$\beta_h = -\Delta\theta^{(a)}\sin(2\theta_B) - \tfrac{1}{2}\chi_0(1-\gamma) \quad . \tag{10.35}$$

Attention has to be paid to the fact that the total solution is a sum of complex field amplitudes and therefore it is vital to use expressions for the perfect crystal amplitudes $D_0^{(p)}$ and $D_h^{(p)}$ with the correct phases.

In figure 10.3 we show the results of the calculation of the rocking curve for Bragg reflection from 10 μm thick InSb (111) (angle of asymmetry $\varphi = 5°$, $\lambda = 1.5406$ Å, σ-pol.) distorted by coherent longitudinal acoustic phonons propagating perpendicularly to the crystal surface with spatial frequency $|\mathbf{q}| = 10^8$ rad/m, temporal frequency $\Omega = 3.75 \times 10^{11}$ rad/s and strain amplitude $\eta_0 = 0.1\%$. The agreement between the perturbative solution, and the full numerical

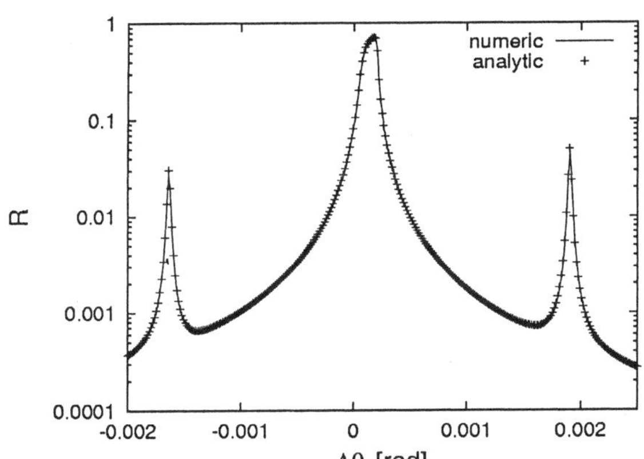

Figure 10.3: Comparison between the full numerical and the perturbative solution for Bragg diffraction from a single crystal in the presence of a coherent acoustic phonon for the conditions stated in the text.

solution of equation (10.14) is excellent – indeed the difference between the two solutions lies within the accuracy of the numerical algorithms employed and cannot be seen on the scale of the graph.

One advantage of such an analytic solution is the ability to immediately see the influence of certain physical parameters. For example, the explicit time-dependence represented by the time derivatives $\partial_t D_0$ and $\partial_t D_h$, which lead to the Ω/ω-terms in A_0 and A_h, only plays a role if the phonon phase velocity is a non-negligible fraction of the speed of light, as, for example, in the presence of polaritons. The major effect of these terms on the rocking curve is the angular shift of the satellites, either away or towards the centre of the rocking curve. The direction of shift depends on whether the phonon (polariton) has a velocity component parallel or anti-parallel in the direction of the reciprocal lattice vector $\mathbf{G}_h$. If the component is anti-parallel, the satellites will move towards the central peak, if it is parallel they will move away. This effect can be understood in terms of the simple wave vector matching scheme introduced in section 3. The structures with reciprocal lattice vectors $\mathbf{G}_h \pm \mathbf{q}$ move now at a speed level that relativistic Doppler effects come into play. The structure with $\mathbf{G}_h + \mathbf{q}$ moves towards the crystal surface and sees a blue-shifted incident X-ray wave, the structure with $\mathbf{G}_h - \mathbf{q}$ moves into the crystal bulk and thus sees a red-shift. In order to quantify this effect one has to fulfil the Laue condition in the rest-frame of each structure and then perform a Lorentz transformation into the rest-frame of the crystal. The Laue condition in the respective rest frames

$$\mathbf{k}' - \mathbf{k}'_0 = \mathbf{G}'_h + \mathbf{q}' \quad , \quad \mathbf{k}'' - \mathbf{k}''_0 = \mathbf{G}''_h - \mathbf{q}'' \tag{10.36}$$

Lorentz transformed results in

$$\delta\mathbf{K}^{\pm} = \frac{2}{c^2} \frac{\mathbf{q} \pm \mathbf{G}_h}{|\mathbf{q} \pm \mathbf{G}_h|^2} \Omega(\omega - \Omega) \quad , \tag{10.37}$$

with the Doppler shift $\delta\mathbf{K}^{\pm} \stackrel{\text{def}}{=} \mathbf{K} - (\mathbf{G}_h \pm \mathbf{q})$ of the scattering vector $\mathbf{K} \stackrel{\text{def}}{=} \mathbf{k} - \mathbf{k}_0$. Under the assumption that $|\mathbf{q}| \ll |\mathbf{G}_h|$ this leads to the angular shift

$$\delta\theta^{\pm} = \pm 2 \tan\theta_B \frac{\Omega(\omega - \Omega)}{c^2 G_h^2} \tag{10.38}$$

of the satellites. A more illustrative, but only qualitative way of looking at the problem, is from the viewpoint of the incident X-ray wave. If the phonon moves at least partially into the same direction, the X-rays see a longer phonon wavelength, because by the time the X-rays have traveled e.g. from one amplitude maximum to another, the second one has moved further away from where the X-rays intercepted the first. Correspondingly, the X-rays will see a shorter phonon wavelength if the phonon moves towards the incident X-rays.

The analytic solution includes an arbitrary propagation direction of the phonon. The numerical solution of this case is computationally expensive, because the full spatially two-dimensional Takagi-Taupin equation has to be solved rather than a one-dimensional reduced variant for only depth dependent distortion profiles [282, 283]. The analytic solution is in principle far simpler in such cases, and may be of use in a study of the proposed phonon-Bragg switch [250].

Finally the perturbative approach has the great advantage of allowing a direct decomposition of a measured rocking curve into contributions from different phonon modes. Because the differential equation (10.27) for $\delta\mathbf{D}'_h = \mathbf{D}'_h - \mathbf{D}_h^{(p)}$ is linear, the contributions from different modes superimpose. This allows to use e.g. a least squares algorithm to fit the modulus of a sum of those analytical solutions for a set of relevant phonon modes to a measured rocking curve, directly revealing the modes' occupation numbers. The fit is, however, somewhat ambiguous for lack of phase information in the measured rocking curve.

The perturbative solution becomes less accurate with increasing strain amplitude, as the the realization of the basic assumption $|\delta\chi_h| \ll |\chi_h|$ gets worse. The assumption works well for strains up to a few tens of a percent, depending on the location of the satellites, i.e. the spatial phonon frequencies $|\mathbf{q}|$. In our studies we have found that the perturbative solution is reasonable if the maximum intensity of the satellites does not exceed 10% of that of the main peak. Care must also be taken in using the perturbative solution when the sidebands are not well separated in angle from the main peak. This will occur for phonons with wavelengths comparable to or longer than an extinction depth. In this case the incident X-rays will not interact with many wavelengths of the periodic disturbance, and the solutions to the equations for the main peak cannot be decoupled from those of the sidebands. This can be seen in figure 10.4, where we show the perturbative and numerical solutions for a very low frequency acoustic phonon – evidently, use of the perturbative solution is not safe when the angular separation of the sidebands from the main peak is of order a rocking curve width. Another case where the perturbative solution becomes inaccurate

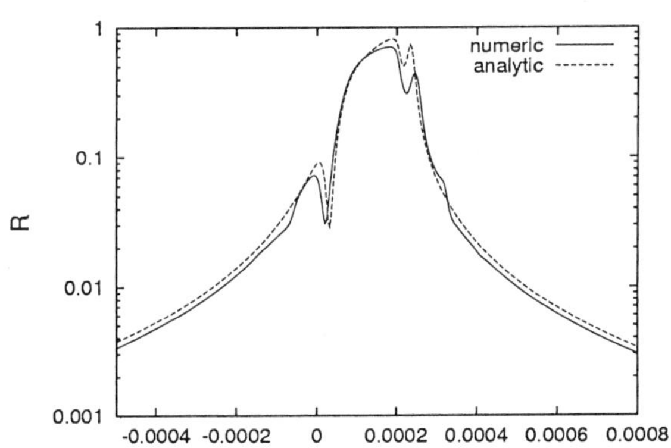

Figure 10.4: Comparison between the full numerical and the perturbative solution for Bragg diffraction from a single crystal in the presence of a very low frequency coherent acoustic phonon with $|\mathbf{q}| = 5 \times 10^6$ rad/m, $\Omega = 1.875 \times 10^{10}$ rad/s, and $\eta_0 = 0.01\%$. Otherwise the parameters are the same as in figure 10.3.

is when the crystal is thin enough to show Laue oscillations. In this case the deviation is of the order of the Laue oscillation amplitude.

6. Numerical solution of the generalised Takagi-Taupin equation

For those situations where perturbation theory is not valid the equations must be solved numerically. If the distortions are not evolving rapidly in time (such that the time dependence of the wave fields can be ignored), then the same numerical algorithms as have been used previously for the classical Takagi-Taupin equation can be employed. For the rocking curves presented in this section, a fourth order Runge-Kutta algorithm with variable step-size [284] has been applied to the one-dimensional reduced variant of the problem [282, 283]. Figures 10.5, 10.6 and 10.7 show the simulated effects of single mode coherent phonons, acoustic and optical, on the rocking curve of α-quartz (010). All three

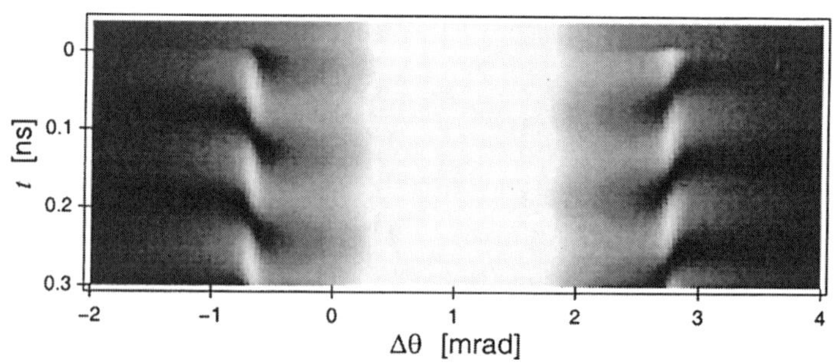

Figure 10.5: Effects of single mode coherent acoustic phonons on the rocking curve of α-quartz (010) for the conditions stated in the text.

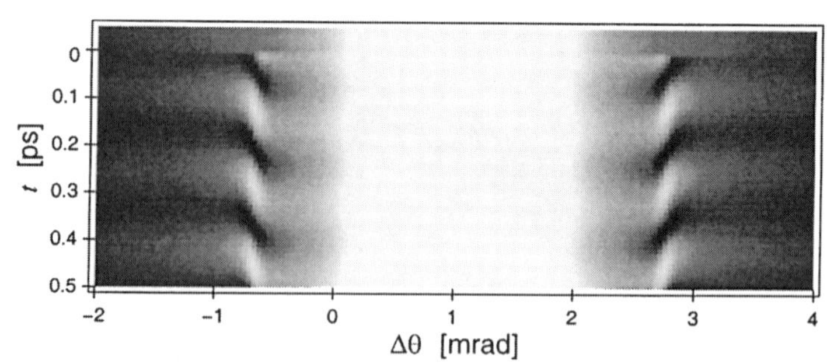

Figure 10.6: Effects of single mode coherent optical phonons with linear polarisation on the rocking curve of α-quartz (010) for the conditions stated in the text.

simulations were done for σ-polarised X-rays of 7.1255 Å wavelength Bragg reflected from a 30 μm thick slab of α-quartz with 45° angle of asymmetry ([010] tilted towards [001]). In all three cases the wave vector was given by $|\mathrm{q}| = 1.414 \times 10^7$ rad/s in the direction parallel to the surface normal. Figure

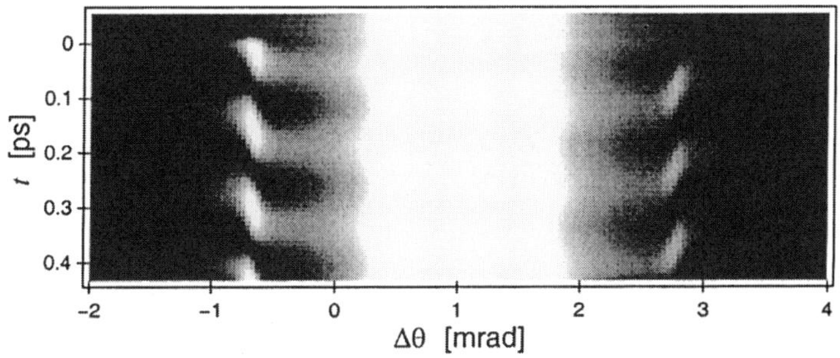

Figure 10.7: Effects of single mode coherent optical phonons with elliptic polarisation on the rocking curve of α-quartz (010) for the conditions stated in the text.

10.5 shows the effects of the acoustic A-mode with frequency $\Omega = 57 \times 10^9$ rad/s, figure 10.6 of the linearly polarised optical A-mode with $\Omega = 37.9 \times 10^{12}$ rad/s and figure 10.7 of the elliptically polarised optical A-mode with $\Omega = 45.18$ rad/s. The phonons have been modelled with the help of UNISOFT, a program package for lattice dynamical calculations [285]. The simulation of α-quartz is based on a simple Born-von Karman force constant model included as an example in UNISOFT.

Both acoustic and optical phonons generate sidebands to the main peak of the rocking curve oscillating with the phonon frequency. There is a phase difference of π between these sidebands in the case of the acoustic mode, but no phase differences for the linearly polarised optical mode. This behaviour is not entirely unexpected, since optical phonons may change the modulus of the susceptibility coefficient χ'_h, whereas low-frequency acoustic phonons only the phase – equivalent to the lattice spacing – thus shifting reflectivity to other angles rather than changing the integral value. Phase shifts between both satellites can be observed for optical phonons, too, as demonstrated in figure 10.7, since optical phonons can alter both the modulus and the phase of χ'_h .

By comparing figure 10.7 with figures 10.5 and 10.6, it becomes obvious that the phonon polarisation type affects the symmetry of the sidebands. Both linearly polarised modes generate sidebands of equal amplitude, whereas the elliptically polarised mode generates sidebands with different amplitudes. This can be understood by considering the expansion of the modulated susceptibility coefficient χ'_h in equation (10.19) and the expressions for the sideband amplitudes in equation (10.31). $\delta\chi_h^+$ is directly responsible for the sideband on one side[2] and $\delta\chi_h^-$ for the other. In the case of linearly polarised phonons $\delta\chi_h^- = \delta\chi_h^+$. Only elliptically polarised phonons allow $\delta\chi_h^- \neq \delta\chi_h^+$, leading to the observed asymmetry.

[2]Which side, depends on the direction of $\mathbf{q}$.

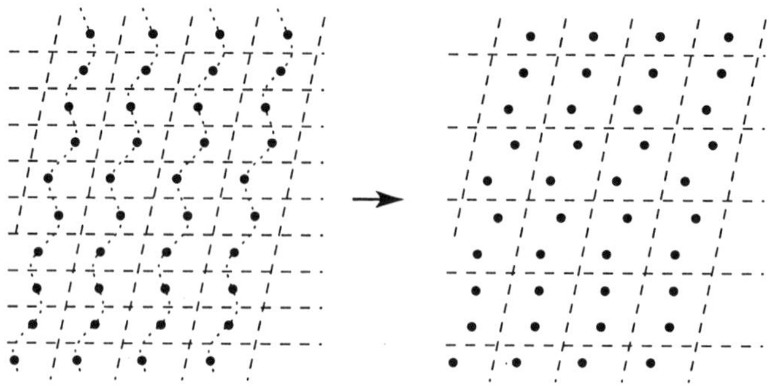

Figure 10.8: If two adjacent unit cells are combined to one, a high frequency acoustic phonon close to the Brillouin zone boundary can be transformed into a low-frequency optical phonon.

Different types of coherent phonons obviously lead to well distinguishable features of the rocking curve. This makes TRXD (time resolved X-ray diffraction) a potentially very powerful tool for their investigation.

7. High spatial frequency phonons

In principle, the generalised Takagi-Taupin equation can be used to model phonons with wave vectors up to the Brillouin zone boundary, as long as the phonon amplitude is much smaller than the lattice period. However the practical difficulty exists, that close to the Brillouin zone boundary the step size of the numerical integration routine would have to to be extremely small – only a fraction of a lattice period. To model the diffraction from a macroscopic crystal the computational costs would be immense. In this section we discuss a method to overcome this difficulty, which in fact makes the modelling of diffraction close to the Brillouin zone boundary as simple as in the Brillouin zone centre.

This method makes use of the fact, that these phonons, whether acoustic or optical, show a phase difference close to π between adjacent unit cells. Therefore, by combining two adjacent unit cells to a new "super cell", as shown in figure 10.8 for the case of one atom per unit cell, an acoustic phonon can be transformed into an optical phonon of much smaller spatial frequency $q_{\mathrm{eff}} = q_{\mathrm{BZ}} - q$, where q_{BZ} stands for the spatial frequency exactly on the Brillouin zone boundary. Another way of looking at this is to imagine that we could consider alternate atoms of the original crystal to be infinitesimally different in terms of their scattering power – clearly this should not alter the physically observed scattering. However, by considering alternate atoms to be different, even by a small amount, we have effectively doubled the period of the crystal lattice, and halved the dimensions of the Brillouin zone, resulting in a splitting of the original acoustic phonon dispersion curve into acoustic and optical branches. A schematic diagram of the reduction of the Brillouin zone is shown in figure 10.9.

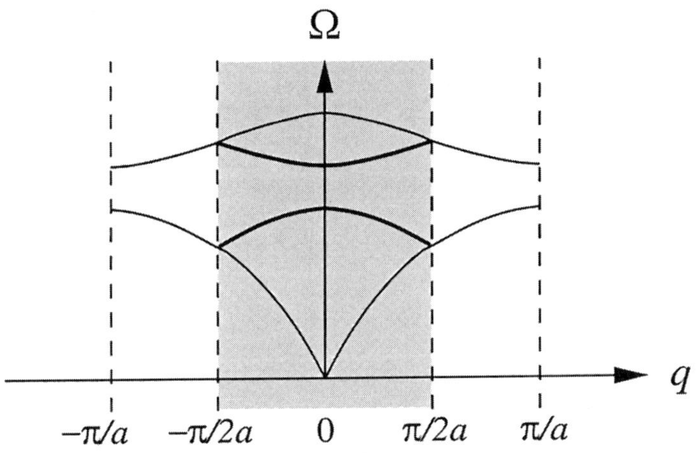

Figure 10.9: Reduction of the Brillouin zone, if two neighbour cells are combined to one. The phonon dispersion curves for $q > \pi/2a$ have to be folded back into the reduced zone, i.e. effectively doubling the number of modes in correspondence with the doubled number of atoms per unit cell. For example the acoustic branch for $q > \pi/2a$ becomes a new optical branch.

Phonons with wave vectors close to the Brillouin zone boundary of the original unit cell are transformed to the centre of the reduced zone. This approach to modelling high spatial frequency phonons in an efficient manner is by no means limited to acoustic phonons or crystals with a monatomic basis. An acoustic or optical phonon in a crystal with non-primitive basis can similarly be transformed into optical phonon close to the zone centre of a reduced Brillouin zone. The approach outlined above may become clearer when we translate this transformation into the formal language of mathematics.

The dissection of the susceptibility into primitive susceptibilities introduced in equation (10.2) can be transformed into:

$$\chi(\mathbf{r},t) = \sum_{N=1}^{2M} \chi_N^{(\mathrm{r})}\left(\mathbf{r} - \mathbf{r}_N - \mathbf{u}_N(\mathbf{r},t)\right) \quad , \tag{10.39}$$

where $\chi_N^{(\mathrm{r})}(\mathbf{r})$ stands for the "reduced primitive" susceptibility which represents only every other atom in the Nth imaginary crystal with primitive basis. Obviously the atomic coordinates $\mathbf{r}_N$ for $N > M$ are given by $\mathbf{r}_{M+N} = \mathbf{r}_N + \mathbf{a}_1$, where $\mathbf{a}_1$ is the lattice basis vector between the unit cells forming the super cell. The "reduced primitive" susceptibility can be expanded in a Fourier series:

$$\chi(\mathbf{r},t) = \sum_{N=1}^{2M} \sum_h \frac{1}{V_{\mathrm{sc}}} \int_{\mathrm{sc}} dV' \, \chi_N^{(\mathrm{r})}(\mathbf{r}') \, e^{i\mathbf{g}_h \cdot \mathbf{r}'} e^{-i\mathbf{g}_h \cdot (\mathbf{r} - \mathbf{r}_N - \mathbf{u}_N(\mathbf{r},t))} \quad . \tag{10.40}$$

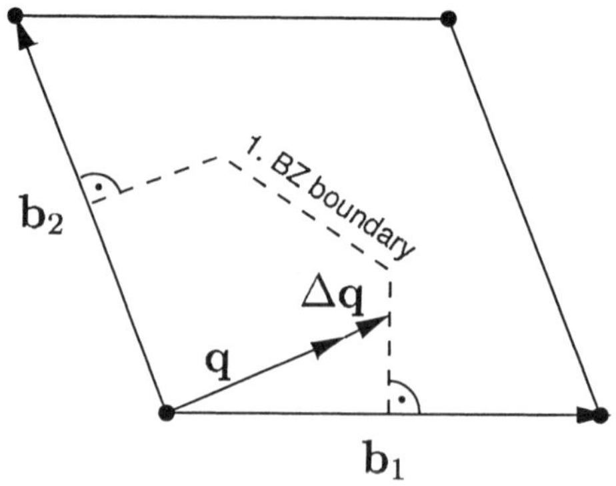

Figure 10.10: Dissection of a phonon wave vector close to the Brillouin zone boundary.

Here $\mathbf{g}_h$ stands for the reciprocal lattice vector $\mathbf{g}_h \stackrel{\text{def}}{=} h_1\mathbf{B}_1 + h_2\mathbf{B}_2 + h_3\mathbf{B}_3$ with $\mathbf{B}_1 = \mathbf{b}_1/2$, $\mathbf{B}_2 = \mathbf{b}_2$, $\mathbf{B}_3 = \mathbf{b}_3$ where $\mathbf{b}_1$, $\mathbf{b}_2$ and $\mathbf{b}_3$ are the reciprocal basis vectors of the original unit cell. The period of the reciprocal lattice in $\mathbf{b}_1$-direction has halved since the period of $\chi_N^{(\mathrm{r})}(\mathbf{r})$ in $\mathbf{a}_1$ direction is twice the period of $\chi_N^{(\mathrm{p})}(\mathbf{r})$. Accordingly the integral stretches over the super cell, thus $V_{\mathrm{sc}} = 2V_{\mathrm{uc}}$.

The phonon displacement fields $\mathbf{u}_N(\mathbf{r}, t)$ are given by equation (10.17). It turns out to be helpful to split up the phonon wave vector $\mathbf{q}$ into the vector $\mathbf{q}_{\mathrm{BZ}}$ exactly on the Brillouin zone boundary and a small difference $\Delta\mathbf{q} = \mathbf{q}_{\mathrm{BZ}} - \mathbf{q}$, as illustrated in figure 10.10. Then the displacement field falls into a high and a low (spatial) frequency part:

$$\mathbf{u}_N(\mathbf{r}, t) = \frac{1}{2}\mathbf{p}_N\; e^{-i\mathbf{q}_{\mathrm{BZ}}\cdot\mathbf{r}}\; e^{i(\Omega t + \Delta\mathbf{q}\cdot\mathbf{r})} + \mathrm{c.c.} \tag{10.41}$$

The high spatial frequency part $\exp(-i\mathbf{q}_{\mathrm{BZ}} \cdot \mathbf{r})$ has the same periodicity in $\mathbf{a}_1$-direction as the lattice with the enlarged unit cell, i.e. it is cell independent. If one further assumes that it is constant over the region over which atom N extends,[3] the spatial coordinate $\mathbf{r}$ can be replaced by the local atomic coordinate vector $\mathbf{r}_N$. What remains of the high frequency part then, is a constant phase factor $\exp(-i\phi_N)$ with $\phi_N \stackrel{\text{def}}{=} \mathbf{q}_{\mathrm{BZ}} \cdot \mathbf{r}_N$:

$$\mathbf{u}_N(\mathbf{r}, t) = \frac{1}{2}\mathbf{p}_N\; e^{i(\Omega t + \Delta\mathbf{q}\cdot\mathbf{r} - \phi_N)} + \mathrm{c.c.} \quad . \tag{10.42}$$

[3]This assumption has to be made also for other reasons. Without it, $\mathbf{u}_N(\mathbf{r}, t)$ would describe deformations of single atoms. Spatial phonon frequencies $q > \pi/a$ wouldn't be equivalent to $q' = q - \pi/a$ anymore.

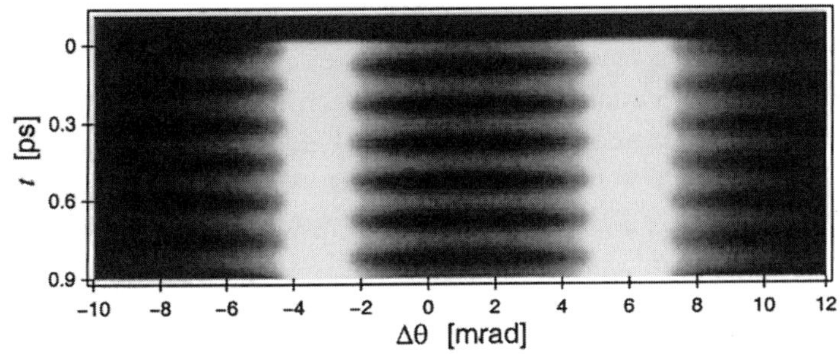

Figure 10.11: Effects of coherent single mode phonons with a wave vector close to the Brillouin zone boundary on the rocking curve of quartz $(00\frac{3}{2})$ under the conditions stated in the text.

Substituting this expression into equation (10.40) leads to:

$$\chi(\mathbf{r},t) = \sum_h \chi_h'^{\,(\mathrm{hf})}(\mathbf{r},t)\; e^{-i\mathbf{g}_h\cdot\mathbf{r}} \quad , \tag{10.43}$$

with the modulated susceptibility coefficient

$$\chi_h'^{\,(\mathrm{hf})} = \sum_{N=1}^{2M} \chi_{Nh}^{(\mathrm{r})} e^{i\mathbf{g}_h\cdot\mathbf{r}_N}\; e^{i\mathbf{g}_h\cdot\{\mathbf{p}_N \exp(i[\Omega t+\Delta\mathbf{q}\cdot\mathbf{r}-\phi_N])/2+\mathrm{c.c.}\}} \tag{10.44}$$

and $\chi_{Nh}^{(\mathrm{r})} \stackrel{\mathrm{def}}{=} V_{\mathrm{sc}}^{-1}\int_{\mathrm{sc}} dV'\; \chi_N^{(\mathrm{r})}(\mathbf{r}')\exp(i\mathbf{g}_h\cdot\mathbf{r}')$.

This approach to modelling the diffraction from coherent phonons not only at zone centre but also close to the original Brillouin zone boundary could be extremely useful. It is this region that often contains interesting physics: for example the soft modes associated with many phase transitions are either close to the Brillouin zone centre (*ferroelectric* phase transitions) or the Brillouin zone boundary (*anti-ferroelectric*).

One can expect that these high frequency distortions generate effects far away from the rocking curve centre. It turns out that peaks appear in the middle between two diffraction orders of the same set of lattice planes. As they can't be considered as a supplementary feature of neither of these orders, it is regarded as an independent diffraction peak with half-integral Miller indices.

Figure 10.11 shows the reflected intensity of α-quartz $(00\frac{3}{2})$ for single mode phonons with a wave vector only $\Delta q = 1.414 \times 10^7$ rad/m away from the Brillouin zone boundary. As previous simulations it is for a 30 μm thick slab of α-quartz with 45° angle of asymmetry ([001] tilted towards [010]), but now for single mode coherent phonons with $q = 0.8222 \times 10^{10}$ rad/m and $\Omega = 21.49 \times 10^{12}$ rad/s propagating parallel to the crystal surface normal probed with σ-polarised X-rays of 7.1255 Å wavelength in Bragg geometry. The first thing that catches the eye is the fact, that there is no main line, just two satellites. The reason for this is that one is effectively looking at a forbidden reflection. Remember that the $(00\frac{3}{2})$-reflection has been generated artificially, by combining two identical unit cells to one. It can be shown easily that, because the new cell consists of two identical halves, these additional reflections are forbidden. Thus figure 10.11 actually shows the lifting of the ban of a forbidden reflection due to optical phonons. A second closer look reveals, that the satellite intensities are modulated with twice the phonon frequency, unlike the satellites of non-forbidden reflections which are modulated with the phonon frequency itself. Once more the perturbation theory from section 5 gives the explanation. Multiplying the perturbative solution from equation (10.33) with its complex conjugate gives

$$\begin{aligned} D_h D_h^* &= |D_h^{(\mathrm{p})}|^2 + |d_h^+|^2 + |d_h^-|^2 \\ &+ 2\Re\Big\{D_h^{(\mathrm{p})} d_h^{-*} \exp(-i[\Omega t - \mathbf{q}\cdot\mathbf{r}])\Big\} \\ &+ 2\Re\Big\{D_h^{(\mathrm{p})} d_h^{+*} \exp(i[\Omega t - \mathbf{q}\cdot\mathbf{r}])\Big\} \\ &+ 2\Re\Big\{d_h^{+} d_h^{-*} \exp(2i[\Omega t - \mathbf{q}\cdot\mathbf{r}])\Big\} \quad , \end{aligned} \tag{10.45}$$

which is proportional to the diffracted intensity. In the case of a forbidden reflection the perfect crystal amplitude of the diffracted wave, $D_h^{(\mathrm{p})}$, is zero, thus the 1st, 4th and 5th term vanish equally. The only time dependent term which remains then is the 6th term, which obviously oscillates with frequency 2Ω. In the case of a non-forbidden reflection this term is usually too weak to be observed.

Half integral peaks, i.e. X-ray reflection maxima with half-integer Miller indices, have been observed already in laser irradiated α-quartz [286], though with insufficient temporal resolution to see any phonon oscillations.

It has been demonstrated that the presented generalisation of the Takagi-Taupin theory not only is applicable to a new class of distortions – including optical phonons – but also overcomes the restriction of the classical theory to low spatial frequency distortions.

Chapter 11

NONLINEAR RESPONSE FUNCTIONS FOR X-RAY LASER PULSES

P. Zambianchi
Advanced Photon Source, Argonne National Laboratory, Argonne, IL 60439, USA
zambianc@aps.anl.gov

Abstract Response functions are macroscopic quantities that encode relevant information about interacting systems. In particular, they can provide the strength of scattering and nonlinear relaxation processes. Although nonlinear *optical* susceptibilities have been extensively studied, both theoretically and phenomenologically, nonlinear response functions of atomic and molecular systems at x-rays wavelengths have received little attention. This scenario is likely to change with the advent of the 4th-generation light sources, the x-ray free-electron lasers (XFELs), where nonlinear effects may become observable from the interaction of these ultrahigh laser pulses and matter. In this paper, we provide a general semiclassical formalism to calculate nonlinear response functions for x-rays, which avoids the usual dipole approximations well suited for optical frequencies. Quick estimates suggest that, for x-rays frequencies, resonant nonlinear effects for atoms and in crystalline media are rather small, ranging from $\chi^{(2)} \approx 10^{-20} - 10^{-14} esu$ to $\chi^{(3)} \approx 10^{-32} - 10^{-19} esu$. From these values, cross sections are calculated.

1. Introduction

Since its beginnings in the sixties, nonlinear optics has motivated intense theoretical and experimental research, and now it has become a well-established field. On the other hand, nonlinear effects at x-rays wavelengths (~ 1Å) remain a relatively unexplored domain. Although the first theoretical and experimental investigations in nonlinear x-ray diffraction appeared almost 30 years ago [126, 287, 288, 289, 127, 87, 290, 291, 111], the lack of a fully coherent, high-intensity source for x-rays has motivated little further research in nonlinear x-rays physics [292, 293, 113]. With the proposal of linac-based synchrotron radiation sources, namely, the Linear Coherent Light Source (LCLS) [294] and TESLA [295] x-ray free-electron laser projects (XFELs), this scenario is about to change. Operating in the self-amplified spontaneous emission regime

(SASE) close to 1 Å, these machines will provide pulses as short as 230 fs of fully transverse coherent radiation whose peak brilliance, 10^{33} $photons/(s\ .1\% - bw\ mrad^2\ mm^2)$, is several orders of magnitude more intense than that of the current synchrotron sources [294]. At these extraordinary intensities, it is expected that the x-ray counterparts of nonlinear optical effects, such as harmonic generation and parametric down conversion, will be observable. Sure enough, although Compton scattering, thermal diffuse scattering and the *linear* effects in general are expected to be dominant, nonlinear phenomena arising from subpicosecond pulses can provide valuable spectroscopic information-inner shell processes in condensed matter and femtochemistry among others [294]-and, therefore, must be investigated thoroughly.

Motivated by this new promising regime, we developed a perturbative, semi-classical formalism to calculate nonlinear response functions for x-rays. Estimates [296] suggest that for *unfocused* XFEL pulses the strong-field regime will not be achieved for inner shell electrons of intermediate-Z materials, and, therefore, the perturbative approach remains valid. The approach shown here is limited to tightly bound electrons, including x-ray resonances. In this case, the steep charge density gradients near the nuclei can lead to strong anharmonicities in the binding potentials, which are expressed through higher order susceptibilities. With outer shell electrons whose binding energy is much less than the x-ray photon energy, the order-of-magnitude estimate in ch. 2, sec. 1 shows that nonlinear x-ray optical effects due to anharmonic potentials are much weaker than those due to the Lorentz force.

Our calculations make use of perturbation series expansion in powers of the classical vector potential $\vec{A}$ to generalize the linear response theory into the nonlinear regime, thus obtaining the corresponding nonlinear tensorial response coefficients. The key difference in our formalism is that it avoids the usual dipole approximation in the interaction Hamiltonian, which is well suited for much longer (optical) wavelengths. In dealing with light whose wavelength is comparable to atomic sizes, a basic treatment that makes use of this dipole approximation as a starting point becomes questionable. Instead, our semi-classical formalism makes use of the minimal coupling $(\vec{p} - (e/c)\vec{A})^2$ [297] describing the interaction between quantized matter and classical radiation.

The presentation of this paper is as follows. In the next section, we present the general formalism and the perturbative calculation of the linear and non-linear response functions up to third order in the vector potential. Section III deals with estimates of various nonlinear processes. These estimates can be done with different degrees of sophistication and accuracy, e.g., the use of hydrogenic functions of the atomic transitions, full set of selection rules, Hartree-Fock (many-body) formalism, etc. Instead, in order to obtain quick numerical estimates of nonlinear effects involving an x-ray laser and matter, we have followed the simplest approach, which is the estimate of the *scaling* of the matrix elements, in particular contributions from dipole transitions. In this spirit, the numerical results only provide estimates of the order of magnitude of the nonlinear processes. We stress that our formalism is *general*; the *examples* provided here are particular cases (scaling of matrix elements from dipole transitions). From the estimates of section III, we calculate the corresponding cross sections of nonlinear processes that might be observable at 4th-generation light sources, in particular nonlinear atomic spectroscopy involving x-rays plus an optical laser (four-wave mixing). The conclusion and final remarks are presented in Section IV.

2. Nonlinear response functions: general formalism

In generalizing the linear response theory into the nonlinear regime, the basic formalism developed here makes use of wave functions instead of density matrices. This is justified if we assume that pure states are to evolve into pure states upon interacting with the external fields. For a more general treatment that takes into account mixed states and decoherence effects see ref. [298].

We start with the Hamiltonian that describes the interaction between electromagnetic fields and matter [297] in the Coulomb gauge, $\nabla \cdot \vec{A}(\vec{r}, t) = 0$:

$$H_{int} = H_1 + H_2, \tag{11.1}$$

where

$$H_1 = -\frac{e}{c} \int d^3\vec{r}\vec{j}(\vec{r}) \cdot \vec{A}(\vec{r}, t) \tag{11.2}$$

$$H_2 = \frac{e^2}{2mc^2} \int d^3\vec{r}\rho(\vec{r})\vec{A}^2(\vec{r}, t), \tag{11.3}$$

where $j(\vec{r})$ and $\rho(\vec{r})$, given by

$$\rho(\vec{r}) = \sum_i \delta(r - \vec{r}_i) \tag{11.4}$$

and

$$\vec{j}(\vec{r}) = \frac{1}{2} \sum_i \left[\frac{\vec{p}_i}{m} \delta(\vec{r} - \vec{r}_i) + \delta(\vec{r} - \vec{r}_i) \frac{\vec{p}_i}{m} \right] \tag{11.5}$$

are the electronic current and charge density operators, respectively. In the presence of the electromagnetic field, the true electric current density operator $e\vec{J}(\vec{r}, t)$ is given by [297]:

$$e\vec{J}(\vec{r}, t) = e\vec{j}(\vec{r}) - \frac{e^2}{mc} \vec{A}(\vec{r}, t)\rho(\vec{r}), \tag{11.6}$$

The high-intensity incident laser pulse $\vec{A}(\vec{r},t)$, with a given divergence and bandwidth, will be treated classically. Expanding it into its Fourier components

$$\vec{A}(\vec{r},t) = \int_{-\infty}^{+\infty} \frac{d^4k}{(2\pi)^4} \vec{A}(\omega,\vec{k}) e^{i\vec{k}\cdot\vec{r}-i\omega t}, \tag{11.7}$$

with $d^4k \equiv d\omega d^3\vec{k}$ and the reality of $\vec{A}(\vec{r},t)$ implies $\vec{A}^*(\omega,\vec{k}) = \vec{A}(-\omega,-\vec{k})$. Without loss of generality, we will assume that at $t = 0$ our system is found in the ground state and that it will evolve in time remaining close to zero temperature. If this last assumption is met, the density matrix approach is reduced to that of wave functions.

The theory of response functions establishes a proportionality between the external electromagnetic field and the induced polarization or, equivalently, the proportionality between the external vector potential and the induced current [299]. Therefore, the calculation of the expected value of the current density operator, $\langle e\vec{J}(\vec{r},t)\rangle \equiv \langle\Psi(t)|e\vec{J}(\vec{r},t)|\Psi(t)\rangle$, expanded in powers of the vector potential provides the tensorial response functions coefficients $R_{ijk...}(\vec{r},t)$ (see appendix). From the expression for $\langle e\vec{J}(\vec{r},t)\rangle$:

$$\left\langle e\vec{J}(\omega,\vec{k})\right\rangle = \int dtd^3r \left\langle e\vec{J}(\vec{r},t)\right\rangle e^{-i\vec{k}\cdot\vec{r}+i\omega t}. \tag{11.8}$$

For example, in the linear case,

$$\left\langle e\vec{J}(\omega,\vec{k})\right\rangle = \int \frac{d^3\vec{k}}{(2\pi)^3} R_{ij}(\omega,\vec{k},\vec{k}_1) A_j(\omega,\vec{k}_1), \tag{11.9}$$

where

$$R_{ij}(k,\vec{k}_1) \equiv -\frac{e^2}{mc}\langle 0|\rho(\vec{k}-\vec{k}_1)|0\rangle\delta_{ij} + \frac{e^2}{c}\sum_n \frac{\langle 0|j_i(\vec{k})|n\rangle\langle n|j_j(-\vec{k}_1)|0\rangle}{E_n - E_0 - \hbar\omega} \tag{11.10}$$

The first term on the right-hand side is responsible for the Thomson scattering and the second term gives the anomalous scattering.

In the general case, the relation between the nonlinear response functions $R^{(n)}$ and the corresponding nonlinear susceptibilities $\chi^{(n)}$ is

$$\chi^{(n)}(\omega,\vec{k};\omega_1,\vec{k}_1,\omega_2,\vec{k}_2,...,\omega_n,\vec{k}_n) = \tag{11.11}$$
$$i^{1-n}\frac{c^n}{\omega_1\omega_2...\omega_n\omega} R^{(n)}(\omega,\vec{k};\omega_1,\vec{k}_1,\omega_2,\vec{k}_2,...,\omega_n,\vec{k}_n).$$

In the linear case, $n = 1$ and $\omega_1 = \omega$ (elastic scattering):

$$\chi_{ij}(\omega,\vec{k};\vec{k}_1) = \frac{c}{\omega^2} R_{ij}(\omega,\vec{k};\vec{k}_1). \tag{11.12}$$

As an example of a quadratic process, consider the current

$$\langle eJ_i(\omega,\vec{k})\rangle^{(2)} \propto R^{(2)}_{ijk}(\omega,\vec{k})A_j(\omega_1,\vec{k}_1)A_k(\omega_2,\vec{k}_2). \quad (11.13)$$

In the semiclassical description of this quadratic process (harmonic generation, eq. (11.13)), the combination of the monochromatic incident fields A_j and A_k polarizes the medium, which, in turn, emits radiation with frequency $\omega = \omega_1+\omega_2$, as given by $R_{ijk}(\omega,\vec{k})$. In quantum language, two incident photons ω_1 and ω_2 originate an outgoing photon ω, with $\hbar\omega = \hbar\omega_1 + \hbar\omega_2$. Moreover, the intensity of nonlinear processes is a function proportional to $\Delta\vec{K} \equiv \vec{k}_2 - \vec{k}_1 - \vec{k}$: for perfect phase matching (constructive interference), $\Delta\vec{K} = 0$, the intensity the process achieves its maximum. In particular, when crystalline media are present there will be an "edge enhancement" [135] when $\Delta\vec{K} = \vec{k}_2 - \vec{k}_1 - \vec{k} = \vec{H}$ and $\vec{k}$, $\vec{k}_1$ are collinear, where $\vec{H}$ is a reciprocal lattice vector. In this case, the crystalline medium itself supplies the extra $\vec{H}$ for the total momentum conservation. The previous expression represents a nonlinear generalization of the well known Bragg equation of x-ray diffraction. In order to obtain an upper bound value for nonlinear processes, perfect frequency and phase matching have been assumed throughout our calculations. The expressions for the nonlinear response functions up to 3rd order in the vector potential for free atoms can be found in the appendix, where only resonant terms have been kept.

3. Applications

We start by evaluating the linear response function and the corresponding susceptibility for x-rays. Away from resonances, i.e., neglecting the contribution from the anomalous scattering, the dipole contribution to the Thomson scattering (see appendix) is

$$R^{(1)} \sim \frac{Ze^2}{mc}, \quad (11.14)$$

and the corresponding susceptibility is

$$\chi^{(1)}(\omega) = \frac{c}{\omega^2}R^{(1)}. \quad (11.15)$$

Typical values for x-rays are $\lambda \sim 1\text{Å}(\hbar\omega \sim 12.4keV)$. The generalization of the nonlinear susceptibility to crystalline structure will be done in a forthcoming paper [300]. However, a crude estimate can be obtained if we divide the atomic nonlinear susceptibility by the volume of the crystal unit cell.[1] It is expected

[1] In the theory of x-ray dynamical diffraction, the linear susceptibility is defined as $\chi(\vec{r}) = \sum_{\vec{H}} \chi_{\vec{H}} \exp(i\vec{H}\cdot\vec{r})$, where $\vec{H}$ is a reciprocal lattice vector and $\chi_{\vec{H}}$ is proportional to the Fourier transform of the charge density. In this paper, we are estimating $\chi^{(n)}_{\vec{H}}$ by the nonlinear susceptibility $\chi^{(n)}_C$, eq. (11.16).

that the estimates for crystalline structure will be smaller than the real values, since we have replaced *crystalline* matrix elements by *atomic* matrix elements [300]. For crystalline media with lattice parameter $a \sim 5$ Å and atomic number $Z \sim 30$, we obtain

$$\chi_C^{(1)} \sim \frac{1}{V_C} \chi^{(1)} \sim 1.8 \times 10^{-7}, \tag{11.16}$$

where V_C is the volume of the unit cell. This estimate gives a reasonably good approximation for the linear susceptibility in crystalline media. For nonlinear processes, our estimates are as follows.

a) Quadratic processes. Assuming a monochromatic, collimated incident beam, $\vec{k}_1 = \vec{k}_2 = \vec{k}_0$, the resonant quadratic response that describes second harmonic generation is (see appendix):

$$R_{ijk}^{(2)}(\omega = 2\omega_0, \vec{k}) = -\frac{e^3}{2mc^2} \sum_n \frac{\langle 0|j_i(\vec{k})|n\rangle\langle n|\rho^\dagger(2\vec{k}_0)|0\rangle}{E_n - E_0 - 2\hbar\omega_0} \delta_{ij}. \tag{11.17}$$

In order to obtain a quick estimate, we will consider the contribution arising from dipole transitions only. In this case, the matrix elements are:

$$\langle n|\rho^\dagger(2\vec{k}_0)|0\rangle = i2\vec{k}_0 \cdot \langle n|\vec{R}|0\rangle \tag{11.18}$$

and

$$\langle n|j_i(\vec{k})|0\rangle = -i(\frac{E_n - E_0}{\hbar})\langle n|R_i|0\rangle, \quad (E_n - E_0)/\hbar \equiv 2\omega_0. \tag{11.19}$$

The dipole matrix elements scale with the linear dimension of the atomic system for Coulomb interactions, that is, $|\langle 0|\vec{R}|n\rangle| \sim e^2/\hbar\omega_{n0}$. In addition, if only one strong resonance with linewidth Γ is excited, e.g., a 1s electron is promoted to a vacant atomic level near continuum after absorbing two x-ray photons, then

$$R^{(2)}(\omega = 2\omega_0, \vec{k}) \sim \alpha^2 r_e \frac{ec}{\Gamma}, \tag{11.20}$$

where $r_e = 2.8 \times 10^{-13}\ cm$ is classical electron radius and $\alpha \sim 1/137$ is the fine structure constant. Using $\Gamma \sim 5\ eV$, the quadratic susceptibility for one atom is

$$\chi^{(2)}(\omega = 2\omega_0) \sim \frac{c^2}{2\omega_0^3} R^{(2)}(2\omega_0) \tag{11.21}$$

or

$$\chi^{(2)} \sim 1.8 \times 10^{-42}\ cm^5\ stC^{-1}, \tag{11.22}$$

and for a crystal,[2]

$$\chi_C^{(2)} \sim \frac{1}{V_C}\chi^{(2)} \tag{11.23}$$

$$\chi_C^{(2)} \sim 1.5 \times 10^{-20} cm^2 stC^{-1}, \tag{11.24}$$

where stC means the Gaussian unit of statCoulomb.

By inspection, we see that all the expressions of the quadratic responses bear the same functional dependence. Therefore, for parametric down converted x-ray photons, $\hbar\omega_x \rightarrow \hbar\omega_x/2 + \hbar\omega_x/2$, the atomic susceptibility is

$$\chi^{(2)} \sim 1.4 \times 10^{-41}\ cm^5\ stC^{\ -1}, \tag{11.25}$$

and for a crystalline medium

$$\chi_C^{(2)} \sim 1.2 \times 10^{-19} cm^2 stC^{-1}. \tag{11.26}$$

We can expect an increase in the previous susceptibilities if the double resonant denominator in $R^d{}_{jik}$ can be excited with the emission of one x-ray photon plus one optical photon. Then, the additional contribution to parametric down conversion processes is

$$\chi^{(2)}(\omega_{op}) \sim \frac{c^2}{\omega^2{}_x\omega_{op}} R^{(2)}(\omega_{op}). \tag{11.27}$$

Making $\hbar\omega_{mn} \sim 10eV$ (optical frequencies, ω_{op}), we obtain

$$\chi^{(2)} \sim 3.4 \times 10^{-36} cm^5 stC^{-1}, \tag{11.28}$$

and for a crystal

$$\chi_C^{(2)} \sim 2.7 \times 10^{-14}\ cm^2\ stC^{\ -1}, \tag{11.29}$$

which represents a significant increase-up to five orders of magnitude-from the previous single resonant term.

b) Nonlinear correction to the atomic scattering factor. An important question that arises is how important, if at all, are the nonlinear corrections to the atomic scattering factor (which is proportional to $\chi^{(1)}$), the x-ray counterpart of the optical Kerr effect. If they are appreciable, one can no longer express the scattering factor simply as the Fourier transform of the charge density, and the interpretation of scattering data has to be modified accordingly. Fortunately, our estimates show that this nonlinear correction can be safely neglected.

[2] The idea of having expressions and estimates of nonlinear susceptibilities of crystalline media is to develop a theory of x-ray nonlinear diffraction (both, kinematical and dynamical cases) in analogy to the linear case. The expressions of $\chi_C^{(n)}$ will be given in a forthcoming paper [300].

Assuming again a plane wave, monochromatic incident beam on a "fixed" atom (neglect recoil), the nonlinear correction to the elastic scattering comes from $\chi^{(3)}$ (see appendix):

$$\langle eJ_i(\omega_0,\vec{k})\rangle^{(3)} = \frac{e^4}{2m^2c^3}\sum_n \left[\frac{\langle 0|\rho(2\vec{k})|n\rangle\langle n|\rho^\dagger(2\vec{k}_0)|0\rangle}{E_n - E_0 - 2\hbar\omega_0} + \right.$$

$$\left. + \frac{\langle 0|\rho^\dagger(2\vec{k}_0)|n\rangle\langle n|\rho(2\vec{k})|0\rangle}{E_n^* - E_0 + 2\hbar\omega_0}\right] \delta_{jk}\delta_{li}\, A_j(\omega_0,\vec{k}_0)A_k(\omega_0,\vec{k}_0)A_i^*(\omega_0,\vec{k}). \tag{11.30}$$

There are two general cases of interest:
i) Away from resonances and if $2\hbar\omega_0 \gg E_n - E_0$, typical for optical elements (low-Z materials, $Z < 10$) used in x-ray diffraction, then, after expanding the denominators, we get

$$\langle eJ_i(\omega_0,\vec{k})\rangle^{(3)} = -\frac{e^4}{2m^2c^3}\frac{k_0^2}{(\hbar\omega_0)^2}\sum_n (E_n - E_0)\langle 0|\hat{k}\cdot\vec{R}|n\rangle\langle n|\hat{k}_0\cdot\vec{R}|0\rangle$$

$$\delta_{jk}\delta_{li}\, A_j(\omega_0,\vec{k}_0)A_k(\omega_0,\vec{k}_0)A_i^*(\omega_0,\vec{k}), \tag{11.31}$$

where we have used the dipole expansion again. If $\hat{k} = \hat{k}_0$ (forward scattering) and making use of the dipole sum rule [297], the above expression is reduced to

$$\langle eJ_i(\omega_0,\vec{k})\rangle^{(3)} = \frac{-e^4}{2m^2c^3}\frac{1}{(\hbar\omega_0)^2}Z\frac{(\hbar k_0)^2}{2m}\delta_{jk}\delta_{li}\, A_j(\omega_0,\vec{k}_0)A_k(\omega_0,\vec{k}_0)A_i^*(\omega_0,\vec{k}). \tag{11.32}$$

Therefore,

$$R^{(3)}_{jkli}(\omega_0,\vec{k})A_j(\omega_0,\vec{k}_0)A_k(\omega_0,\vec{k}_0) = -\frac{e^4}{2m^2c^3}\frac{1}{(\hbar\omega_0)^2}Z\frac{(\hbar k_0)^2}{2m}\delta_{jk}\delta_{li}A_j(\omega_0,\vec{k}_0)A_k(\omega_0,\vec{k}_0). \tag{11.33}$$

As an important check of the validity of the perturbation expansion, we note that

$$\frac{R^{(3)}(\omega_0,\vec{k})\,.\,|A(\omega_0,\vec{k}_0)|^2}{R^{(1)}} \sim 6.2\times 10^{-13}, \tag{11.34}$$

where $|A(\omega_0)|^2 = (8\pi c/\omega_0^2)P$. In the above expression, we have used $P = 1.5\times 10^{14}\ W/cm^2$ from the (peak) power density of a LCLS pulse (unfocused beam). The third-order susceptibility is

$$\chi^{(3)} \sim 1.6\times 10^{-54}\ cm^7\ stC^{-2}, \tag{11.35}$$

for a crystal

$$\chi_C^{(3)} \sim 1.3 \times 10^{-32}\ cm^4\ stC^{\ -2}. \tag{11.36}$$

Therefore, the nonlinear correction of the atomic scattering factor can be safely neglected for optical components.

ii) If only one resonance is excited, $2\hbar\omega_0 = E_n - E_0$ for a particular n, then

$$R^{(3)}_{jkli}(\omega_0, \vec{k}) A_j(\omega_0, \vec{k}_0) A_k(\omega_0, \vec{k}_0) \;=\; -4i\frac{e^4}{m^2c^3}({k_0}^2/\Gamma)\ \langle 0|\hat{k}\cdot\vec{R}|n\rangle\langle n|\hat{k}_0\cdot\vec{R}|0\rangle\delta_{jk}\delta_{li}\ A_j(\omega_0, \vec{k}_0)A_k(\omega_0, \vec{k}_0), \tag{11.37}$$

or

$$R^{(3)}(\omega_0, \vec{k})|A(\omega_0, \vec{k}_0)|^2 \sim 4\frac{\alpha r_e^2}{\hbar c}\frac{mc^2}{\Gamma}\frac{R^{(1)}}{Z}\ |A(\omega_0, \vec{k}_0)|^2, \tag{11.38}$$

which gives for $Z \sim 30$

$$\frac{R^{(3)}(\omega_0, \vec{k})\ .\ A^2}{R^{(1)}(\omega_0, \vec{k})} \sim 1.7 \times 10^{-12}. \tag{11.39}$$

The resonant susceptibility is

$$\chi^{(3)} \sim 1.3 \times 10^{-53}\ cm^7\ stC^{\ -2}, \tag{11.40}$$

and for a crystal

$$\chi_C^{(3)} \sim 1.1 \times 10^{-31}\ cm^4 stC^{\ -2}. \tag{11.41}$$

The above result shows that, even for high-Z materials where resonances can be excited with incoming x-rays, the nonlinear corrections arising from dipole transitions to the elastic scattering and, therefore, beam self-focusing and defocusing, are negligible.

c) X-ray + optical mixing. An interesting application to nonlinear atomic spectroscopy is the mixing of a x-ray laser pulse with optical lasers (four-wave mixing). More specifically, we are interested in the following processes:[3]

$$\hbar\omega_{x-ray} + \hbar\omega_{optical} \longrightarrow \hbar\omega'_{x-ray} + \hbar\omega'_{optical}. \tag{11.42}$$

In the following estimates, we consider processes where multiple (double and triple) atomic resonances can be achieved by the appropriate incident photon energies, that is, an inner shell electron gets promoted to near continuum

[3] For a third-order process involving three incident beams, creating an outgoing signal (coherent x-ray Raman scattering), see [301].

available levels by the combination of the x-ray and optical incoming (outgoing) photons (see also [302]). Combining a Ti:Sapphire laser [40] with x-ray photons from the LCLS and making the same approximations from the previous section, we obtain from the double-resonant term the susceptibility

$$\chi^{(3)}(\omega_x) = \frac{c^3}{{\omega_x}^2\,\omega_{op}^2}\,R^{(3)}(\omega_x), \tag{11.43}$$

or

$$\chi^{(3)} \sim 1.4 \times 10^{-43}\ cm^7\ stC^{\ -2} \tag{11.44}$$

and for a crystalline medium

$$\chi_C^{(3)} \sim 1.2 \times 10^{-21}\ cm^4\ stC^{\ -2}. \tag{11.45}$$

In addition, if another atomic resonance can be achieved, then the contribution from this triple resonant term is:

$$R^{(3)} \sim \alpha^6 \frac{\hbar^2 c^3}{\Gamma_l \Gamma_m \Gamma_n}, \tag{11.46}$$

which implies

$$\chi^{(3)} \sim 5.3 \times 10^{-41}\ cm^7\ stC^{\ -2} \tag{11.47}$$

for one atom and

$$\chi_C^{(3)} \sim 4.3 \times 10^{-19}\ cm^4\ stC^{\ -2} \tag{11.48}$$

for a crystal, where $\hbar\omega_{nm}$ and $\hbar\omega_{ml}$ are the optical photons. This estimate shows that a significant enhancement (about two orders of magnitude) in the susceptibility of four-wave mixing involving x-rays and optical lasers and is to be expected from multiple atomic resonances.

4. Event rates and cross sections

The previous numerical estimates of nonlinear χ play a central role in the calculation of the event rates for the corresponding nonlinear processes. This calculation will be carried out in this section. In what follows, we will still treat the incident laser fields classically. This is justified by their large occupation number and the relatively weak conversion (depletion) of the incident beams by nonlinear process. However, all fields *created* by nonlinear processes (such as biphotons-signal and idler-in parametric down conversion) have no classical analogue and, therefore, must be treated quantum mechanically. Although we could have used a fully quantized treatment describing matter and radiation fields, the semiclassical approach was preferred because it provides a straightforward calculation of important spectroscopic quantities.

Following the usual theoretical approach in nonlinear quantum optics, the Hamiltonian describing the nonlinear interaction between electromagnetic fields (S: "signal", I: "idler" and P: "pump") and matter is described by [135, 303]

$$H = \frac{1}{3}\int d^3\vec{r}\chi^{(2)}_{ijk}E^P_i \hat{E}^S_j \hat{E}^I_k. \tag{11.49}$$

In general, the above Hamiltonian should take into account nonlocal effects in space and time. The locality of eq. (11.49) is a consequence of the long-wavelength regime (dipole approximation). The use of the Hamiltonian (11.49) to calculate x-ray scattering is not contradictory, though, for we will make use here of $\chi^{(2)}$ in the dipole approximation from the previous section.

The signal and idler quantized fields are given by

$$\vec{E}(\vec{r}) = \sum_{\alpha=1,2}\int \frac{d^3\vec{k}}{(2\pi)^3}\sqrt{\frac{\hbar\omega}{2V}}\left[ia_\alpha(\vec{k})\hat{\epsilon}_\alpha e^{i\vec{k}\cdot\vec{r}} - i{a^\dagger}_\alpha(\vec{k})\hat{\epsilon}^*_\alpha e^{-i\vec{k}\cdot\vec{r}}\right], \tag{11.50}$$

and the classical pump field is

$$\vec{E}(\vec{r}) = \sum_{\alpha=1,2}\int \frac{d^3\vec{k}}{(2\pi)^3}\frac{i\omega}{c}\left[A_\alpha(\vec{k})\hat{\epsilon}_\alpha e^{i\vec{k}\cdot\vec{r}} - {A^*}_\alpha(\vec{k})\hat{\epsilon}^*_\alpha e^{-i\vec{k}\cdot\vec{r}}\right]. \tag{11.51}$$

For parametric down conversion processes, the Hamiltonian is reduced to

$$H = \frac{-i\hbar}{3c}\frac{1}{2V}\int \frac{d^3\vec{k}_P d^3\vec{k}_S d^3\vec{k}_I}{(2\pi)^9}\chi^{(2)}\omega_P\sqrt{\omega_S\omega_I}A_{\alpha P}(\vec{k}_P){a^\dagger}_{\alpha_S}(\vec{k}_S){a^\dagger}_{\alpha_I}(\vec{k}_I)$$
$$+H.C., \tag{11.52}$$

where $\chi^{(2)} \equiv \chi^{(2)}_{ijk}{\epsilon^i}_{\alpha_P}{\epsilon^{*j}}_{\alpha_S}{\epsilon^{*k}}_{\alpha_I}$ and $H.C.$ stands for "Hermitian conjugated." The transition rate from the initial state $|i\rangle = |N_{\vec{k}_S,\alpha_S} = 0; N_{\vec{k}_I,\alpha_I} = 0\rangle$ to $|f\rangle = |N_{\vec{k}'_I,{\alpha'}_I} = 1; N_{\vec{k}'_S,{\alpha'}_S} = 1\rangle$ can be computed from the Fermi Golden Rule

$$\Gamma_{i\to f} = \frac{2\pi}{\hbar}\left|\langle f|H|i\rangle\right|^2\delta(E_i - E_f). \tag{11.53}$$

Without loss of generality, let us assume a monochromatic, collimated and linearly polarized incident field. The total rate of production of signal-idler photons has to be summed over the total number of signal-idler final states,

$$\Gamma = \frac{1}{2}\int V\frac{d^3{\vec{k}'}_S}{(2\pi)^3}V\frac{d^3{\vec{k}'}_I}{(2\pi)^3}\Gamma_{i\to f}. \tag{11.54}$$

Integrating the above expression over the idler and assuming perfect phase matching, the production rate rate of the signal per unit energy per unit solid angle is

$$\langle \frac{d\Gamma_S}{d\Omega_S} \rangle_t = \frac{c}{576\pi^4} \frac{1}{(\hbar c)^9} (\hbar\omega_P)^2 (\hbar\omega_S)^3 (\hbar\omega_P - \hbar\omega_S)^3 |\chi^{(2)}|^2 \langle |A_{\vec{k}_P}|^2 \rangle_t, \tag{11.55}$$

and since $|A_{\vec{k}}|^2 = (2\pi\hbar c^2/\omega_P) N_{\vec{k}}$, where $N_{\vec{k}}$ is the occupation number of the incident laser pulse, the cross section for parametric down conversion is

$$\frac{d\sigma}{d\Omega}^{(2)} = \frac{1}{288\pi^3} \frac{1}{(\hbar c)^7} (\hbar\omega_P)(\hbar\omega_S)^3 (\hbar\omega_P - \hbar\omega_S)^3 \, |\chi^{(2)}|^2. \tag{11.56}$$

If $\hbar\omega (= 12.4 \; keV) \rightarrow \hbar\omega'(= 6.2 \; keV) + \hbar\omega'(= 6.2 \; keV)$ and making use of the previous nonlinear susceptibility, the result is

$$\frac{d\sigma}{d\Omega}^{(2)} \sim 1.3 \times 10^{-26} P^2 \; cm^2 \tag{11.57}$$

where P^2 is the polarization factor. Compared with the Thomson cross section

$$\frac{d\sigma}{d\Omega} = r_e{}^2 \, |f|^2 \, \sin^2\theta \tag{11.58}$$

making $Z \sim 30$, we have

$$\frac{d\sigma}{d\Omega} \sim 5.7 \times 10^{-23} \, \sin^2\theta \; cm^2 \tag{11.59}$$

That is, the parametric down conversion cross section is roughly 4 orders of magnitude smaller than the Thomson cross section. This result is to be compared with those of ch. 5, where, for a gas of free electrons, the down conversion cross section is about nine orders of magnitude smaller than the linear scattering. This is consistent with our results being based upon atomic resonances.

The differential cross section for the analogous process but involving a down converted *optical* photon instead, $\hbar\omega(= 12.4 \; keV) \rightarrow \hbar\omega'(= 12.39 \; keV) + \hbar\omega'(= 10eV)$, is

$$\frac{d\sigma}{d\Omega}^{(2)} \sim 2.6 \times 10^{-23} P^2 cm^2, \tag{11.60}$$

which now is comparable to the Thomson cross section.

5. Conclusions

Motivated by the unique opportunities that XFELs are to offer in the near future, in particular in the novel nonlinear x-ray regime, we have developed a systematic formalism for calculating nonlinear response functions consisting of a semiclassical, perturbative expansion of matter-radiation interactions. Our estimates of the nonlinear susceptibilities and cross sections have shown the magnitude of nonlinear processes-second as harmonic generation, down conversion and x-ray + optical mixing-for XFELs. In particular, we have shown that nonlinear corrections to the atomic scattering factor can be safely neglected. In calculating nonlinear response functions, this paper represents a first step toward a theory of x-rays nonlinear diffraction. Once our estimates have shown that nonlinear effects are not to be neglected, a second step involving a detailed theory of x-rays nonlinear diffraction, either in the kinematical or dynamical regimes, is in place. This will done in a future paper.

6. Acknowledgments

The author is indebted to Dr. B. Adams, Dr. S. Mukamel and Dr. A. Caticha for many valuable discussions. This work was supported by the U.S. Department of Energy, Basic Energy Sciences, Office of Science, under Contract No. W-31-109-Eng-38.

7. Appendix

The state $|\psi(t)\rangle$ of the system at time t is given by the usual perturbation series expansion [304],

$$|\psi(t)\rangle = \sum_n C_n(t) e^{-iE_n t/\hbar}|n\rangle, \tag{11.61}$$

where the coefficients are

$$C_n(t) = C_n{}^{(0)}(t) + C_n{}^{(1)}(t) + C_n{}^{(2)}(t) + ... \tag{11.62}$$

with $C_n^{(0)}(t) = \delta_{n0}$ and

$$C_n{}^{(1)}(t) = -\frac{i}{\hbar}\int_{t_0}^{t} dt' \langle n|H_{int}(t')|0\rangle e^{i\omega_{n0}t'} \tag{11.63}$$

and

$$C_n^{(2)}(t) = (-\frac{i}{\hbar})^2 \sum_m \int_{t_0}^{t} dt' \int_{t_0}^{t'} dt'' e^{i\omega_{nm}t'} \langle n|H_{int}(t')|m\rangle\, e^{i\omega_{m0}t''} \langle m|H_{int}(t'')|0\rangle \tag{11.64}$$

and so forth. Using the minimal coupling Hamiltonian in the above expressions and keeping only the single- and double-resonant terms in the expansion, the induced current up to 3rd order in the vector potential is:

$$\langle eJ_i(\omega,\vec{k})\rangle = \langle eJ_i(\omega,\vec{k})\rangle^{(1)} + \langle eJ_i(\omega,\vec{k})\rangle^{(2)} + \langle eJ_i(\omega,\vec{k})\rangle^{(3)}, \tag{11.65}$$

where

$$\langle eJ_i(\omega,\vec{k})\rangle^{(1)} = \int \frac{d^3\vec{k}_1}{(2\pi)^3} R_{ij}(\omega,\vec{k},\vec{k}_1) A_j(\omega,\vec{k}_1) \tag{11.66}$$

$$\begin{aligned}\langle eJ_i(\omega,\vec{k})\rangle^{(2)} &= \int \frac{d^4k_1 d^3\vec{k}_2}{(2\pi)^7}\Big[\\ & R^a_{ijk}(\omega,\vec{k},\omega_1,\vec{k}_1,\vec{k}_2)A_j(\omega_1,\vec{k}_1)A_k(\omega-\omega_1,\vec{k}_2)+ \\ &+R^b_{jik}(\omega,\vec{k},\omega_1,\vec{k}_1,\vec{k}_2)A_j(\omega-\omega_1,\vec{k}_2)A_k(\omega_1,\vec{k}_1)+ \\ &+R^c_{jik}(\omega,\vec{k},\omega_1,\vec{k}_1,\vec{k}_2)A^*_j(\omega_1-\omega,\vec{k}_2)A_k(\omega_1,\vec{k}_1)+ \\ &+R^d_{jik}(\omega,\vec{k},\omega_1,\vec{k}_1,\vec{k}_2)A^*_j(\omega_1,\vec{k}_1)A_k(\omega+\omega_1,\vec{k}_2)+ \\ &+R^e_{ijk}(\omega,\vec{k},\omega_1,\vec{k}_1,\vec{k}_2)A_j(\omega-\omega_1,\vec{k}_2)A_k(\omega_1,\vec{k}_1)\Big]\end{aligned} \quad (11.67)$$

and

$$\begin{aligned}\langle eJ_i(\omega,\vec{k})\rangle^{(3)} &= \int \frac{d^4k_1 d^4k_2 d^3\vec{k}_3}{(2\pi)^{11}}\Big[\\ & R^a_{jkli}(k,k_1,k_2,\vec{k}_3)A_j(\omega_2,\vec{k}_2)A_k(\omega-\omega_1-\omega_2,\vec{k}_3)A_l(\omega_1,\vec{k}_1) \\ &+R^b_{jkli}(k,k_1,k_2,\vec{k}_3)A^*_j(\omega_2,\vec{k}_2)A^*_k(-\omega+\omega_1-\omega_2,\vec{k}_3)A_l(\omega_1,\vec{k}_1) \\ &+R^c_{ijkl}(k,k_1,k_2,\vec{k}_3)A_j(\omega-\omega_1-\omega_2,\vec{k}_3)A_k(\omega_1,\vec{k}_1)A_l(\omega_2,\vec{k}_2) \\ &+R^d_{ijkl}(k,k_1,k_2,\vec{k}_3)A_j(\omega_1,\vec{k}_1)A_k(\omega_2,\vec{k}_2)A_l(\omega-\omega_1-\omega_2,\vec{k}_3) \\ &+R^e_{ijkl}(k,k_1,k_2,\vec{k}_3)A_j(\omega_1,\vec{k}_1)A_k(\omega_2,\vec{k}_2)A_l(\omega-\omega_1-\omega_2,\vec{k}_3) \\ &+R^f_{jikl}(k,k_1,k_2,\vec{k}_3)A^*_j(\omega_1,\vec{k}_1)A^*_k(\omega_2,\vec{k}_2)A_l(\omega+\omega_1-\omega_2,\vec{k}_3) \\ &+R^g_{jikl}(k,k_1,k_2,\vec{k}_3)A^*_j(\omega_1,\vec{k}_1)A_k(\omega_2,\vec{k}_2)A_l(\omega+\omega_1-\omega_2,\vec{k}_3) \\ &+R^h_{ijkl}(k,k_1,k_2,\vec{k}_3)A_j(\omega_1,\vec{k}_1)A^*_k(\omega_2,\vec{k}_2)A^*_l(-\omega+\omega_1-\omega_2,\vec{k}_3) \\ &+R^i_{jkil}(k,k_1,k_2,\vec{k}_3)A^*_j(\omega_2,\vec{k}_2)A^*_k(-\omega+\omega_1-\omega_2,\vec{k}_3)A_l(\omega_1,\vec{k}_1) \\ &+R^j_{jkil}(k,k_1,k_2,\vec{k}_3)A^*_j(\omega_2,\vec{k}_2)A_k(\omega-\omega_1+\omega_2,\vec{k}_3)A_l(\omega_1,\vec{k}_1) \\ &+R^k_{jkil}(k,k_1,k_2,\vec{k}_3)A_j(\omega_2,\vec{k}_2)A_k(\omega-\omega_1-\omega_2,\vec{k}_3)A_l(\omega_1,\vec{k}_1) \\ &+R^l_{kjil}(k,k_1,k_2,\vec{k}_3)A^*_j(\omega_2,\vec{k}_2)A^*_k(-\omega-\omega_1-\omega_2,\vec{k}_3)A^*_l(\omega_1,\vec{k}_1)\Big].\end{aligned} \quad (11.68)$$

By inspection, therefore, the response functions are:
Linear response:

$$R_{ij}(k, \vec{k}_1) \equiv -\frac{e^2}{mc}\langle 0|\rho(\vec{k}-\vec{k}_1)|0\rangle\delta_{ij} + \frac{e^2}{c}\sum_n \frac{\langle 0|j_i(\vec{k})|n\rangle\langle n|j_j(-\vec{k}_1)|0\rangle}{E_n - E_0 - \hbar\omega}, \tag{11.69}$$

where the first and second terms on the LHS above correspond to the Thomson and anomalous scattering, respectively.
Quadratic response:

$$R^a_{ijk}(k, k_1, \vec{k}_2) \equiv -\frac{e^3}{2mc^2}\sum_n \frac{\langle 0|j_i(\vec{k})|n\rangle\langle n|\rho(-\vec{k}_1-\vec{k}_2)|0\rangle}{E_n - E_0 - \hbar\omega}\delta_{jk}$$

$$R^b_{jik}(k, k_1, \vec{k}_2) \equiv -\frac{e^3}{mc^2}\sum_n \frac{\langle 0|\rho(\vec{k}-\vec{k}_1)|n\rangle\langle n|j_j(-\vec{k}_2)|0\rangle}{E_n - E_0 - \hbar(\omega-\omega_1)}\delta_{ik}$$

$$R^c_{jik}(k, k_1, \vec{k}_2) \equiv -\frac{e^3}{mc^2}\sum_n \frac{\langle 0|j_j^\dagger(-\vec{k}_2)|n\rangle\langle n|\rho(\vec{k}-\vec{k}_1)|0\rangle}{E_n^* - E_0 - \hbar(\omega_1-\omega)}\delta_{ik}$$

$$R^d_{jik}(k, k_1, \vec{k}_2) \equiv \frac{e^3}{c^2}\sum_{m,n} \frac{\langle 0|j_j^\dagger(-\vec{k}_1)|m\rangle\langle m|j_i(\vec{k})|n\rangle\langle n|j_k(-\vec{k}_2)|0\rangle}{(E_m^* - E_0 - \hbar\omega_1)(E_n - E_0 - \hbar(\omega+\omega_1))}$$

$$R^e_{ijk}(k, k_1, \vec{k}_2) \equiv \frac{e^3}{c^2}\sum_{m,n} \frac{\langle 0|j_i(\vec{k})|m\rangle\langle m|j_j(-\vec{k}_2)|n\rangle\langle n|j_k(-\vec{k}_1)|0\rangle}{(E_n - E_0 - \hbar\omega_1)(E_m - E_0 - \hbar\omega)}$$

Cubic response:

$$R^a_{jkli}(k, k_1, k_2, \vec{k}_3) \equiv \frac{e^4}{2m^2c^3}\sum_n \frac{\langle 0|\rho(\vec{k}-\vec{k}_1)|n\rangle\langle n|\rho(-\vec{k}_2-\vec{k}_3)|0\rangle}{E_n - E_0 - \hbar(\omega-\omega_1)}\delta_{jk}\delta_{li}$$

$$R^b_{jkli}(k, k_1, k_2, \vec{k}_3) \equiv \frac{e^4}{2m^2c^3}\sum_n \frac{\langle 0|\rho^\dagger(-\vec{k}_2-\vec{k}_3)|n\rangle\langle n|\rho(\vec{k}-\vec{k}_1)|0\rangle}{E_n^* - E_0 - \hbar(-\omega+\omega_1)}\delta_{jk}\delta_{li}$$

$$R^c_{ijkl}(k, k_1, k_2, \vec{k}_3) \equiv$$

$$\equiv -\frac{e^4}{2mc^3}\sum_{m,n} \frac{\langle 0|j_i(\vec{k})|n\rangle\langle n|j_j(-\vec{k}_3)|m\rangle\langle m|\rho(-\vec{k}_1-\vec{k}_2)|0\rangle}{(E_m - E_0 - \hbar\omega_1 - \hbar\omega_2)(E_n - E_0 - \hbar\omega)}\delta_{kl}$$

$$R^d_{ijkl}(k, k_1, k_2, \vec{k}_3) \equiv$$

$$\equiv -\frac{e^4}{2mc^3}\sum_{m,n} \frac{\langle 0|j_i(\vec{k})|n\rangle\langle n|\rho(-\vec{k}_2-\vec{k}_3)|m\rangle\langle m|j_j(-\vec{k}_1)|0\rangle}{(E_m - E_0 - \hbar\omega_1)(E_n - E_0 - \hbar\omega)}\delta_{kl}$$

$$R^{e}_{ijkl}(k, k_1, k_2, \vec{k}_3) \equiv$$
$$\equiv \frac{e^4}{c^3} \sum_{m,n,l} \frac{\langle 0|j_i(\vec{k})|n\rangle\langle n|j_j(-\vec{k}_1)|m\rangle\langle m|j_k(-\vec{k}_2)|l\rangle\langle l|j_l(-\vec{k}_3)|0\rangle}{(E_l - E_0 - \hbar(\omega-\omega_1-\omega_2))(E_m - E_0 - \hbar(\omega-\omega_1))(E_n - E_0 - \hbar\omega)}$$

$$R^{f}_{jikl}(k, k_1, k_2, \vec{k}_3) \equiv$$
$$\equiv -\frac{e^4}{2mc^3} \sum_{m,n} \frac{\langle 0|j_j^{\dagger}(-\vec{k}_1)|m\rangle\langle m|j_i(\vec{k})|n\rangle\langle n|\rho(-\vec{k}_2 - \vec{k}_3)|0\rangle}{(E_m^* - E_0 - \hbar\omega_1)(E_n - E_0 + \hbar(\omega + \omega_1))} \delta_{kl}$$

$$R^{g}_{jikl}(k, k_1, k_2, \vec{k}_3) \equiv$$
$$\equiv \frac{e^4}{c^3} \sum_{m,n,l} \frac{\langle 0|j_i^{\dagger}(-\vec{k}_1)|m\rangle\langle m|j_i(\vec{k})|n\rangle\langle n|j_k(-\vec{k}_2)|l\rangle\langle l|j_l(-\vec{k}_3)|0\rangle}{(E_m^* - E_0 - \hbar\omega_1)(E_l - E_0 - \hbar\omega_2)(E_n - E_0 + \hbar(\omega + \omega_1))}$$

$$R^{h}_{ijkl}(k, k_1, k_2, \vec{k}_3) \equiv$$
$$\equiv -\frac{e^4}{2mc^3} \sum_{m,n} \frac{\langle 0|\rho^{\dagger}(-\vec{k}_2 - \vec{k}_3)|m\rangle\langle m|j_i(\vec{k})|n\rangle\langle n|j_j(-\vec{k}_1)|0\rangle}{(E_m^* - E_0 + \hbar(\omega - \omega_1))(E_n - E_0 - \hbar\omega_1)} \delta_{kl}$$

$$R^{i}_{jkil}(k, k_1, k_2, \vec{k}_3) \equiv$$
$$\equiv \frac{e^4}{c^3} \sum_{m,n,l} \frac{\langle 0|j_j^{\dagger}(-\vec{k}_2)|l\rangle\langle l|j_k^{\dagger}(-\vec{k}_3)|m\rangle\langle m|j_i(\vec{k})|n\rangle\langle n|j_l(-\vec{k}_1)|0\rangle}{(E_l^* - E_0 - \hbar\omega_2)(E_m^* - E_0 + \hbar(\omega - \omega_1))(E_n - E_0 - \hbar\omega_1)}$$

$$R^{j}_{jkil}(k, k_1, k_2, \vec{k}_3) \equiv$$
$$\equiv -\frac{e^4}{mc^3} \sum_{m,n} \frac{\langle 0|j_j^{\dagger}(-\vec{k}_2)|m\rangle\langle m|\rho(\vec{k} - \vec{k}_1)|n\rangle\langle n|j_k(-\vec{k}_3)|0\rangle}{(E_m^* - E_0 - \hbar\omega_2)(E_n - E_0 - \hbar(\omega - \omega_1 + \omega_2))} \delta_{il}$$

$$R^{k}_{jkil}(k, k_1, k_2, \vec{k}_3) \equiv$$
$$\equiv -\frac{e^4}{mc^3} \sum_{m,n} \frac{\langle 0|\rho(\vec{k} - \vec{k}_1)|m\rangle\langle m|j_j(-\vec{k}_2)|n\rangle\langle n|j_k(-\vec{k}_3)|0\rangle}{(E_n - E_0 - \hbar(\omega - \omega_1 - \omega_2))(E_m - E_0 - \hbar(\omega - \omega_1))} \delta_{il}$$

$$R^{l}_{kjil}(k, k_1, k_2, \vec{k}_3) \equiv$$
$$\equiv -\frac{e^4}{mc^3} \sum_{m,n} \frac{\langle 0|j_k^{\dagger}(-\vec{k}_3)|m\rangle\langle m|j_j^{\dagger}(-\vec{k}_2)|n\rangle\langle n|\rho(\vec{k} + \vec{k}_1)|0\rangle}{(E_n^* - E_0 + \hbar(\omega + \omega_1 + \omega_2))(E_m^* - E_0 + \hbar(\omega + \omega_1))} \delta_{il}$$

References

[1] J.D. Jackson. *Classical Electrodynamics*. John Wiley and Sons, New York, 1975.

[2] R.F. Service. Battle to become the next-generation x-ray source. *Science*, 298:1356–1358, 2002.

[3] D. von der Linde, K. Sokolowski-Tinten, C. Blome, C. Dietrich, P. Zhou, A. Tarasevitch, A. Cavalleri, C.W. Siders, C.P.J. Barty, J. Squier, K.R. Wilson, I. Uschmann, and E. Förster. Generation and application of ultrashort x-ray pulses. *Laser and Particle Beams*, 19:15–22, 2001.

[4] A.A. Hauer and G.A. Kyrala. Laser-plasma x-ray emission: its creation, diagnosis, and applications in transient diffration. In J.R. Helliwell and P.M. Rentzepis, editors, *Time-Resolved Diffraction*, chapter 3. Oxford University Press, Great Clarendon Street, Oxford OX2 6DP, UK, 1997.

[5] P. Antoine, A. L'Huillier, and M. Lewenstein. Attosecond pulse trains using high-order harmonics. *Phys. Rev. Lett.*, 77:1234–1237, 1996.

[6] F.R. Elder, A.M. Gurewitsch, R.V. Langmuir, and H.C. Pollock. Radiation from electrons in a synchrotron. *Phys. Rev.*, 71:829–829, 1947.

[7] F. R. Elder, R. V. Langmuir, and H. C. Pollock. Radiation from electrons accelerated in a synchrotron. *Phys. Rev.*, 74:52–56, 1948.

[8] J. Schwinger. On the classical radiation of accelerated electrons. *Phys. Rev.*, 75:1912–1925, 1949.

[9] ESRF, BP 220, F-38043 Grenoble Cedex, France. *Performances*, 2002. http://www.esrf.fr/machine/performances/parameters.html.

[10] Advanced Photon Source (APS), Argonne National Laboratory, 9700 S. Cass Ave., Argonne, IL 60439, USA. *APS Operations Division, User Operations*, 2002. http://www.aps.anl.gov/xfd/operations/welcome.html.

[11] A.A. Zholents and M.S. Zolotorev. Femtosecond x-ray pulses of synchrotron radiation. *Phys. Rev. Lett.*, 76:912–915, 1996.

[12] R.W. Schoenlein, S. Chattopadhyay, H.H.W. Chong, T.E. Glover, P.A. Heimann, W.P. Leemans, C.V. Shank, A. Zholents, and M. Zolotorev. Generation of femtosecond x-ray pulses via laser-electron beam interaction. *Appl. Phys. B*, 71:1–10, 2000.

[13] R.W. Schoenlein, A.H. Chin, H.H.W. Chong, R.W. Falcone, T.E. Glover, P.A. Heimann, S.L. Johnson, A.M. Lindenberg, C.V. Shank, A.A. Zholents, and M.S. Zolotorev. Ultrafast x-ray science at the advanced light source. *Synchrotron Radiation News*, 14:20–27, 2001.

[14] R.B. Palmer. Interaction of relativistic particles and free electromagnetic waves in the presence of a static helical magnet. *J. Appl. Phys.*, 43:3014–3023, 1972.

[15] D. Bilderback, I. Bazarov, K. Finkelstein, S. Gruner, G. Krafft, L. Merminga, H. Padamsee, Q. Shen, C. Sinclair, M. Tigner, and R. Talman. New energy recovery linac source of synchrotron x-rays. *Synchrotron Radiation News*, 14:12–21, 2001.

[16] SLAC, 2575 Sand Hill Road, Menlo Park, CA 94025, USA. *LCLS, Linear Coherent Light Source*, 2002. http://www-ssrl.slac.stanford.edu/lcls.

[17] Deutsches Elektronen-Synchrotron DESY, Notkestr. 85, 22607 Hamburg, Germany. *TESLA*, 2002. http://tesla.desy.de.

[18] DESY, Notkestraße 85, D-22607 Hamburg, Germany. *The TESLA Test Facility (TTF) at DESY*, 2002. http://tesla.desy.de/TTF_intro.html.

[19] BESSY, Albert-Einstein-Str. 15, D-12489 Berlin, Germany. *FEL*, 2002. http://www.bessy.de/lab_profile/01.FEL/index.php.

[20] Massachusetts Institute of Technology, 21 Manning Ave., Middleton, MA 01949, USA. *Bates Linear Accelerator Center*, 2002. http://mitbates.mit.edu/index2.stm.

[21] T. Zwart, M. Farkhondeh, E. Ihloff, R. Milner, S. Sobczynski, C. Tschalar, J. van der Laan, F. Wang, A. Zolfaghari, and D.E. Moncton. The MIT Bates x-ray laser project. In *Proceedings of the FEL 2002 conference*, 2002.

[22] Spring8, Japan Synchrotron Radiation Research Institute (JASRI) 1-1-1 Kouto Mikazuki-cho Sayo-gun Hyogo 679-5198 Japan. *SCSS Soft-X-Ray FEL Project*, 2002. http://www-xfel.spring8.or.jp/SCSS.htm.

[23] G.E. Moore. Cramming more components onto integrated circuits. *Electronics*, 38, 1965.

[24] A. Bambini, A. Renieri, and S. Stenholm. Classical theory of the free-electron laser in a moving frame. *Phys. Rev. A*, 19:2013–2025, 1979.

[25] P. Meystre and M. Sargent III. *Elements of quantum optics*. Springer, Berlin, Heidelberg, New York, 1991.

[26] J.M.J. Madey. Relationship between mean radiated energy, mean square radiated energy and spontaneous power spectrum in a power series expansion of the equations of motion in a free-electron laser. *Il Nuovo Cimento B*, 50:64–88, 1978.

[27] D.E. Nikonov, Y.V. Rostovtsev, and G. Sussmann. Madey's and Liouville's theorems related to free-electron lasers without inversion. *Phys. Rev. E*, 57:3444–3453, 1998.

[28] R.H. Dicke. Coherence in spontaneous radiation processes. *Phys. Rev.*, 93:99–110, 1954.

[29] G. Materlik and T. Tschentscher, editors. *TESLA technical design report, part V*. DESY, 2001. http://tesla.desy.de/new_pages/TDR_CD/PartV/fel.html.

[30] V. A. Bordovitsyn. Maximal acceleration, zitterbewegung, and spin. *Nucl. Instrum. Methods A*, 47:7–10, 2001.

[31] I.V. Smetanin. Quantum limit in x-ray fel theory. In *Free Electron Lasers 1996, Proceedings of the Eighteenth International Free Electron Laser Conference, Rome, Italy, August 26-31, 1996*, Amsterdam, Lausanne, New York, Oxford, Shannon, Singapore, Tokyo, 1997. Elsevier Science B.V.

[32] E.M. Belenov, S.V. Grigor'ev, A.V. Nazarkin, and I.V. Smetanin. X-ray free-electron laser in the quantum regime. *J. Exp. Theo. Phys. (JETP)*, 78:431–435, 1994.

[33] M. Orszag, W. Becker, and M.O. Scully. Theory of a correlated-emission free-electron laser. *Phys. Rev. A*, 36:1310–1315, 1987.

[34] D.E. Nikonov, M.O. Scully, and G. Kurizki. Free-electron laser without inversion via transverse momentum discrimination. *Phys. Rev. E*, 54:6780–6787, 1996.

[35] G. Kurizki, M.O. Scully, and C. Keitel. Free-electron lasing without inversion by interference of momentum states. *Phys. Rev. Lett.*, 70:1433–1436, 1993.

[36] B. Sherman, G. Kurizki, D.E. Nikonov, and M.O. Scully. Universal classical mechanism of free-electron lasing without inversion. *Phys. Rev. Lett.*, 75:4602–4605, 1995.

[37] A.I. Artemiev, M.V. Fedorov, Y.V. Rostovtsev, G. Kurizki, and M.O. Scully. Free-electron laser without inversion: Gain optimization and implementation scheme. *Phys. Rev. Lett.*, 85:4510–4513, 2000.

[38] Y.I. Salamin and F.H.M. Faisal. Harmonic generation by superintense light scattering from relativistic electrons. *Phys. Rev. A*, 54:4383–4395, 1996.

[39] B.C. Stuart, M.D. Perry, J. Miller, G. Tietbohl, S. Herman, J.A. Britten, C. Brown, D. Pennington, V. Yanovsky, and K. Wharton. 125-TW Ti:sapphire/Nd:glass laser system. *Opt. Lett.*, 22:242–244, 1997.

[40] K. Yamakawa, M. Aoyama, S. Matsuoka, T. Kase, Y. Akahane, and H. Takuma. 100-TW sub-20-fs Ti:sapphire laser system operating at a 10-Hz repetition rate. *Opt. Lett.*, 23:1468–1470, 1998.

[41] M.D. Perry, D. Pennington, B.C. Stuart, G. Tietbohl, J.A. Britten, C. Brown, S. Herman, B. Golick, M. Kartz, J. Miller, H.T. Powell, M. Vergino, and V. Yanovsky. Petawatt laser pulses. *Opt. Lett.*, 24:160–162, 1999.

[42] M. Hillery and L.D. Mlodinov. Quantization of electrodynamics in nonlinear dielectric media. *Phys. Rev. A*, 30:1860–1865, 1984.

[43] N. Bloembergen. *Nonlinear Optics*. World Scientific, Singapore, 1965.

[44] J. Frenkel. Die Bewegung eines freien Elektrons im Felde ebener elektromagnetischer Wellen. *Z. Phys.*, 32:27–31, 1925.

[45] T.W.B. Kibble. Frequency shift in high-intensity Compton scattering. *Phys. Rev.*, 138:740–753, 1963.

[46] J.H. Eberly and A. Sleeper. Trajectory and mass shift of a classical electron in a radiation pulse. *Phys. Rev.*, 176:1570–1576, 1968.

[47] D.M. Wolkow. Über eine Klasse von Lösungen der Diracschen Gleichung. *Z. Phys.*, 94:250–260, 1935.

[48] E. Esarey, P. Sprangle, and J. Krall. Laser acceleration of electrons in vacuum. *Phys. Rev. E*, 52:5443–5453, 1995.

[49] Y.I. Salamin and C.H. Keitel. Subcycle high electron acceleration by crossed laser beams. *Appl. Phys. Lett.*, 77:1082–1084, 2000.

[50] Y.I. Salamin, G.R. Mocken, and C.H. Keitel. Relativistic electron dynamics in intense crossed laser beams: Acceleration and Compton harmonics. *Phys. Rev. E*, 67:016501, 2003.

[51] P.M. Woodward. A method for Calculating the field over a plane aperture required to produce a given polar diagram. *J. IEEE*, 93:1554–1558, 1946.

[52] J.D. Lawson. *IEEE Trans. Nucl. Sci.*, 26:4217–, 1979.

[53] E.S. Sarachik and G.T. Schappert. Classical theory of the scattering of intense laser radiation by free electrons. *Phys. Rev. D*, 1:2738–2752, 1970.

[54] W. Scheid and H. Hora. On electron acceleration by plane transverse electromagnetic pulses in vacuum. *Laser and Particle Beams*, 7:315–322, 1989.

[55] V.G. Bagrov, D.M. Gitman, and A.V. Jushin. Solutions for the motion of an electron in electromagnetic fields. *Phys. Rev. D*, 12:3200–3202, 1975.

[56] Y.I. Salamin and F.H.M. Faisal. Relativistic free-electron dynamics and light-emission spectra in the simultaneous presence of a superintense laser field and a strong uniform magnetic field. *Phys. Rev. A*, 58:3221–3234, 1998.

[57] Y.I. Salamin and F.H.M. Faisal. Ultrahigh electron acceleration and Compton emission spectra in a superintense laser pulse and a uniform axial magnetic field. *Phys. Rev. A*, 61:043801, 2000.

[58] J.G. Woodworth, M.N. Kreisler, and A.K. Kerman. A free-wave accelerator. *AIP Conf. Proc.*, 356:378–388, 1996.

[59] T. Tajima and J.M. Dawson. Laser electron accelerator. *Phys. Rev. Lett.*, 43:267–270, 1979.

[60] T. Tajima. High energy laser plasma accelerators. *Laser and Particle Beams*, 3:351–413, 1985.

[61] F. Dorchies, F. Amiranoff, S. Baton, D. Bernard, B. Cros, D. Descamps, F. Jacquet, V. Malka, J.R. Marques, G. Matthieussant, P. Mine, A. Modena, P. Mora, J. Morillo, Z. Najmudin, and A. Solodov. Electron acceleration in laser wakefield experiment at ecole polytechnique. *Laser and Particle Beams*, 17:299–305, 1999.

[62] K. Nakajima. Particle acceleration by ultraintense laser interactions with beams and plasmas. *Laser and Particle Beams*, 18:519–528, 2000.

[63] V.A. Balakirev, V.I. Karas, I.V. Karas, and V.D. Levchenko. Plasma wake-field excitation by relativistic electron bunches and charged particle acceleration in the presence of external magnetic field. *Laser and Particle Beams*, 19:597–604, 2001.

[64] M. Lax, W.H. Louisell, and W.B. McKnight. From Maxwell to paraxial wave optics. *Phys. Rev. A*, 11:1365–1370, 1975.

[65] M.O. Scully and M.S. Zubairy. Simple laser accelerator: Optics and particle dynamics. *Phys. Rev. A*, 44:2656–2663, 1991.

[66] E.D. Courant, C. Pellegrini, and W. Zakowicz. High-energy inverse free-electron-laser accelerator. *Phys. Rev. A*, 32:2813–2823, 1985.

[67] A. van Steenbergen, J. Gallardo, J. Sandweiss, and J.-M. Fang. Observation of energy gain at the BNL inverse free-electron-laser accelerator. *Phys. Rev. Lett.*, 77:2690–2693, 1996.

[68] Y.C. Huang, D. Zheng, W.M. Tulloch, and R.L. Byer. Proposed structure for a crossed-laser beam, GeV per meter gradient, vacuum electron linear accelerator. *Appl. Phys. Lett.*, 68:753–755, 1996.

[69] B. Hafizi, E. Esarey, and P. Sprangle. Laser-driven acceleration with Bessel beams. *Phys. Rev. E*, 55:3539–3545, 1997.

[70] T.W.B. Kibble. Mutual refraction of electrons and photons. *Phys. Rev.*, 150:1060–1069, 1966.

[71] T.W.B. Kibble. Refraction of electron beams by intense electromagnetic waves. *Phys. Rev. Lett.*, 16:1054–1056, 1966.

[72] G.V. Stupakov and M.S. Zolotorev. Ponderomotive laser acceleration and focusing in vacuum for generation of attosecond electron bunches. *Phys. Rev. Lett.*, 86:5274–5277, 2001.

[73] B.W. Batterman and H. Cole. Dynamical diffraction of x rays by perfect crystals. *Rev. Mod. Phys.*, 36:681–716, 1964.

[74] A. Authier. *Dynamical Theory of X-Ray Diffraction*. Oxford University Press, Great Clarendon Press, Oxford OX2 6DP, Great Britain, 2001.

[75] T.J. Davis. X-ray propagation and diffraction in distorted, anisotropic, dielectric and magnetic crystals. *Acta Cryst. A*, 51:354–360, 1995.

[76] P. Penning and D. Polder. Anomalous transmission of x-rays in elastically deformed crystals. *Philips Res. Rept.*, 16:419–440, 1961.

[77] N. Kato. Pendellösung fringes in distorted crystals i. fermat's principle for bloch waves. *J. Phys. Soc. Japan,*, 18:1785–1791, 1963.

[78] N. Kato. Pendellösung fringes in distorted crystals ii. application to two-beam cases. *J. Phys. Soc. Japan,*, 19:67–77, 1964.

[79] N. Kato. Pendellösung fringes in distorted crystals iii. application to homogeneously bent crystals. *J. Phys. Soc. Japan,*, 19:971–985, 1964.

[80] U. Bonse. Theorie der Ausbreitung von Röntgen-Wellenfeldern im schwach deformierten Kristallgitter. *Z. Phys.*, 177:385–423, 1964.

[81] U. Bonse and W. Graeff. Propagation of x-ray wavefield beams in a slightly deformed crystal. *Z. Naturforschung*, 28a:558–564, 1973.

[82] S. Takagi. Dynamical theory of diffraction applicable to crystals with any kind of small distortion. *Acta Cryst.*, 15:1311–1312, 1962.

[83] S. Takagi. A dynamical theory of diffraction for a distorted crystal. *J. Phys. Soc. Jpn.*, 26:1239–1253, 1969.

[84] D. Taupin. Théorie dynamique de la diffraction des rayons x par les cristeaux déformés. *Bull. Soc. Franc. Miner. Crist.*, 87:469–511, 1964.

[85] B. Adams. Extended Takagi-Taupin theory for strongly scattering and strongly deformed crystals. *Surface investigation*, 12:415–420, 1997.

[86] B.W. Adams. Dynamical diffraction of x-rays under conditions of a rapidly changing structure factor: Theory and possible applications for femtosecond x-rays. *SPIE Proc.*, 4500:89–100, 2002.

[87] I. Freund. Nonlinear diffraction. *Phys. Rev. Lett.*, 21:1404–1406, 1968.

[88] T. Fukamachi, R. Negishi, S. Zhou, M. Yoshizawa, and T. Kawamura. Extinction effect and Borrmann effect of resonant dynamical scattering in the Bragg case. *Acta Cryst. A*, 58:552–558, 2002.

[89] B. Adams. Bau eines Röntgeninterferometers. diploma thesis, University of Munich, 1989.

[90] G. Borrmann and G. Hildebrandt. Röntgen-Wellenfelder in grossen Kalkspatkristallen und das Wirken einer Deformation. *Z. Naturforschung A*, 11:585–587, 1956.

[91] J.P. Sutter, E.E. Alp, M.Y. Hu, P.L. Lee, H. Sinn, W. Sturhahn, T.S. Toellner, G. Bortel, and R. Colella. Multiple-beam x-ray diffraction near exact backscattering in silicon. *Phys. Rev. B*, 63:094111, 2001.

[92] H. Römer. *Theoretische Optik.* VCH, Weinheim, New York, Basel, Cambridge, 1994.

[93] M. v. Laue. Die Energieströmung bei Röntgenstrahl-Interferenzen in Kristallen. *Acta Cryst.*, 5:619–625, 1952.

[94] M. v. Laue. Der Teilchenstrom bei Raumgitterinterferenzen von Materiewellen. *Acta Cryst.*, 6:217–217, 1953.

[95] N. Kato. The flow of x-rays and material waves in ideally perfect single crystals. *Acta Cryst.*, 11:885–887, 1958.

[96] W. Kossel, V. Loeck, and H. Voges. . *Z. Phys.*, 94:139–, 1935.

[97] T. Gog, D. Bahr, and G. Materlik. Kossel diffraction in perfect crystals: X-ray standing waves in reverse. *Phys. Rev. B*, 51:6761–6764, 1995.

[98] M.J. Bedzyk and G. Materlik. Two-beam dynamical diffraction solution of the phase problem: A determination with x-ray standing-wave fields. *Phys. Rev. B*, 32:6456–6463, 1985.

[99] K. Hirano, K. Kanzaki, M. Mikami, M. Miura, K. Tamasaku, T. Ishikawa, and S. Kikuta. Tunable-wavelength production of circularly polarized x-rays with a perfect-crystal quarter-wave plate. *J. Appl. Cryst.*, 25:531–535, 1992.

[100] Yu. A. Kravtsov and Yu. I. Orlov. *Geometrical Optics of Inhomogeneous Media.* Springer, Berlin Heidelberg New York, 1990.

[101] F. Balibar. Généralisation de la notion de champs d'ondes en théorie dynamique. *Acta Cryst. A*, 25:650–658, 1969.

[102] A. Authier and F. Balibar. Création de nouveaux champs d'ondes généralisés dus à la présence d'un objet diffractant à l'intérieur d'un cristal parfait. ii. Cas d'un défaut isolé. *Acta Cryst. A*, 26:647–654, 1970.

[103] D. Hilbert R. Courant. *Methoden der Mathematischen Physik II.* Springer, Berlin Heidelberg New York, 1968.

[104] F. Balibar, Y. Epelboin, and C. Malgrange. Study of the new wave fields appearing in crystals deformed by large thermal gradients. *Acta Cryst. A*, 31:836–840, 1975.

[105] W.F. Ames. *Numerical Methods for Partial Differential Equations.* Thomas Nelson and Sons, Ltd., London, 1969.

[106] P.V. Petrashen. The physical foundations of the computer simulation of x-ray traverse topographs. *Acta Cryst. A*, 36:287–295, 1980.

[107] C.A.M. Carvalho and Y. Epelboin. Simulation of x-ray topographs: a new method to calculate the diffracted field. *Acta Cryst. A*, 49:460–467, 1993.

[108] F. Balibar, F.N. Chukhovskii, and C. Malgrange. Dynamical x-ray propagation: a theoretical approach to the creation of new wave fields. *Acta Cryst. A*, 39:387–399, 1983.

[109] J. Kulda. A novel approach to dynamical neutron diffraction by a deformed crystal. *Acta Cryst. A*, 40:120–126, 1984.

[110] I.R. Entin et al. X-ray acoustic resonance in a perfect silicon crystal. *Sov. Phys. Solid State,*, 20:754–756, 1978.

[111] P. Eisenberger and S.L. McCall. X-ray parametric conversion. *Phys. Rev. Lett.*, 26:684–688, 1971.

[112] Y. Yoda, T. Suzuki, X.-W. Zhang, K. Hirano, and S. Kikuta. Parametric down conversion of x-rays. *J. Synchrotron Rad.*, 5:980–982, 1998.

[113] B. Adams, P. Fernandez, W.-K. Lee, G. Materlik, D.M. Mills, and D.V. Novikov. Parametric down conversion of x-ray photons. *J. Synchrotron Rad.*, 7:81–88, 2000.

[114] B. Adams, Y. Nishino, D.V. Novikov, G. Materlik, and D.M. Mills. Parametric down conversion of x-rays, recent experiments. *Nucl. Instrum. Methods A*, 467:1019–1020, 2001.

[115] R. Dejus M. Sanchez del Rio. *program XOP*, 2002. http://www.esrf.fr/computing/scientific/xop.

[116] Ioffe Physico-Technical Institute, 26 Polytekhnicheskaya, St. Petersburg 194021, Russian Federation. *GaP - Gallium Phosphide, Basic Parameters at 300 K*, 2002. http://www.ioffe.rssi.ru/SVA/NSM/Semicond/GaP/basic.html.

[117] I.S. Gradshtein and I.M. Ryzhik. *Tables of Series, Products, and Integrals*. Harri Deutsch, Thun, 1981.

[118] S.D. Shastri, P. Zambianchi, and D.M. Mills. Femtosecond x-ray dynamical diffraction by perfect crystals. *Proc. SPIE*, 4143:69–77, 2001.

[119] S.D. Shastri, P. Zambianchi, and D.M. Mills. Dynamical diffraction of ultrashort x-ray free-electron laser pulses. *J. Synchrotron Rad.*, 8:1131–1135, 2001.

[120] P. Zambianchi, S.D. Shastri, and D.M. Mills. Numerical simulations of dynamically diffracted ultrashort x-ray free-electron laser pulses. *Proc. SPIE*, 4500:75–88, 2001.

[121] W. Graeff. Short x-ray pulses in a Laue-case crystal. *J. Synchrotron Rad.*, 9:82–85, 2002.

[122] W. Graeff. Time dependence of the polarization of short x-ray pulses after crystal reflection. *J. Synchrotron Rad.*, 9:293–297, 2002.

[123] M.F. DeCamp, D.A. Reis, P.H. Bucksbaum, B. Adams, J.M. Caraher, R. Clarke, C.W.S. Conover, E.M. DuFresne, R. Merlin, V. Stoica, and J.K. Wahlstrand. Coherent control of pulsed x-ray beams. *Nature*, 413:825–828, 2001.

[124] D.A. Reis, M.F. DeCamp, P.H. Bucksbaum, R. Clarke, E. Dufresne, M. Hertlein, R. Merlin, R. Falcone, H. Kapteyn, M.M. Murnane, J. Larsson, T. Missalla, and J.S. Wark. Probing impulsive strain propagation with x-ray pulses. *Phys. Rev. Lett.*, 86:3072–3075, 2001.

[125] A. Hauer and S.J. Burns. Observation of an x-ray shuttering mechanism utilizing acoustic interruption of the Borrmann effect. *Appl. Phys. Lett.*, 27:524–526, 1975.

[126] B.F. Levine and I. Freund. Parametric down conversion of x-rays. *Opt. Comm.*, 1:419–422, 1970.

[127] I. Freund and B.F. Levine. Parametric conversion of x rays. *Phys. Rev. Lett.*, 23:854–857, 1969.

[128] S. Friberg, C.K. Hong, and L. Mandel. Measurement of time delays in the parametric production of photon pairs. *Phys. Rev. Lett.*, 54:2011–2013, 1985.

[129] C.K. Hong, Z.Y. Ou, and L. Mandel. Measurement of subpicosecond time intervals between two photons by interference. *Phys. Rev. Lett.*, 59:2044–2046, 1987.

[130] R. Rüffer and A.I. Chumakov. Nuclear resonance beamline at esrf. *Hyperfine Interactions*, 97/98:589–604, 1996.

[131] J.E. Field. *The properties of natural and synthetic DIAMOND*. Academic Press, Ltd., 24/28 Oval Road, London NWI 7DX, 1992.

[132] Röntec AG, Schwarzschildstr. 12, 12489 Berlin, Germany. *X-Flash 1001 detector*, 2002. http://www.roentec.de.

[133] Drukker International BV, P.O. Box 119, 5430 AC Cuijk, Holland, 1999.

[134] B.W. Adams. Characterization of diamond samples. Technical report, DESY, HASYLAB, 1999.

[135] D.A. Kleinman. Theory of optical parametric noise. *Phys. Rev.*, 174:1027–1041, 1968.

[136] E.M. Dufresne, D.A. Arms, S.B. Dierker, R. Clarke, Y. Yacoby, J. Pitney, B. MacHarrie, and R. Pindak. Design and performance of a stable first crystal mount for a cryogenically cooled Si monochromator at the advanced photon source. *Rev. Sci. Instrum.*, 73:1511–1513, 2002.

[137] B. Adams. A time- and energy-resolving, multichannel event logger. *Rev. Sci. Instrum.*, 74:1128–1134, 2003.

[138] P.G. Kwiat, A.M. Steinberg, R.Y. Chiao, P.H. Eberhard, and M.D. Petroff. High-efficiency single-photon detectors. *Phys. Rev. A*, 48:r867–r870, 1993.

[139] B.W. Adams, M.F. DeCamp, E.M. Dufresne, and D.A. Reis. Picosecond Laser-pump, x-ray probe spectroscopy of GaAs. *Rev. Sci. Instrum.*, 73:4150–4156, 2002.

[140] P. Lechner, S. Eckbauer, R. Hartmann, S. Krisch, D. Hauff, R. Richter, H. Soltau, L. Strüder, C. Fiorini, E. Gatti, A. Longoni, and M. Sampietro. Silicon drift detectors for high resolution room temperature x-ray spectroscopy. *Nucl. Instrum. Methods A*, 377:346–351, 1996.

[141] C. Fiorini, A. Longoni, R. Hartmann, P. Lechner, and L. Strüder. Silicon drift detectors with on-chip electronics for x-ray spectroscopy. *J. X-ray Sc. Tech.*, 7:117–129, 1997.

[142] B.W. Adams. Characterization of the Röntec detector no, 290. Technical report, DESY, HASYLAB, 1998.

[143] S. Techert, F. Schotte, and M. Wulff. Picosecond x-ray diffraction probed transient structural changes in organic solids. *Phys. Rev. Lett.*, 86:2030–2033, 2001.

[144] R. Neutze, R. Wouts, S. Techert, J. Davidsson, M. Kocsis, A. Kirrander, F. Schotte, and M. Wulff. Visualizing photochemical dynamics in solution through picosecond x-ray scattering. *Phys. Rev. Lett.*, 87:195508, 2001.

[145] B. Adams. Proposal for a femtosecond x-ray detector. *Nucl. Instrum. Methods A*, 459:339–346, 2001.

[146] A.N. Artemiev, M.M. Vsevolodov, D.P. Grechukin, A.V. Zabelin, V.N. Kosyakov, S.V. Romanov, and A.A. Soldatov. About systematic studying of chemical shifts in x-ray emission lines with the use of synchrotron radiation. *Nucl. Instrum. Methods A*, 359:266–269, 1995.

[147] D.A. Arms, E.M. Dufresne, R. Clarke, S.B. Dierker, N.R. Pereira, and D. Foster. Refractive optics using lithium metal. *Rev. Sci. Instrum.*, 73:1492–1494, 2002.

[148] Canberra, Inc., 800 Research Parkway, Meriden, CT 06450, USA. *specifications of model 2025 AFT research amplifier*, 2001.

[149] M.C. Beard, G.M. Turner, and C.A. Schmuttenmaer. Transient photoconductivity in GaAs as measured by time-resolved terahertz spectroscopy. *Phys. Rev. B*, 62:15764–15777, 2000.

[150] G. Decker. APS accelerator division, personal communication, 2002.

[151] unpublished results of esrf beam time mi-431, 2000.

[152] K. Hämäläinen, D.P. Siddons, J.B. Hastings, and L.E. Berman. Elimination of the inner-shell lifetime broadening in x-ray absorption spectroscopy. *Phys. Rev. Lett.*, 67:2850–2853, 1991.

[153] M.O. Krause and J.H. Oliver. Natural width of atomic K and L levels, K_α x-ray lines and several KLL Auger lines. *J. Phys. Chem. Ref. Data*, 8:329–338, 1979.

[154] M. Hase, K. Ishioka, M. Kitajima, K. Ushida, and S. Hishita. Dephasing of coherent phonons by lattice defects in bismuth films. *Appl. Phys. Lett.*, 76:1258–1260, 2000.

[155] K. Ishioka, M. Hase, M. Kitajima, and K. Ushida. Ultrafast carrier and phonon dynamics in ion-irradiated graphite. *Appl. Phys. Lett.*, 78:3965–3967, 2001.

[156] M. Hase, M. Kitajima, S. Nakashima, and K Mizoguchi. Dynamics of coherent anharmonic phonons in Bismuth using high density photoexcitation. *Phys. Rev. Lett.*, 88:067401, 2002.

[157] C. Rose-Petruck, R. Jimenez, T. Guo, A. Cavalleri, C.W. Siders, F. Raksi, J.A. Squier, B.C. Walker, K.R. Wilson, and C.P.J. Barty. Picosecond-milliångström lattice dynamics measured by ultrafast x-ray diffraction. *Nature*, 398:310–312, 1999.

[158] A.M. Lindenberg, I. Kang, S.L. Johnson, T. Missalla, P.A. Heimann, Z. Chang, J. Larsson, P.H. Bucksbaum, H.C. Kapteyn, H.A. Padmore,

R.W. Lee, J.S. Wark, and R.W. Falcone. Time-resolved x-ray diffraction from coherent phonons during a laser-induced phase transition. *Phys. Rev. Lett.*, 84:111–114, 2000.

[159] K. Sokolowski-Tinten, J. Solis, J. Bialkowski, J. Siegel, C.N. Afonso, and D. von der Linde. Dynamics of ultrafast phase changes in amorphous GeSb films. *Phys. Rev. Lett.*, 81:3679–3682, 1998.

[160] J.P. Callan, A.M.-T. Kim, C.A.D. Roeser, E. Mazur, J. Solis, J. Siegel, C.N. Afonso, and J.C.G. de Sande. Ultrafast laser-induced phase transitions in amorphous GeSb films. *Phys. Rev. Lett.*, 86:3650–3653, 2001.

[161] H.O. Jeschke, M.E. Garcia, and K.H. Bennemann. Microscopic analysis of the laser-induced femtosecond graphitization of diamond. *Phys. Rev. B*, 60:r3701–r3704, 1999.

[162] A. Cavalleri, C. Tóth, C.W. Siders, J.A. Squier, F. Ráksi, P. Forget, and J.C. Kieffer. Femtosecond structural dynamics in VO_2 during an ultrafast solid-solid phase transition. *Phys. Rev. Lett.*, 87:237401, 2001.

[163] C.V. Shank, R. Yen, and C. Hirlimann. Femtosecond-time-resolved surface structural dynamics of optically excited silicon. *Phys. Rev. Lett.*, 51:900–902, 1983.

[164] H.W.K. Tom, G.D. Aumiller, and C.H. Brito-Cruz. Time-resolved study of laser-induced disorder of Si surfaces. *Phys. Rev. Lett.*, 60:1438–1441, 1988.

[165] D.H. Reitze, H. Ahn, and M.C. Downer. Optical properties of liquid carbon measured by femtosecond spectroscopy. *Phys. Rev. B*, 45:2677–2693, 1992.

[166] K. Sokolowski-Tinten, J. Bialkowski, and D. von der Linde. Ultrafast laser-induced order-disorder transitions in semiconductors. *Phys. Rev. B*, 51:14186–14198, 1995.

[167] A. Leitenstorfer, S. Hunsche, J. Shah, M.C. Nuss, and W.H. Knox. Ultrafast high-field transport in semiconductors. *Physica B*, 272:348–352, 1999.

[168] R.W. Schoenlein, S. Chattopadhyay, H.H.W. Chong, T.E. Glover, P.A. Heimann, C.V. Shank, A.A. Zholents, and M.S. Zolotorev. Generation of femtosecond pulses of synchrotron radiation. *Science*, 287:2237–2240, 2000.

[169] C.W. Siders, A. Cavalleri, K. Sokolowski-Tinten, Cs. Tóth, T. Guo, M. Kammler, M. Horn von Hoegen, K.R. Wilson, D. von der Linde,

and C.P.J. Barty. Detection of nonthermal melting by ultrafast x-ray diffraction. *Science*, 286:1340–1342, 1999.

[170] A. Rousse, C. Rischel, S. Fourmaux, I. Uschmann, S. Sebban, G. Grillon, P. Balcou, E. Förster, J.P. Geindre, P. Audebert, J.C. Gauthier, and D. Hulin. Non-thermal melting in semiconductors measured at femtosecond resolution. *Nature*, 410:65–68, 2001.

[171] A. Rousse, C. Rischel, and J.-C. Gauthier. Colloquium: Femtosecond x-ray crystallography. *Rev. Mod. Phys.*, 73:17–31, 2001.

[172] M.S. Sherwin. Terahertz optics of semiconductor nanostructures near and far from equilibrium. *Current Opinion in Solid State & Materials Science*, 3:191–, 1998.

[173] A. Perez, P. Melinon, V. Dupuis, P. Jensen, B. Prevel, J. Touaillon, L. Bardotti, C. Martet, M. Treilleux, M. Broyer, M. Pellarin, J.L. Vaille, B. Palpant, and J. Lerme. Cluster assembled materials: A novel class of nanostructured solids with original structures and properties. *J. Phys. D*, 30:709–721, 1997.

[174] M. Terrones, Wen Kuang Hsu, H. W. Kroto, and D.R.M. Walton. Nanotubes: A revolution in materials science and electronics. In A. Hirsch, editor, *Fullerenes and Related Structures, Topics in Current Chemistry 199*. Springer, 1999.

[175] F.P. Bundy, W.A. Bassett, R.S. Weathers, R.J. Hemley, H.K. Mao, and A.F. Goncharov. The pressure-temperature phase and transformation diagram for Carbon, updated through 1994. *Carbon*, 34:141–153, 1996.

[176] S.I. Anisimov, B.L. Kapeliovich, and T.L. Perelman. *J. Exp. Theo. Phys. (JETP)*, 39:375–, 1974.

[177] M. Togaya. Pressure dependences of the melting temperature of graphite and the electrical resistivity of liquid Carbon. *Phys. Rev. Lett.*, 79:2474–2477, 1997.

[178] J.M. Glosli and F.H. Ree. Liquid-liquid phase transformation in carbon. *Phys. Rev. Lett.*, 82:4659–4662, 1999.

[179] P. Stampfli and K.H. Bennemann. Theory for the instability of the diamond structure of Si, Ge, and c induced by a dense electron-hole plasma. *Phys. Rev. B*, 42:7163–7173, 1990.

[180] P. Stampfli and K. Bennemann. Dynamical theory of the laser-induced lattice instability of silicon. *Phys. Rev. B*, 46:10686–10692, 1992.

[181] P. Stampfli and K.H. Bennemann. Time dependence of the laser-induced femtosecond lattice instability of Si and GaAs: Role of longitudinal optical distortions. *Phys. Rev. B*, 49:7299–7305, 1994.

[182] R.F.W. Herrmann, J. Gerlach, and E.E.B. Campbell. Ultrashort pulse laser ablation of Silicon: An MD simulation study. *Appl. Phys. A*, 66:35–42, 1998.

[183] G. Galli, R.M. Martin, R. Car, and M. Parrinello. Carbon: The nature of the liquid state. *Phys. Rev. Lett.*, 63:988–991, 1989.

[184] P.L. Silvestrelli, A. Alavi, M. Parrinello, and D. Frenkel. Ab initio molecular dynamics simulation of laser melting of Silicon. *Phys. Rev. Lett.*, 77:3149–3152, 1996.

[185] A. De Vita, G. Galli, A. Canning, and R. Car. A microscopic model for surface-induced diamond-to-graphite transitions. *Nature*, 379:523–526, 1996.

[186] P.L. Silvestrelli and M. Parrinello. Ab initio molecular dynamics simulation of laser melting of graphite. *J. Appl. Phys.*, 83:2478–2483, 1998.

[187] C.Z. Wang, K.M. Ho, M.D. Shirk, and P.A. Molian. Laser-induced graphitization on a diamond (111) surface. *Phys. Rev. Lett.*, 85:4092–4095, 2000.

[188] H.O. Jeschke, M.E. Garcia, and K.H. Bennemann. Theory for laser-induced ultrafast phase transitions in Carbon. *Appl. Phys. A*, 69:s49–s53, 1999.

[189] H.O. Jeschke, M.E. Garcia, and K.H. Bennemann. Time-dependent energy absorption changes during ultrafast lattice deformation. *J. Appl. Phys.*, 91:18–, 2002.

[190] H.O. Jeschke, M.E. Garcia, M. Lenzner, J. Bonse, J. Krüger, and W. Kautek. Laser ablation thresholds of Silicon for different pulse durations: Theory and experiment. *Appl. Surf. Sc.*, 197:839–844, 2002.

[191] M.E. Garcia and H.O. Jeschke. in press. *Appl. Surf. Sc.*, 2003.

[192] H.O. Jeschke, M.E. Garcia, and K.H. Bennemann. Theory for the ultrafast ablation of graphite films. *Phys. Rev. Lett.*, 87:015003, 2001.

[193] H.O. Jeschke and M.E. Garcia. Theoretical description of the ultrafast ablation of diamond and graphite: dependence of thresholds on pulse duration. *Appl. Surf. Sc.*, 197:107–113, 2002.

[194] H.O. Jeschke, M.E. Garcia, and J.A. Alonso. Nonthermal fragmentation of C_{60}. *Chem. Phys. Lett.*, 352:154–162, 2002.

[195] J.S. Graves and R.E. Allen. Response of GaAs to fast intense laser pulses. *Phys. Rev. B*, 58:13627–13633, 1998.

[196] T. Dumitrica, J.S. Graves, and R.E. Allen. Second-order susceptibility from a tight-binding hamiltonian. *Phys. Rev. B*, 58:15340–15343, 1998.

[197] T. Dumitrica and R.E. Allen. Time evolution of second order susceptibility in GaAs following a fast intense laser pulse. *Sol. St. Comm.*, 113:653–656, 2000.

[198] R.E. Allen, T. Dumitrica, and B. Torralva. *Semiconductors and Semimetals*, 67:315–, 2001.

[199] P. Blanc, P. Audebert, F. Fallies, J.P. Geindre, J.C. Gauthier, A. Dos Santos, A. Mysyrowicz, and A. Antonetti. Phase dynamics of reflected probe pulses from sub-100-fs laser-produced plasmas. *J. Opt. Soc. Am. B*, 13:118–124, 1996.

[200] A. Yariv. *Quantum Electronics*. John Wiley & Sons, 1975.

[201] H.C. Andersen. Molecular dynamics simulations at constant pressure and/or temperature. *J. Chem. Phys.*, 72:2384–2393, 1980.

[202] M. Parrinello and A. Rahman. Crystal structure and pair potentials: A molecular-dynamics study. *Phys. Rev. Lett.*, 45:1196–1199, 1980.

[203] C.L. Cleveland. New equations of motion for molecular dynamics systems that change shape. *J. Chem. Phys.*, 89:4987–4993, 1988.

[204] J.R. Ray. Molecular dynamics equations of motion for systems varying in shape and size. *J. Chem. Phys.*, 79:5128–5130, 1983.

[205] R.M. Wentzcovitch. Invariant molecular-dynamics approach to structural phase transitions. *Phys. Rev. B*, 44:2358–2361, 1991.

[206] T.A. Halgren. The representation of van der Waals (vdw) interactions in molecular mechanics force fields: Potential form, combination rules, and vdw parameters. *J. Am. Chem. Soc.*, 114:7827–7843, 1992.

[207] J. Tersoff. New empirical approach for the structure and energy of covalent systems. *Phys. Rev. B*, 37:6991–7000, 1988.

[208] F.H. Stillinger and T.A. Weber. Computer simulation of local order in condensed phases of silicon. *Phys. Rev. B*, 31:5262–5271, 1985.

[209] C.H. Xu, C.Z. Wang, C.T. Chan, and K.M. Ho. A transferable tight-binding potential for carbon. *J. Phys. Cond. Mat.*, 4:6047–6054, 1992.

[210] M. E. Garcia and H. O. Jeschke. to be published.

[211] W.H. Knox, D.S. Chemla, G. Livescu, J.E. Cunningham, and J.E. Henry. Femtosecond carrier thermalization in dense Fermi seas. *Phys. Rev. Lett.*, 61:1290–1293, 1988.

[212] R.M. Wentzcovitch, J.L. Martins, and P.B. Allen. Energy versus free-energy conservation in first-principles molecular dynamics. *Phys. Rev. B*, 45:11372–11374, 1992.

[213] M. Weinert and J.W. Davenport. Fractional occupations and density-functional energies and forces. *Phys. Rev. B*, 45:13709–13712, 1992.

[214] D.W. Heermann. *Computer Simulation Methods in Theoretical Physics*. Springer-Verlag, Berlin Heidelberg, 1990.

[215] P.A. Egelstaff. *An introduction to the liquid state, 2nd ed.* Clarendon Press, Oxford, 1992.

[216] A. Cavalleri, K. Sokolowski-Tinten, D. von der Linde, I. Spagnolatti, M. Bernasconi, G. Benedek, A. Podestà, and P. Milani and. Generation of the low-density liquid phase of carbon by non-thermal melting of fullerite. *Europhys. Lett.*, 57:281–287, 2002.

[217] I. Manika, J. Maniks, R. Pokulis, J. Kalnacs, and D. Erts. Polymerization and damage of C_{60} single crystals under low fluency laser irradiation. *Phys. Stat. Sol. A*, 188:989–998, 2001.

[218] A. Gambirasio, M. Bernasconi, G. Benedek, and P.L. Silvestrelli. Ab initio simulation of laser-induced transformations in fullerite. *Phys. Rev. B*, 62:12644–12647, 2000.

[219] A. Cavalleri, C.W. Siders, F.L.H. Brown, D.M. Leitner, C. Toth, J.A. Squier, C.P.J. Barty, K.R. Wilson, K. Sokolowski-Tinten, M. Horn von Hoegen, D. von der Linde, and M. Kammler. Anharmonic lattice dynamics in germanium measured with ultrafast x-ray diffraction. *Phys. Rev. Lett.*, 85:586–589, 2000.

[220] A.H. Chin, R.W. Schoenlein, T.E. Glover, P. Balling, W.P. Leemans, and C.V. Shank. Ultrafast structural dynamics in inSb probed by time-resolved x-ray diffraction. *Phys. Rev. Lett.*, 83:336–339, 1999.

[221] V. Srajer, T. Teng, T. Ursby, C. Pradervand, Z. Ren, S. Adachi, W. Schildkamp, D. Bourgeois, M. Wulff, and K. Moffatt. Photolysis of the carbon

monoxide complex of myoglobin: Nanosecond time-resolved crystallography. *Science*, 274:1726–1729, 1996.

[222] B. Perman, V. Srajer, Z. Ren, T. Teng, C. Pradervand, T. Ursby, D. Bourgeois, F. Schotte, M. Wulff, R. Kort, K. Hellingwerf, and K. Moffat. Energy transduction on the nanosecond time scale: Early structural events in a xanthopsin photocycle. *Science*, 279:1946–1950, 1998.

[223] M. Gruebele, J. Sabelko, R. Ballew, and J. Ervin. Laser temperature jump induced protein refolding. *Acc. Chem. Res.*, 31:699–707, 1998.

[224] M.C.R. Shastry, J.M. Sauder, and H. Roder. Kinetic and structural analysis of submillisecond folding events in cytochrome c. *Acc. Chem. Res.*, 31:717–725, 1998.

[225] A.M. Lindenberg, I. Kang, S.L. Johnson, R.W. Falcone, P.A. Heimann, Z. Chang, R. W. Lee, and J.S. Wark. Coherent control of phonons probed by time-resolved x-ray diffraction. *Opt. Lett.*, 27:869–871, 2002.

[226] R. Keller, T. Renner, and D.J. Massoletti. Electron beam diagnostics using synchrotron radiation at the Advanced Light Source. *AIP Conf. Proc.*, 390:240–247, 1997.

[227] Z. Chang, A. Rundquist, J. Zhou, M.M. Murnane, H.C. Kapteyn, X. Liu, B. Shan, J. Liu, L. Niu, M. Gong, and X. Zhang. Demonstration of a sub-picosecond x-ray streak camera. *Appl. Phys. Lett.*, 69:133–135, 1996.

[228] B.L. Henke, J.P. Knauer, and K. Premaratne. Characterization of x-ray photocathodes in the 0.1-10-kev photon energy region. *J. Appl. Phys.*, 52:1509–1520, 1981.

[229] B.E. Warren. *X-ray diffraction*. Addison Wesley Pub. Co., Reading, MA, 1969.

[230] W.H. Zachariasen. *Theory of x-ray diffraction in crystals*. Dover Publications Inc., New York, 1994.

[231] M. Holt, Z. Wu, H. Hong, P. Zschack, P. Jemian, J. Tischler, H. Chen, and T.-C. Chiang. Determination of phonon dispersions from x-ray transmission scattering: The example of Silicon. *Phys. Rev. Lett.*, 83:3317–3320, 1999.

[232] W.H. Zachariasen. A theoretical study of the diffuse scattering of x-rays by crystals. *Phys. Rev.*, 57:597–602, 1940.

[233] E. Burkel. *Inelastic scattering of x-rays with very high energy resolution.* Springer-Verlag, Berlin,Heidelberg, 1991.

[234] C. Thomsen, H.T. Grahn, H.J. Maris, and J. Tauc. Surface generation and detection of phonons by picosecond light pulses. *Phys. Rev. B*, 34:4129–4138, 1986.

[235] O. Madelung, editor. *Semiconductors–Basic Data.* Springer-Verlag, Berlin, Heidelberg, 1996.

[236] J.J. Baumberg, D.A. Williams, and K. Köhler. Ultrafast acoustic phonon ballistics in semiconductor heterostructures. *Phys. Rev. Lett.*, 78:3358–3361, 1997.

[237] O.B. Wright. Ultrafast nonequilibrium stress generation in gold and silver. *Phys. Rev. B*, 49:9985–9988, 1994.

[238] M. Nisoli, S. De Silvestri, A. Cavalleri, and A.M. Malvezzi. Coherent acoustic oscillations in metallic nanoparticles generated with femtosecond optical pulses. *Phys. Rev. B*, 55:r13424–r13427, 1997.

[239] J.F. Vetelino, S.P. Gaur, and S.S. Mitra. Debye-waller factor for zinc-blende-type crystals. *Phys. Rev. B*, 5:2360–2366, 1972.

[240] F.A. Lindemann. Über die Berechnung Molecularer Eigenfrequenzen. *Z. Phys.*, 11:609–612, 1910.

[241] E.D. Potter, J.L. Herek, S. Pedersen, Q. Liu, and A.H. Zewail. Femtosecond laser control of a chemical reaction. *Nature*, 55:66–68, 1992.

[242] R.J. Gordon, L. Zhu, and T. Seideman. Coherent control of chemical reactions. *Acc. Chem. Res.*, 32:1007–1016, 1998.

[243] R.J. Levis, G.M. Menkir, and H. Rabitz. Selective bond dissociation and rearrangement with optimally tailored, strong-field laser pulses. *Science*, 292:709–713, 2001.

[244] A.M. Weiner, D.E. Leaird, G.P. Wiederrecht, and K.A. Nelson. Femtosecond pulse sequences used for optical manipulation of molecular motion. *Science*, 247:1317–1319, 1990.

[245] T. Dekorsy, W. Kutt, T. Pfeifer, and H. Kurz. Coherent control of LO-phonon dynamics in opaque semiconductors by femtosecond laser pulses. *Europhys. Lett.*, 23:223–228, 1993.

[246] R. Bartels, S. Backus, E. Zeek, L. Misoguti, G. Vdovin, I.P. Christov, M.M. Murnane, and H.C. Kapteyn. Shaped-pulse optimization of coherent emission of high-harmonic soft x-rays. *Nature*, 406:164–166, 2000.

[247] D. Meshulach, D. Yelin, and Y. Silberberg. Adaptive real-time femtosecond pulse shaping. *J. Opt. Soc. Am.*, 15:1615–1619, 1998.

[248] A. Bartels, T. Dekorsky, H. Kurz, and K Köhler. Coherent control of acoustic phonons in semiconductor superlattices. *Appl. Phys. Lett.*, 72:2844–2846, 1998.

[249] Ü. Özgür, C.-W. Lee, and H.O. Everitt. Control of coherent acoustic phonons in semiconductor quantum wells. *Phys. Rev. Lett.*, 86:5604–5607, 2001.

[250] P.H. Bucksbaum and R. Merlin. The phonon Bragg switch: A proposal to generate sub-picosecond x-ray pulses. *Sol. St. Comm.*, 111:535–539, 1999.

[251] C. Thomsen, J. Strait, Z. Vardeny, H.J. Maris, J. Tauc, and J.J. Hauser. Coherent phonon generation and detection by picosecond light pulses. *Phys. Rev. Lett.*, 53:989–982, 1984.

[252] H.-Y. Hao and H.J. Maris. Study of phonon dispersion in Silicon and Germanium at long wavelengths using picosecond ultrasonics. *Phys. Rev. Lett.*, 84:5556–5559, 2000.

[253] H.-Y. Hao and H.J. Maris. Experiments with acoustic solitons in crystalline solids. *Phys. Rev. B*, 64:064302, 2001.

[254] H.-Y. Hao and H.J. Maris. Dispersion of the long-wavelength phonons in Ge, Si, GaAs, quartz, and sapphire. *Phys. Rev. B*, 63:224301, 2001.

[255] N.V. Chigarev, D.Y. Paraschuk, X.Y. Pan, and V.E. Gusev. Coherent phonon emission in the supersonic expansion of photoexcited electron-hole plasma in Ge. *Phys. Rev. B*, 61:15837–15840, 2000.

[256] S.A. Akhmanov and V.E. Gusev. Laser excitation of ultrashort acoustic pulses: New possibilities in solid-state spectroscopy, diagnostics of fast processes, and nonlinear acoustics. *Sov. Phys. Usp.*, 35:153–191, 1992.

[257] P. Chen, I.V. Tomov, and P.M. Rentzepis. Time resolved heat propagation in a gold crystal by means of picosecond x-ray diffraction. *J. Chem. Phys.*, 104:10001–10007, 1996.

[258] C. Rischel, A. Rousse, I. Uschmann, P.-A. Albouy, J.-P. Geindre, P. Audebert, J.-C. Gauthier, E. Förster, J.-L. Martin, and A. Antonetti. Femtosecond time-resolved x-ray diffraction from laser-heated organic films. *Nature,*, 390:490–492, 1997.

[259] A. Cavalleri, C.W. Siders, C. Rose-Petruck, R. Jimenez, C. Tóth, J.A. Squier, C.P.J. Barty, K.R. Wilson, K. Sokolowski-Tinten, M. Horn von Hoegen, and D. von der Linde. Ultrafast x-ray measurement of laser heating in semiconductors: Parameters determining the melting threshold. *Phys. Rev. B*, 63:193306, 2001.

[260] K. Sokolowski-Tinten, C. Blome, C. Dietrich, A. Tarasevitch, M. Horn von Hoegen, D. von der Linde, A. Cavalleri, J. Squier, and M. Kammler. Femtosecond x-ray measurement of ultrafast melting and large acoustic transients. *Phys. Rev. Lett.*, 87:225701, 2001.

[261] Matthew F. DeCamp. *Seeing sound: Dynamical effects in ultrafast x-ray diffraction.* PhD thesis, Univesity of Michigan, 2002.

[262] C.V. Shank and D.H. Auston. Parametric coupling in an optically excited plasma in Ge. *Phys. Rev. Lett.*, 34:479–481, 1975.

[263] K. Sokolowski-Tinten, J. Bialkowski, M. Boing, A. Cavalleri, and D. von der Linde. Thermal and nonthermal melting of gallium arsenide after femtosecond laser excitation. *Phys. Rev. B*, 58:r11805–r11808, 1998.

[264] O.B. Wright, B. Perrin, O. Matsuda, and V.E. Gusev. Ultrafast carrier diffusion in gallium arsenide probed with picosecond acoustic pulses. *Phys. Rev. B*, 64:081202, 2001.

[265] H.M. Bizek, editor. *The Advanced Photon Source List of Parameters.* Argonne National Labs, 1996.

[266] D. Atwood. *Soft X-rays and Extreme Ultraviolet Radiation: Principles and Applications.* Cambridge University Press, 1999.

[267] A. Baron. A fast, convenient, x-ray detector. *Nucl. Instrum. Methods A*, 400:124–132, 1997.

[268] S. Backus, C.G. Durfee III, M.M. Murnane, and H.C. Kapteyn. High power ultrafast lasers. *Rev. Sci. Instrum.*, 69:1207–1223, 1998.

[269] D. Strickland and G. Mourou. Compression and amplification of chirped optical pulses. *Opt. Comm.*, 56:219–221, 1985.

[270] David R. Lide, editor. *Handbook of Chemistry and Physics.* CRC press, 1996.

[271] C.R. Wie, T.A. Tombrello, and T. Vreeland, Jr. Dynamical x-ray diffraction from nonuniform crystalline films: Application to x-ray rocking curve analysis. *J. Appl. Phys.*, 59:3743–3746, 1986.

[272] N.C. Woolsey and J.S. Wark. Modeling of time resolved x-ray diffraction from laser-shocked crystals. *J. Appl. Phys.*, 81:3023–3037, 1997.

[273] K.D. Liss, A. Magerl, R. Hock, A. Remhof, and B. Waibel. Towards a new (q, t) regime by time-resolved x-ray diffraction: Ultra-sound excited crystals as an example. *Europhys. Lett.*, 40:369–374, 1997.

[274] B. E. Warren. *X-Ray Diffraction.* Dover Publications Inc., 1990.

[275] A. Authier, S. Lagomarsino, and B.K. Tanner, editors. *X-ray and Neutron Dynamical Diffraction: Theory and Applications.* NATO ASI Series, 1996.

[276] S.D. LeRoux, R. Colella, and R. Bray. X-ray-diffraction studies of acoustoelectrically amplified phonon beams. *Phys. Rev. Lett.*, 35:230–234, 1975.

[277] S.D. LeRoux, R. Colella, and R. Bray. Effect of acoustoelectric phonons on anomalous transmission of x rays. *Phys. Rev. Lett.*, 37:1056–1059, 1976.

[278] P. Sondhauss and J.S. Wark. Extension of the time-dependent dynamical diffraction theory to 'optical phonon'-type distortions: application to diffraction from coherent acoustic and optical phonons. *Acta Cryst. A*, 59:7–13, 2003.

[279] F.N. Chukhovskii and E. Förster. Time-dependent x-ray Bragg diffraction. *Acta Cryst. A*, 51:668–672, 1995.

[280] J.S. Wark and H. He. Subpicosecond x-ray diffraction. *Laser and Particle Beams*, 12:507–513, 1994.

[281] J.S. Wark and R.W. Lee. Simulations of femtosecond x-ray diffraction from unperturbed and rapidly heated single crystals. *J. Appl. Cryst.*, 32:692–703, 1999.

[282] J. Gronkowski. Propagation of x-rays in distorted crystals under dynamical diffraction. *Phys. Rpt.*, 206:1–41, 1991.

[283] B. Klar and F. Rustichelli. Dynamical neutron diffraction by ideally curved crystals. *Il Nuovo Cimento B*, 13b:249–271, 1973.

[284] W.H. Press, S.A. Teukolsky, W.T. Vetterling, and B.P. Flannery. *Numerical recipes in C: the art of scientific computing.* Cambridge Univ. Press, 2 edition, 1992.

[285] G. Eckold, M. Stein-Arsic, and H.J. Weber. Unisoft - a program package for lattice dynamical calculations. *J. Appl. Cryst.*, 20:134–139, 1987.

[286] L.D. Chapman, S.N. Ehrlich, and N.M. Lazarz. A time-resolved x-ray scattering technique for obervation of non-equilibrium phonons. *Rev. Sci. Instrum.*, 61:86–89, 1990.

[287] I. Freund and B.F. Levine. Optically modulated x-ray diffraction. *Phys. Rev. Lett.*, 25:1241–1245, 1970.

[288] I. Freund and B.F. Levine. Resonant dispersion of the nonlinear atomic scattering factor. *Opt. Comm.*, 3:101–104, 1971.

[289] I. Freund and B.F. Levine. Surface effects in the nonlinear interaction of x-ray and optical fields. *Phys. Rev. B*, 8:3059–3060, 1973.

[290] I. Freund. Nonlinear x-ray diffraction. determination of valence electron charge ditributions. *Chem. Phys. Lett.*, 12:583–588, 1972.

[291] P.M. Eisenberger and S.L. McCall. Mixing of x-ray and optical photon. *Phys. Rev. A*, 3:1145–1151, 1971.

[292] H. Danino and I. Freund. Parametric down conversion of x rays into the extreme ultraviolet. *Phys. Rev. Lett.*, 46:1127–1130, 1981.

[293] T. Takahashi, T. Ishikawa, and S. Kikuta. Generation of the x-ray second harmonic under the dynamical diffraction condition. *Nucl. Instrum. Methods A*, 246:768–771, 1986.

[294] SLAC, 2575 Sand Hill Road, Menlo Park, CA 94025, USA. *Linac Coherent Light Source (LCLS) Design Study Report*, 1998. http://www.slac.stanford.edu/pubs/slacreports/slac-r-521.html.

[295] DESY, Notkestr. 85, D-22607 Hamburg, Germany. *Free Electron Laser*, 2002. http://www-hasylab.desy.de/facility/fel/parameters.

[296] M.A. Kornberg, A.L. Godunov, S. Itza-Ortiz, D.L. Ederer, J.H. McGuire, and L. Young. Interaction of atomic systems with x-ray free-electron lasers. *J. Synchrotron Rad.*, 9:298–303, 2002.

[297] G. Baym. *Lectures on Quantum Mechanics*. W.A. Benjamin, New York, Amsterdam, 1969.

[298] S. Tanaka, V. Chernayk, and S. Mukamel. Time-resolved x-ray spectroscopies: Nonlinear response functions and Liouville-space pathways. *Phys. Rev. A*, 63:063405, 2001.

[299] P.N. Butcher and D. Cotter. *The elements of nonlinear optics*. Cambridge University Press, Cambridge (UK), New York, 1990.

[300] P. Zambianchi. Interaction between x-ray free-electron laser pulses and matter: Nonlinear susceptibilities. *in preparation*, 2003.

[301] S. Tanaka and S. Mukamel. Coherent x-ray Raman spectroscopy: A nonlinear local probe for electronic excitations. *Phys. Rev. Lett.*, 89:043001, 2002.

[302] P. Kalman. Laser-assisted x-ray absorption near the absorption edge. *Phys. Rev. A*, 38:5458–5460, 1988.

[303] L. Mandel and E. Wolf. *Optical Coherence and Quantum Optics.* Cambridge University Press, Cambridge, New York, 1995.

[304] J.J. Sakurai. *Modern Quantum Mechanics.* Addison-Wesley publishing Co., USA, 1985.

Notation

Terminology